Mechanical Effects of Welding

International Union of Theoretical and Applied Mechanics

L. Karlsson, L.-E. Lindgren,
M. Jonsson (Eds.)

Mechanical Effects of Welding

IUTAM Symposium, Luleå/Sweden
June 10-14, 1991

Springer-Verlag
Berlin Heidelberg New York
London Paris Tokyo
Hong Kong Barcelona Budapest

Prof. Lennart Karlsson
Prof. Lars-Erik Lindgren
Prof. Mikael Jonsson

Luleå University of Technology
Div. of Computer Aided Design
951 87 Luleå
Sweden

ISBN-13: 978-3-642-84733-2 e-ISBN-13: 978-3-642-84731-8
DOI: 10.1007/978-3-642-84731-8

Library of Congress Cataloging-in-Publication Data
IUTAM Symposium (1991 : Luleå, Sweden)
Mechanical effects of welding : IUTAM Symposium, Luleå, Sweden, 1991
L. Karlsson, L.-E. Lindgren, M. Jonsson, eds.
At head of title: International Union of Theoretical and Applied Mechanics.
ISBN-13: 978-3-642-84733-2

1. Welding-Congresses. 2. Residual stresses-Congresses. 3. Deformations (Mechanics)-Congresses.
I. Karlsson, L. (Lennart). II. Lindgren, L.-E. (Lars-Erik). III. Jonsson, M. (Mikael).
IV. International Union of Theoretical and Applied Mechanics. V. Title.
TS227.I95 1991 671.5'20422--dc20 92-7092

Softcover reprint of the hardcover 1st edition 1992

Typesetting: Camera ready by authors

61/3020-5 4 3 2 1 0 – Printed on acid-free paper.

Preface

The International Union of Theoretical and Applied Mechanics (IUTAM) initiated and sponsored an International Symposium on The Mechanical Effects of Welding. The Symposium was held in Luleå, Sweden, 10-14 June 1991. The intention of the Symposium was to gather active scientists in order to assess the current state of the art and future directions.

The field of welding is an area which includes a large number of scientific disciplines, such as materials science, solid mechanics, thermal science, and also mechanical engineering design and production engineering. The intention of the Symposium was to cover the direct mechanical effects of welding and their influence on the in-service behaviour of welded structures.

The Mechanical Effects of Welding is a very appropriate theme for an IUTAM Symposium. Progress in this field requires close interaction between researchers in several disciplines. This is reflected in the topics covered.

The topics of the different sessions were:

◊ Calculations of Temperatures, Strains and Stresses
◊ Residual Stresses and Residual Deformations
◊ Measurements of Residual Strains and Stresses
◊ Effects of Defects and Residual Stresses on Fracture and Fatigue
◊ Effects of Residual Stresses on Creep Deformation
◊ Effects of Residual Deformations and Residual Stresses on Buckling

There were 50 participants from 12 countries at the Symposium. The 28 papers presented at the Symposium are collected in this volume. A Scientific Committee, appointed by the Bureau of IUTAM, selected the participants to be invited and the papers to be presented.

The Scientific Committee consisted of the following members:

Prof L. Karlsson	Luleå University of Technology, Luleå, SWEDEN (Chairman)
Prof U. Dilthey	Welding Institute, Aachen, FRG
Prof J. Goldak	Carleton University, Ottawa, CANADA
Mr J. Hepworth	Powergen, Nottingham, UK
Prof J. Hult	Chalmers University of Technology, Gothenburg, SWEDEN
Dr L. Josefson	Chalmers University of Technology, Gothenburg, SWEDEN
Dr M. Kanninen	Southwest Research Institute, San Antonio, USA
Prof J. B. Leblond	Lab. de Modélisation en Mecanique, Paris, FRANCE
Prof Z. Mroz	IPPT PAN, Warsaw, POLAND
Prof Y. Ueda	Welding Research Institute Osaka University, Osaka, JAPAN

The success of the Symposium would not have been possible without the excellent work of the Local Organizing Committee. Members of the Committee were the staff of the Division of Computer Aided Design at Luleå University of Technology: C. Hardell, H-Å. Häggblad, M. Isaksson, P. Jeppsson, M. Jonsson, L. Karlsson (Chairman), L-E. Lindgren, M. Näsström, M. Oldenburg, A. Stensson, A. Svoboda, S-O. Westberg, L. Wikander and A-K. Wikman. Furthermore it included T. Bergqvist from the University's conference center, CENTEK.

Luleå, October 1991

Lennart Karlsson

List of Participants

Alfredsson Svante, Chalmers University, Division of Solid Mechanics, 412 96 Göteborg, Sweden

Bertram, Lee, Sandia National Laboratory, P O Box 969, Division 8243, Livermore CA 94550, USA

Brown, Stuart, Massachusetts Institute of Technology, Department of Materials Science and Engineering, 77 Massachusetts Avenue, Cambridge, MA 02139, USA

Carlestam, Anders, Swedish Steel, Oxelösund, Box 1000, 613 01 Oxelösund, Sweden

Dahlberg, Lars, The Swedish Plant Inspectorate, Box 49306, 100 20 Stockholm, Sweden

Denis, Sabine, Laboratoire de Metallurgie, Parc de Saurupt, 50442 Cedex, France

Elfström, Bengt-Olof, Luleå University of Technology, Division of Computer Aided Design, 951 87 Luleå, Sweden

Eriksson, Kjell, Luleå University of Technology, Division of Solid Mechanics, 951 87 Luleå, Sweden

Finnie, Iain, University of California, Department of Mechanical Engineering, Berkeley, CA 947 20, USA

Fischer, Franz Dieter, University for Mining and Metal, Institute of Mechanics, Franz Josef Strasse 18, 8700 Leoben, Austria

Goglio, Luca, Politecnico of Torin, Dipartimento di Meccanica, Corso Duca Abruzzi 24, 10123 Torino, Italy

Goldak, John A, Carleton University, Department. of Mechanical and Aerospace Engineering, Ottawa, Ontario, Canada

Häggblad, Hans-Åke, Luleå University of Technology, Division of Computer Aided Design, 951 87 Luleå, Sweden

Johansson, Lars, The Swedish Welding Commission, Box, 5073, 102 42 Stockholm, Sweden

Jonsson, Mikael, Luleå University of Technology, Division of Computer Aided Design, 951 87 Luleå, Sweden

Josefson, Lennart, Chalmers University of Technology, Division of Solid Mechanics, 412 96 Göteborg, Sweden

Karlsson, Lennart, Luleå University of Technology, Division of Computer Aided Design, 951 87 Luleå, Sweden

Kinugawa, Junichi, National Research Institute of Metals, Advanced Materials Processing Division, 3-12, 2-Chome, Nakameguro, Meguroku, Tokyo 153, Japan

Leblond, Jean-Baptiste, Université de Paris VI, Laboratoire de Modélisation en Méchanique, Tour 66, 4 place Jussien, 75005 Paris, France

Lindén, Gunnar, Luleå University of Technology, Division of Computer Aided Design, 951 87 Luleå, Sweden

Lindgren, Lars-Erik, Luleå University of Technology, Division of Computer Aided Design, 951 87 Luleå, Sweden

Lixing, Huo, Tianjin University, Welding Division, Department of Mechanical Engineering, Box 300072, Tianjin, P R China

Murakawa, Hidekazu, Osaka University, Osaka Welding Research Institute, 11-1, Mihogaoka, Ibaraki, Osaka 567, Japan

Nitschke-Pagel, Thomas, Universität des Landes Hessen, Gesamthochschule Kassel, Institute für Werkstofftechnik, Wilhelmshöher Allee 73, 3500 Kassel, Germany

Näsström, Mats, Luleå University of Technology, Division of Computer Aided Design, 951 87 Luleå, Sweden

Ohji, Takayoshi, Osaka University, Department of Welding and Production, 2-1, Yamada-oka, Suita, Osaka, 565, Japan

Oldenburg, Mats, Luleå University of Technology, Division of Computer Aided Design, 951 87 Luleå, Sweden

Orlowski, Cezary, Technical University of Gdansk, Mechanical Technololgy Faculty, 80-952 Gdansk-Wrzeszcz, Poland

Pawlak, Andrzej, Technical University of Czestochowa, Institute of Mechanics and Machine Design, ul Deglera 35, 42-200 Czestochowa, Poland

Rammerstorfer, Franz G, Vienna Technical University, Institute of Light Weight Structures, Gußhausstraße 27-29/317, 1040 Vienna, Austria

Ranta, Arno, Helsinki Institute of Technology, Eerikinkatu 36, 00180 Helsinki, Finland

Ronda, Jacek, University of Cape Town, Department of Applied Mathematics, Private Bag; Rondebosch 7700, South Africa

Saarenheimo, Arja, Research Centre of Finland, Nuclear Engineering Laboratory, P O Box 169, 00181 Helsinki 18, Finland

Samuelsson, Lars Åke, The Swedish Plant Inspectorate, Box 49306 100 20 Stockholm, Sweden

Schwalbe, Karl-Heinz,GKSS Research Centre Geesthacht, Postfach 1160, 2054 Geesthacht, Germany

Segle, Peter, The Swedish Plant Inspectorate, Box 49306, 100 20 Stockholm, Sweden

Seo, Kenji, Himeji Institute of Technology, Department of Mechanical Engineering, 2167 Shosha Himeji, Hyogo 671-22, Japan

Svoboda, Ales, Luleå University of Technology, Division of Computer Aided Design, 951 87 Luleå, Sweden

Tang, Muyao, Xi'an Jiaotong University, Mechanical Engineering Department, 710049 Xi'an, P R China

Ueda, Yukio, Osaka University, Osaka Welding Institute, 11-1, Mihogaoka, Ibaraki, Osaka 567, Japan

Wang, Jianhua, Shanghai Jiao Tong University, Department of Materials Science and Engineering, 1954 Hua Shan Road, Shanghai 200030, P R China

Webster, Peter, University of Salford, Department of Civil Engineering, Salford M5 4WT, United Kingdom

Wikander, Lars, Luleå University of Technology, Division of Computer Aided Design, 951 87 Luleå, Sweden

Wikman, Anna-Karin, Luleå University of Technology, Division of Computer Aided Design, 951 87 Luleå, Sweden

Wohlfahrt, Helmut, TU Braunschweig, Institut für Schweißtechnik und Werkstofftechnologie, Langer Kamp 8, 3300 Braunschweig, Germany

Yao, Tetsuya, Hiroshima University, Department of Naval Architecture and Ocean Engineering, Saijo, Higashi-Hiroshima 724, Japan

Yin, Delan, Southwest Jiao Tong University, Department of Bridge and Underground Engineering, 18005 Nanyuan, Chengdu, Sichuan, 610031, P R China

Yuan, Mingang, Diakin Industries Ltd, Production Engineering Centre, Sakai Plant, 1304 Kanaoka-cho, Sakai, Osaka 591, Japan

Yuan, Zhihua, Technical Research Centre of Finland, Laboratory of Production Engineering, KOT/VTT, P O Box 111, 02151 Espoo, Finland

Öberg, Tero, Technical Research Centre of Finland, Metals Laboratory, P O Box 111, 02151 Espoo, Finland

Contents

CONSTITUTIVE MODELLING OF METALS AT HIGH TEMPERATURES

J. Goldak, A. Oddy, M. Gu, W. Ma, A. Mashaie and E. Hughes: *Coupling Heat Transfer, Microstructure Evolution and Thermal Stress Analysis in Weld Mechanics* 1

S.Brown, P. Kumar and V. Dave: *Very High Homologous Temperature Constitutive Models for Semi-Solid and Solid Metals* 31

F.D. Fischer: *Transformation Plasticity (TRIP) under a Triaxial Stress State* 39

J. Wang, C. Chen and F. Ding: *Research on Welding Stresses during Phase Transformation* 49

S. Denis, D. Farias and A. Simon: *Modelling Phase Transformations for the Calculation of Internal Stresses During Fast Heating and Cooling in Steels* 57

C. Orlowski: *The Analytical Method of Bainite Transformation Evaluation in Microalloyed Steels in the Process of Rapid Cooling* 67

CALCULATIONS OF TEMPERATURES, STRAINS AND STRESSES

J. Ronda and O. Mahrenholtz: *Thermal Problems of Welding* 75

J.M. Bergheau, D. Pont and J.B. Leblond: *Three-Dimensional Simulation of a Laser Surface Treatment through Steady State Computation in the Heat Source's Comoving Frame* 85

R. Parkitny, A. Pawlak and W. Piekarska: *Temperature Fields and Stress States in Welded Tubes of Rectangular Cross Section* 93

RESIDUAL STRESSES AND RESIDUAL DEFORMATIONS

Y. Ueda and M.G. Yuan: *The Characteristics of the Source of Welding Residual Stress (Inherent Strain) and its Application to Measurement and Prediction* 105

Th. Nitschke-Pagel and H. Wohlfahrt: *Residual Stress Distributions after Welding as a Consequence of the Combined Effect of Physical, Metallurgical and Mechanical Sources* 123

MEASUREMENTS OF RESIDUAL STRAINS AND STRESSES

W. Cheng and I. Finnie: *Measurement of Residual Stress Distributions Near the Toe of a Weld between a Bracket and a Plate Using the Crack Compliance Method* 133

P.J. Webster: *The Neutron Strain Scanner: Measurements in Welds* 143

EFFECTS OF DEFECTS AND RESIDUAL STRESSES ON FRACTURE AND FATIGUE

K.-H. Schwalbe: *Effect of Weld Metal Mis-match on Toughness Requirements: Some Simple Analytical Considerations using the Engineering Treatment Model (ETM)* 151

Y. Ueda, H. Murakawa and H. Kimura: *Compressive Strength of Structural Members with Undermatching Weld Joints* 159

K. Seo, F. Nogata and M. Kusaka: *Effect of Mechanical Heterogeneity on the Absorbed Energy of Welded Joint* 167

K. Eriksson: *On the Correlation between Charpy Notch Toughness and Fracture Toughness of Base and Weld Metal* 175

POSTER SESSION

S. Brown, K. Christie and H. Song: *Three Dimensional Finite Element Modeling of Welded Structures* 181

H. Lixing, Z. Yufeng and W. Lijun: *Influence of Weldmatching on Mechanical Behaviours* 189

M. Näsström, L. Wikander, L. Karlsson, L-E. Lindgren and J. Goldak: *Combined Solid and Shell Element Modelling of Welding* 197

T. Ohji, A. Ohkubo and K. Nishiguchi: *Mathematical Modelling of Molten Pool in Arc Welding* 207

T. T. Öberg: *Computation of Temperature Distribution due to Welding in Piping Systems* 215

EFFECTS OF RESIDUAL STRESSES ON CREEP DEFORMATION

L.A. Samuelsson, P. Segle and S.T. Tu: *Design of Weldments in Structures Subjected to Creep* 223

J. Kinugawa, Y. Monma, H. Hongo, M. Yamazaki and T. Watanabe: *Creep Behaviour of 304 Stainless Steel Welded Joints Composed of Two Different 308 Weld Metals* 231

EFFECTS OF RESIDUAL DEFORMATIONS AND RESIDUAL STRESSES ON BUCKLING

F.G. Rammerstorfer, I. Skrna-Jakl and M. Zelezny: *The Influence of Welding Stresses and Distortions on the Stability of Shells of Revolution* 239

T. Yao, P.I. Nikolov and Y. Miyagawa: *Influences of Welding Imperfections on Stiffness of Rectangular Plate under Thrust* 261

D. Yin and S. Qiang: *Influence of Residual Stresses and Deformation on Buckling Behaviour of Plates* 269

K. S. Alfredsson and B.L. Josefson: *Harmonic Response of a Spot Welded Box Beam - Influence of Welding Residual Stresses and Deformations* 277

1. Yin, D
2. Leblond, J-B
3. Goldak, J A
4. Yuan, M
5. Orlowski, C
6. Schwalbe, K-H
7. Brown, S
8. Ohji, T
9. Ronda, J
10. Rammerstorfer, F G
11. Fischer, Franz D
12. Wang, J
13. Kinugawa, J
14. Samuelsson, L Å
15. Bertram, L
16. Ranta, A
17. Denis, S,
18. Josefson, L
19. Alfredsson S,
20. Ueda, Y
21. Wikander, L
22. Segle, P
23. Karlsson, L
24. Carlestam, A
25. Eriksson, K
26. Lixing, H
27. Näsström, M
28. Oldenburg, M
29. Wikman, A-K
30. Johansson, L
31. Elfström, B-O
32. Wohlfahrt, H
33. Yuan, Z
34. Finnie, I
35. Pawlak, A
36. Lindén, G
37. Saarenheimo, A
38. Dahlberg, L
39. Öberg, T
40. Jonsson, M
41. Nitschke-Pagel, T
42. Goglio, L
43. Svoboda, A
44. Webster, P

Coupling Heat Transfer, Microstructure Evolution and Thermal Stress Analysis in Weld Mechanics

John Goldak, Alan Oddy, Moashi Gu,
Weidan Ma, Akbar Mashaie, Edward Hughes
Mechanical & Aerospace Engineering
Carleton University, Ottawa, Canada, K1S 5B6

IUTAM Symposium
Lulea, Sweden, June 10-14, 1991

Abstract

The mechanical behavior of welds is sensitive to the close coupling between heat transfer, microstructure evolution and thermal stress analysis. Since the temperature field computed from a heat transfer analysis can be considered to drive the mechanics of the welding process, the first step is to solve the energy equation usually with FEM. The research issue is decoupling the physics of the arc and weld pool from the energy equation. This is done by modeling the heating effect of the arc. Although the effects of microstructure and stress-strain evolution on heat transfer are not large, the effect of temperature on the microstructure and thermal stress is dominant. In addition, the coupling between microstructure and thermal stress can be strong and subtle. The microstructure evolution is modeled with algebraic equations for thermodynamics and ordinary differential equations for kinetics. The thermal stress analysis involves large strains and large rotations. The most popular constitutive equation has been elasto-plastic. Phase transformations such as the austenite to martensite transformation, can dominate the stress analysis. Since realistic welding problems tend to be truly three dimensional with complex geometry, transient and nonlinear, numerical methods have advantages. However, the computational demands have limited the size of welds that can be analyzed. In the past five years considerable progress has been made in developing numerical methods to solve this coupled problem with increasing speed and accuracy. Major gains have been made with better mesh grading and more efficient solvers. In addition, software engineering has played a major role in managing the complexity of software.

L. Karlsson, L.-E. Lindgren, M. Jonsson (Eds.)
Mechanical Effects of Welding
IUTAM Symposium Luleå/Sweden 1991

Introduction

From industry's viewpoint, the most critical mechanical effects of welding are cracking, distortion, and buckling. The influence of welds on crack propagation, either in stress corrosion cracking, fatigue or fracture is also a concern. To predict and control these effects, it is an advantage to understand and be capable of predicting the macroscopic transient fields of temperature, displacement, strain and stress. In principle, this can be done by solving the equations of continuum mechanics.

In practice however there are many aspects of welding that have made the rigorous analysis of welds challenging. At the macroscopic level a weld can be considered to be a thermo-mechanical problem of computing transient temperature, displacement, stress and strain. At the microscopic level, it can be considered to be a metal physics problem of computing the phase transformations including grain growth, dissolution and precipitation. The equations of continuum mechanics are macroscopic. In the case of welding they typically resolve length scales that range from 1 mm to 10 m. At the next finer length scale, 0.01 μm to 1 mm, microscopic phenomena deal with grains, precipitates, subgrains. This is the realm of the metallurgist. The deformation of a material is sensitive to the structure at this level. At a length scale of 0.1 nm to 0.01 μm, the phenomena are usually described in terms of atoms and their interactions. This is the realm of the metal physicist. Each level of abstraction or spatial resolution has its own set of equations and utilizes parameters from the level below it.

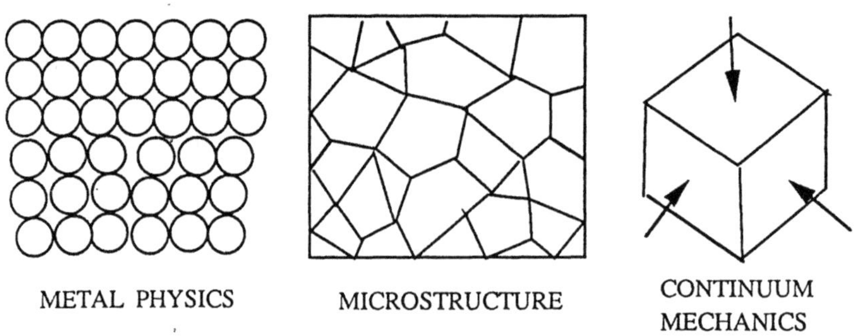

Fig. 1 Length scales in welds range from atomic through microstructure to continuum mechanics.

Realistic welds may involve tens of passes, each of which contribute to the mechanical and metallurgical effects. Interactions between thermal, mechanical, metallurgical and, in the molten pool, chemical and fluid processes, are complex. The difficulties of obtaining relevant material properties, in making experimental measurements to validate predictions and the complexity of the mathematical descriptions have all inhibited progress.

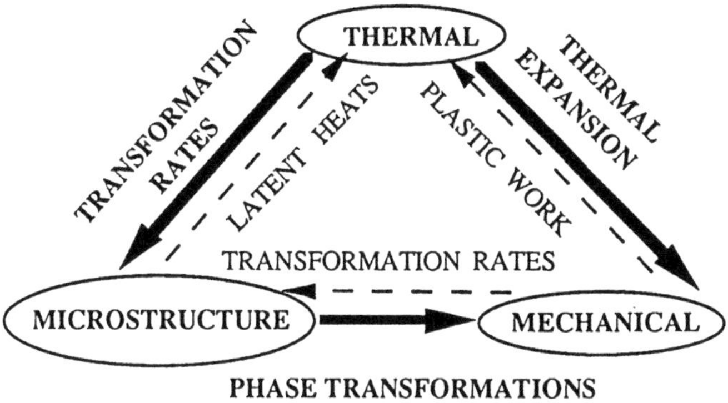

Fig. 2. The dominant couplings in welding are shown with bold lines. The secondary couplings are shown with dotted lines.

The equations of continuum mechanics require the geometry, initial conditions, boundary conditions and constitutive parameters to be specified. If these data are known, the equations can be solved by nonlinear transient finite element analysis (FEA). These equations are most accurately and efficiently solved by organizing or structuring them; i.e., set those interactions that are small to zero. For example, in arc welding the coupling between the hydro-magneto-dynamics of the arc, the fluid mechanics of the weld pool, heat transfer and thermal stress analysis in the solid can and should be limited to the terms that are significant. In this case only the temperature field and possibly solute transportation, couple the weld pool and solid regions. We will consider carefully the couplings that are most effective and seek algorithms to solve these coupled equations.

Inherent difficulties

Foremost among the difficulties has been the problem of adequately describing the essential geometric features of the welded structure while at the same time obtaining sufficient detail around the joint. To describe the heat-affected zone (HAZ) in any detail requires a resolution of at least fractions of a millimeter. Typical welded structures may have dimensions of the order of tens of meters. The important geometric details are frequently of the order of a meter or so. Typical welds may be meters long. Clearly the length scale in each domain of analysis must be coarse enough to make the computations feasible. There is also evidence that decreasing the length scale; i.e. finer mesh, ultimately leads to convergence problems.

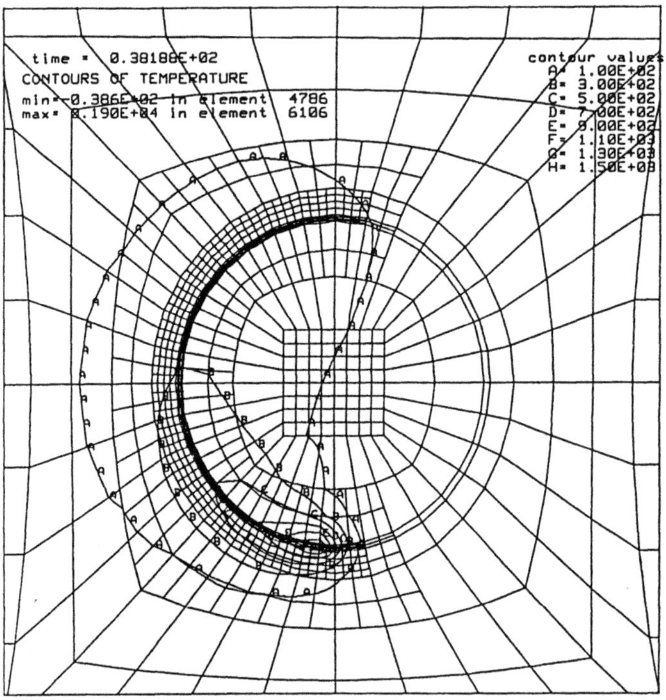

Fig. 3 For welds with complex paths, quasi-stationary steady states may not exist.

The motion of the heat source is an important facet in the analysis of welds. Because of it there are regions at the start and end of any weld where the transient conditions differ significantly from the quasi-steady state conditions in the majority of the weld. For complex joint geometries a quasi-stationary steady state may not exist.

Not only is the metallurgical response of the structure to the thermal cycle complex but there are facets which are still not well understood. With peak temperatures approaching the melting point the metallurgical response will vary rapidly with time and position. The interaction between temperature, stress and microstructure may require the coupled solution of all three aspects. Phase transformations, whose kinetics depend on the stress and thermal history may directly and indirectly affect the stress and thermal history. The direct effects are due to the release of latent heats of transformation, volume changes and transformation plasticity. The indirect effects are due to the effect on material properties. In addition, in multipass welds, tempering of previously transformed material can reduce the hardness and strength while at the same time causing a small volume change.

The occurrence of cracking is, in itself, a complex phenomenon with no single cause or morphology. Cracking may occur because of insufficient ductility at high temperature, excessive hardness, stress and hydrogen content at low temperatures or, during post-weld heat treatment, because of low ductility and non-uniform yield strengths.

The prediction of fracture will require a thorough understanding of the role of local brittle zones [1]. Indeed fracture of welds in general has some unique properties which raise fundamental questions. Fast fracture does not always remain in the most brittle microstructure.

Prediction of the fusion zone size and shape has long been a challenge. The nugget shape is strongly influenced by the flow pattern inside the molten pool. Prediction of such complex flows is a major problem in its own right and a solution is not expected soon.

Validation of predictions of residual stresses by comparison with measured values will be an important step in the industrial acceptance of computational weld mechanics. Measured values however frequently show significant scatter. This is likely to be the fault of the experimental techniques. For instance, even neutron diffraction techniques need to increase their accuracy and precision by a factor of 2 to 3 if residual stress measurements are to be accurate to 30 MPa [2]. Errors on the diffraction measurements lead to anticipated errors of plus or minus 100 MPa. Progress in computational weld mechanics will be difficult without a parallel improvement in experimental techniques.

Fig. 4 Dynamic tear test of multi-weld specimen. Note fracture does not remain between welds or in brittle microstructure. Absorbed energy increases dramatically [1].

Finally, as numerical analysis techniques increase in power the specialized knowledge required to use them has also increased. This has effectively meant that few people interested in weld mechanics have taken advantage of all the available mathematics. It will be increasingly important to design effective user interfaces to hide the complexity from the user where possible.

Brief History of Computational Weld Mechanics

Historically, arc welding began shortly after electrical power became available in the late 1800s. Serious scientific studies date from at least the 1930s. The failure of welded bridges in Europe in the 1930s and the American Liberty ships in World War II, did much to stimulate research in welding in the 1940s. In the USA, the greatest attention was focused on developing fracture mechanics and fracture toughness tests. This could be interpreted as a belief that welding was too complex to analyze and predict and therefore they chose an experimental approach that relied heavily on metallurgical and fracture toughness tests. The steady state heat transfer analysis of Rosenthal was an exception [3].

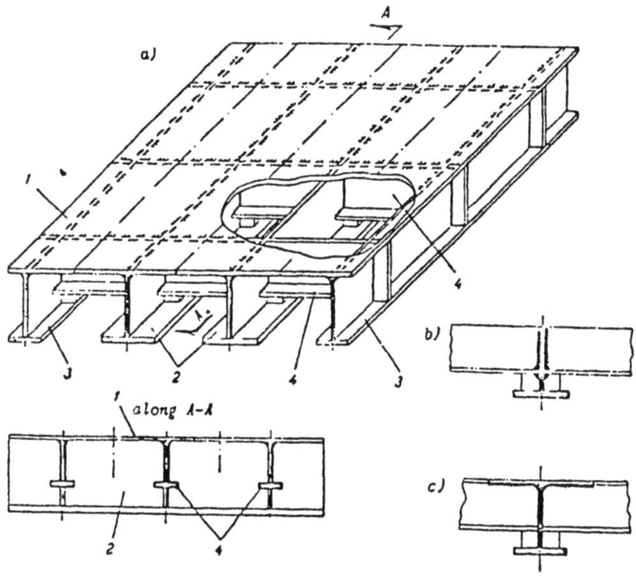

Fig. 5. Structures typical of those analyzed by Okerblom [4].

Russia took a different approach. The books of Okerblom [4] and Vinokurov [5] are a rich record of the analysis of welded structures including multi-pass welds and complex structures. We will not pursue this methodology here because they had to idealize the problem to a greater degree than we are prepared to accept here. The work is admirable and deserves careful study by current researchers.

The current strategy for analyzing welds began in the 1970s with the pioneering work of Hibbitt and Marcal [6], Freidman [7], Masubuchi [8] and Andersson [9]. To reduce the geometric complexity to manageable levels, they used 2D Lagrangian, plane strain, plane stress, or axisymmetric kinematic models. The material was elasto-plastic with temperature dependent properties. The heating effect of the arc was usually modeled as a prescribed flux. Thermal stresses were not computed above some cut-off temperature. Alternative strategies for predicting transverse residual stresses in large multipass welds have reduced the plate to essentially a 1D boundary condition with the weld passes included as a series of layers [10].

The 2D cross-sectional analyses implicitly ignore more than the regions on either end of a weld where the quasi-stationary steady state does not exist. They eliminate the longitudinal interaction with subtle consequences. The 2D plane stress analyses and experimental measurements of Andersson [11] may be the only examinations of this effect. They showed that for stresses, there were surprisingly large regions on the ends where the stresses differed from the values found in the middle.

In the 1980s, Karlsson [12] analyzed the residual stress in welding large plates including the effects of tack welds. Goldak et al [13] introduced the distributed power density heat source model to decouple the weld pool from the solid. Argyris [14] performed a visco-plastic stress analysis of a weld. They first performed an in-plane thermal analysis. Then they coupled this with a cross-sectional thermal analysis and a plane strain stress analysis. Goldak et al published the first 3D transient coupled temperature and stress analysis of a weld [15]. Karlsson et al [16] analyzed the stresses in a girth weld of pipe also using a full 3D transient analysis, coupling temperature and stress. Using shell elements, Lindgren and Karlsson also analyzed the residual stress in the girth weld of a pipe [17].

The addition of filler metal has been modeled by Tekriwal and Mazumder [18]. Basically, the added metal is included in the original mesh. To add metal the elements are turned on. This is accomplished either by bringing them to life or by a penalty method that changes the specific heat in a thermal analysis or the Young's modulus in a stress analysis. This is a relatively crude model of a multipass weld. Ohji et al add filler metal with a weld pool model that includes surface tension, hydrostatic pressure, arc flux and arc pressure but no velocity in the weld pool [19]. This is the most comprehensive weld model short of solving the Navier-Stokes equations in the weld pool.

Multi-pass welds have been analyzed by Ueda [20], Rybicki [21] and Leung [22]. To reduce the cost of separate analyses for each pass, several passes have often been lumped together in different ways. In some, only the last pass in a specific layer is analyzed [20]. In others, it would appear that the volume of the weld deposit of several passes or layers of passes are lumped together and the thermal history of a single pass located in the middle of the deposit is imposed during the stress analysis [21]. Another technique, lumps the thermal histories from several passes together, the temperature at any point, at any instant in time being the greatest value from any pass. All passes in a single layer, except the last, are lumped together. The last pass in any layer is treated separately [22]. Lumping layers together is inadvisable with the possible exception that the lumped layers remain a small proportion of the total thickness. For extremely large numbers of passes even these techniques may not be adequate.

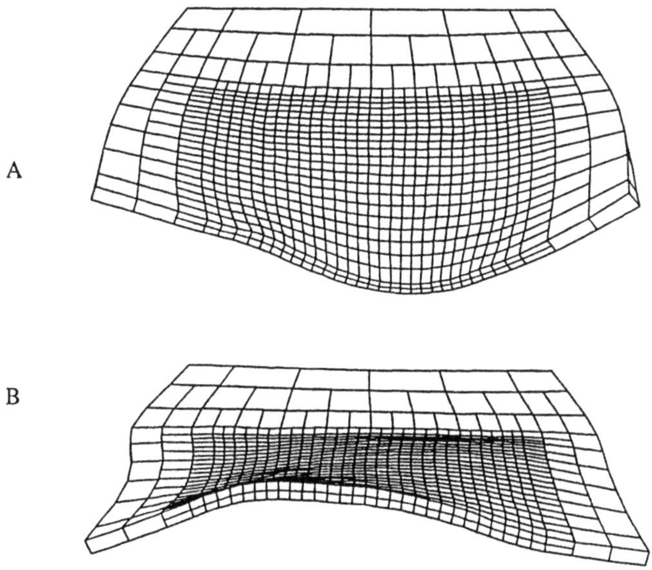

Fig. 6. Deformed shapes of multi-pass, resurfaced, hydroelectric turbine blades; Superimposed residual stress patterns. A: longitudinal patch sequence. B: transverse patch sequence. The distortion is magnified 100 times.

A somewhat more speculative approach was used by the authors to analyze the resurfacing of a thin plate (hydroelectric turbine blade) that involved several hundred passes. Each pass was 10 cm long, groups of 20 passes were done sequentially to cover 10 cm x 10 cm rectangular surface patches. The orientation of each patch, and the patch sequence were varied in order to minimize the deformation. The residual stress pattern of the multipass case was created by superimposing the residual stress pattern of each individual weld and taking the largest value from any individual pass. By building up the stress pattern for each patch and sequentially applying this pattern to the plate it was possible to obtain good qualitative agreement in both the deformed shape and optimum patch sequence. The danger in all these lumping techniques comes from the accuracy with which they model the sequencing effect.

Post-weld heat treatment (PWHT) is a technique frequently used to reduce the residual stresses caused by welding. Two processes are active. The first to occur is a reduction in yield strength during heating. For many steels the yield strength decreases rapidly above approximately 400 ^{o}C. The second is the time dependent creep strain that occurs if the body is held at an elevated temperature. Which effect dominates depends on the temperature and time. The first effect can be included in any thermal stress analysis that includes temperature dependent mechanical properties. The second is more complicated since the constitutive model now contains strain rate dependent terms. The numerical integration of these strain rates must be done with care. There have, nevertheless, been several publications which included both processes [20,23].

A related problem is reheat cracking. If some regions soften faster than others during heating, then the plastic strains may be concentrated in a particular region. Coupled with the fact that the ductility depends on the thermal history and may also not be uniform it becomes apparent that localized material failure may occur. To the authors' knowledge, only one group has ever attempted to predict this phenomenon [24]. A scalar damage factor was used to predict element failure. In the future, more of this type of analysis will be seen especially as continuum damage models improve.

Continuum Mechanics of Welds

The continuum mechanics as described in general texts such as Malvern [25] and Gurtin [26] can and should be applied directly to welds. The aspects of continuum mechanics that are peculiar to welds are the richness of the interactions between various phenomena, the range of temperatures and the range of length scales.

Meshing: Graded, Adaptive and Dynamic:

The need to resolve very fine detail near the weld while including large volumes has been a factor in the authors' pursuit of graded and adaptive meshes. The analysis is performed using the standard methods of nonlinear transient FEM. Near the weld, a fine mesh is needed for adequate accuracy.

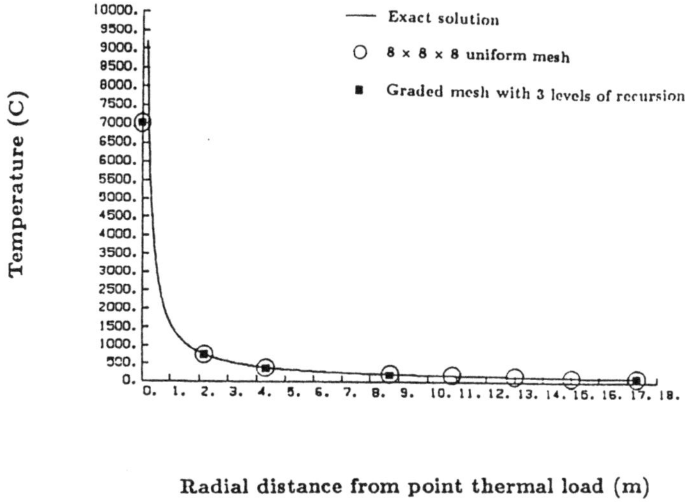

Fig. 7 Temperature distribution in a point loaded semi-infinite body, uniform and recursively graded meshes [27]

Far from the weld a coarse mesh provides adequate accuracy and a fine mesh would be wasteful. McDill et al [27] showed that for a point load on a cube, the use of graded elements would reduce computational complexity of direct solvers from $O(n^7)$ for a uniform mesh to $O(\log_2 n)$ for a graded mesh with no reduction in accuracy (Fig. 7). Here n is the number of degrees of freedom (dofs) along the edge of the cube with the uniform mesh. This grading has done much to make 3D analyses of welds feasible - even on low cost workstations.

The accuracy of any FEA is to a great extent determined during the mesh generation. The results obtained are, in some respects, the least squares projection onto the finite dimensional subspace described by the mesh. Better answers with lower computation effort could be obtained if the mesh were adaptively altered during the analysis. Adaptive mesh management and automatic finite element techniques are a rapidly maturing technology. Dynamic meshing was proposed by Goldak et al. [28], and implemented by

McDill et al. [29]. Gu et al. [30] implemented mixed meshing with brick elements near the weld pool and shell elements far from the weld pool (Fig. 8) . The creation of shell elements and shell-brick transition elements was done automatically. A posteriori error estimates to drive adaptive meshing were developed by Mashaie [31]

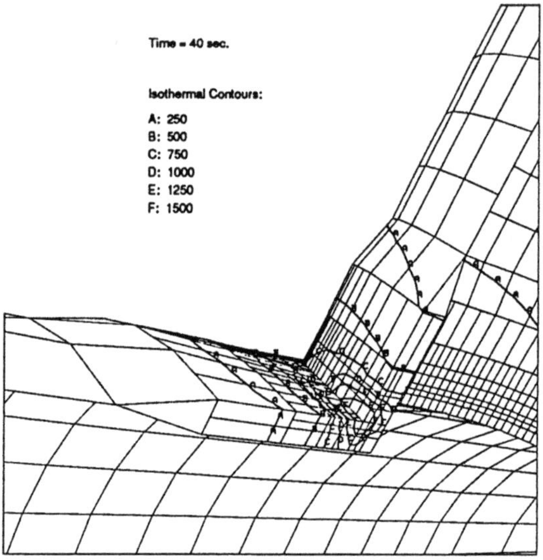

Fig. 8 Mixed meshing with brick elements near the weld pool and shell elements at a distance [30].

Heat Transfer Analysis

Ideally, the heating effect of the arc should be modeled by solving the Boltzmann equations coupled to the fluid mechanics of the weld pool. This is a problem of extreme difficulty from mathematical, numerical, computational and experimental viewpoints. The flow patterns inside the molten pool are the result of the contending forces of gravity, buoyancy, surface tension, electromagnetic force, droplet momentum and surface drag from gas flow. These are in turn affected by the arc voltage, current, shielding and chemistry. Moreover the coupling between the arc and weld pool is strong.

The best known studies of fusion zone size and shape are those of Christensen et al. [32]. Dimensionless nugget characteristics such as the width

$$B' = \frac{Bv}{2\alpha} \tag{1}$$

were related to a dimensionless operating parameter.

$$n = \frac{qv}{4\pi\alpha^2(h_c - h_o)} \tag{2}$$

Since that time there have been other empirical [33] and semi-empirical attempts [34] at predicting fusion zone shape. Others have attempted to predict the shape of the fusion zone without attempting to predict the fluid flow by modifying the thermal analysis [35,36].

In recent years there has been considerable progress towards the numerical prediction of the flow patterns inside molten pool and thus the fusion zone size and shape [37,38]. Practical application of these techniques is hampered somewhat by our ignorance of the conditions inside the arc and material properties at high temperature, especially the effects of trace elements on the surface tension gradient. This is a good example of an increasingly common problem; our analytical capabilities sometimes exceed our knowledge of basic material properties and boundary conditions.

If the size and shape of the weld are known or can be estimated from experiment, the weld pool can be removed from the domain of the energy equation by prescribing the temperature of the solidus line. Alternatively, if the power density distribution in the weld pool is known or can be estimated, it can be prescribed in the energy equation. Either method effectively decouples the arc physics and fluid mechanics of the weld pool from the solid mechanics of arc welds. In a 3D analysis with a Lagrangian frame of reference, prescribing the power density may be more advantageous since it permits large, arbitrary length time steps and non-linear weld paths. Prescribed temperature sources are probably more effective in Eulerian reference frames or for complex weld pool shapes [39]. When models of weld pools and/or arcs are able to predict the size and shape of weld pools, this method will enable them to be coupled simply and efficiently to the solid mechanics of the arc welds.

In solving the heat equation, the thermal conductivity and specific heat are temperature dependent. To date the thermal effects of the latent heat of phase transformations has been modeled by modifying the specific heat or enthalpy as a function of temperature. The transformation rates computed by metal physics to evaluate the rate of evolution of latent heats have not yet been incorporated in the thermal analysis of welds. Since this is being done in castings where the effect of latent heats are dominate [40], this should no longer pose serious difficulty.

Thermal Stress Analysis

The basic equations are the conservation of linear momentum, angular momentum, the constitutive equation and the compatibility or continuity equation. The difficulties arise almost entirely in decomposing the deformation into a translation, rotation and stretch on the one hand and then decomposing the deformation into contributions due to elasticity, plasticity, thermal expansion, transformation plasticity and creep.

Metal Physics of Welds in Steels

At a macroscopic level, the metal physics can be described by specifying the fraction of each phase present, the composition of each phase, and the grain size of each phase. In low alloy steels, the phases of greatest interest are ferrite, pearlite, austenite, bainite, martensite, carbide and liquid. Strictly speaking, pearlite is not a phase in a thermodynamic sense but a microstructure consisting of layers of a ferrite phase and a iron carbide phase. However, in the metallurgical literature the term phase is used to describe pearlite.

Most steels used in welded structures are produced with a ferrite - pearlite microstructure or a ferrite and austenite-martensite constituent. When these steels are heated above the A_{e1} or eutectoid temperature, the pearlite or austenite-martensite constituent rapidly forms austenite with the same composition. Even though the heating rates are high, superheating only amounts to a few tens of degrees, approximately 80 ^oC is common. So the assumption that the transformation occurs at equilibrium can be justified. However, the homogenization of this high carbon austenite is slower. It is a function of the temperature and time. Easterling, Ashby and Li have studied this in detail for laser heat treating surfaces [41].

The growth in the grain size of the austenite that forms can be described as a function of temperature and time. However, growth will not begin until carbo-nitrides such as V and Nb dissolve. Easterling et al. used the equilibrium dissolution temperature for V and Nb to begin grain growth of austenite. TiN does not dissolve but it does coarsen. Ion et al applied Lipschitz, Slyozow and Wagner theory of coarsening to TiN and the Zener theory of grain growth with pinned boundaries to estimate grain growth with TiN [42].

On cooling, the decomposition of austenite begins at the A_{e3} temperature. The first phase to form is ferrite. If possible, pearlite, bainite and martensite will follow. Several strategies have been tried for the prediction of anisothermal phase transformation kinetics. One such is that described by Leblond et al.[43]. The transformation rate is computed from two temperature dependant material parameters, $z_{eq}(T)$ the equilibrium fraction and $\tau(T)$ a characteristic time for the transformation.

$$\dot{z} = \frac{z_{eq}(T) - z}{\tau(T)} \qquad (3)$$

For constant properties the solution of this is:

$$z(t) = z_{eq}\left(1 - e^{\frac{-t}{\tau}}\right) \qquad (4)$$

These parameters may be estimated from CCT diagrams but a distinct CCT diagram is needed for each grain size in each alloy.

Another method, that adopted by the authors, describes the kinetics of austenite decomposition by ordinary differential equations (ODE) based on the work of Kirkaldy et al. [44]. Austenite grain growth is computed according to

$$\frac{dG}{dt} = \frac{1}{2G} k e^{\frac{-Q}{RT}} \qquad (5)$$

Transformation kinetics are computed according to ODEs of the form:

$$\frac{df}{dt} = B(G,T)f^m(1-f)^p \tag{6}$$

The ODEs contain terms which reflect the influence of grain size, undercooling, the alloy and temperature dependence of the diffusivity, and the phase fractions present. This is essentially the model developed by Watt et al. [45]. It was used to compute the microstructures in the HAZ of the problems analyzed in [44].

The hardness of the individual phases can be estimated by empirical equations of the form :

$$H_M = 127 + 949(C) + 27(Si) + 11(Mn) + 8(Ni) + 16(Cr) + 21(\log V) \tag{7}$$

and the hardness of the mixture was estimated using rules of mixtures proposed by [46].

In both methods the effect of stress on the kinetics of austenite decomposition has been neglected. Pearlitic transformation kinetics can be strongly affected by stress and plastic deformation [47].

In multi-pass welds, each successive pass tempers and softens previous passes. It has recently been found that, for some steels at least, the degree of softening during tempering can be quite conveniently described by a single relation [48]

$$\frac{H_o - H}{H_o - H_\infty} = Ae^{\frac{LMP}{B}} \tag{8}$$

In the fusion zone, the microstructure is more complex [49]. On solidification, the liquid transforms to either delta-ferrite or austenite. The liquid solid interface can be planar, columnar or dendritic. The interstitial atoms such as carbon have sufficient time at high temperature for diffusion to homogenize their distribution in the solid but the substitutional atoms do not. The distribution of substitutional atoms in the solid is usually not uniform in the fusion zone. This non-uniform distribution makes the behavior of the fusion zone more complex than that of the HAZ. In addition the strong texture evident in the fusion zone [50] may mean that the common assumption of isotropic thermal and mechanical properties is not valid.

Transformation Plasticity

Transformation plasticity is a phenomenon in which a stressed body undergoing a phase transformation can exhibit permanent deformations which are finite in amount, independent of time, and proportional to the stress, even for stresses much lower than the yield strength.

It has long been known that an additional finite plastic strain occurred when a phase transformation occurred under an applied stress. Greenwood and Johnson [51], and others, showed that the additional, finite, deformation could not be accounted for by creep in either phase. Since the stress was constant and well below the yield strength of all phases present in their experiment, it could not be a classical plastic strain. They argued their experimental results showed, and offered a theory that, the extra plastic strain was caused by the combination of a phase transformation and an applied stress. Leblond presented an elegant theory that invoked a microscopic stress field with spatial frequency terms less than the grain size and a macroscopic stress field with much lower spatial frequency terms [52]. Simply put, the volume change caused by a phase transformation of microscopic volumes can induce local plastic deformation. The interaction between stresses created in these microscopic volumes by the transformation and any macroscopic stress field leads to an irreversible plastic strain. In general transformation plasticity can be described by relations of the form:

$$\dot{\varepsilon}_{ij}^{TrP} = KS_{ij}\dot{z} \tag{9}$$

K depends on the fraction transformed, yield strength and transformation volume change.

Empirical relations describing transformation plasticity have been used extensively as have numerous very crude and unrealistic treatments which artificially lower the yield strength during phase transformations. The first detailed analytical model of transformation plasticity, which specifies the contribution of the rate of a phase transformation to the total strain rate, was formulated elegantly by Leblond [53,54]. By assuming the Hill-Mandel homogenization theorem, Leblond derived formulae for the transformation plastic strain rate that are consistent with fundamental theory of continuum mechanics and agree well with experiment.

The effect of transformation plasticity in welds was analyzed by Dubois et al. [55]. Oddy et al. performed the first 3D thermal stress analysis of a weld including the transformation plasticity that couples the phase transformations to the thermal stress analysis [56]. By analyzing a problem similar to that of Hibbitt and Marcal, they provided a qualitative explanation for the anomalies seen in the experimental stress measurements.

To include phase transformations in the analysis of a weld, there are two problems. The macroscopic response of multi-phase material to micro-stresses is different from homogeneous material. Leblond [52] has shown that such material does not exhibit a distinct yield point even when the transformation rate is zero. This is quite separate from transformation plasticity. When the transformation rate is not zero the additional transformation plasticity strain rate is:

$$\dot{\varepsilon}_{ij}^{TrP} = KS_{ij}\dot{z} \tag{10}$$

This strain rate must be integrated to provide an increment in the strain

$$\Delta\varepsilon_{ij}^{TrP} = \int_{{}^{n}t}^{{}^{n+1}t} \dot{\varepsilon}_{ij}^{TrP}\, dt \tag{11}$$

The strain increment is frequently approximated as

$$\Delta\varepsilon_{ij}^{Trp} \approx {}^{\tau}\dot{\varepsilon}_{ij}^{Trp}\Delta t \tag{12}$$

where

$${}^{n}t < \tau < {}^{n+1}t$$

The integration of the transformation plastic strain rate in this manner can lead to instability unless steps, such as subcycling, are taken.

The first analyses assumed the displacements, rotations and strains were infinitesimal. The first large displacement, large rotation, small strain, 3D stress analysis of a weld may have been performed by Oddy in 1988 [57].

Beyond the purely technical aspects of describing the mechanics in mathematical terms there is the important but often neglected task of implementing the numerical techniques efficiently. This is the domain of software engineering.

In the past, solution of the large matrices has relied on the use of direct solvers. As the number of degrees of freedom (dof) increase, direct solvers are no longer the most efficient. For even what are now fairly modest sized problems, indirect solvers can provide an order of magnitude improvement in the solution time. As problem sizes increase into the range of 10^5 dof for thermal analyses and 10^4 dof for stress problems the advantage of indirect solvers becomes even more pronounced. The authors currently use an element-by-element, preconditioned conjugate gradient solver. Thermal analyses with 10^5 dof are now routinely solved on low-cost workstations in a few hours.

The other major task for software engineering has been in the design of an effective, flexible user interface. In the past, creation of a FEM mesh for a complex problem often required a week of effort by someone proficient in the use of a mesh generator. This and the additional specialized knowledge required to do the thermal and stress analysis of welds effectively has ensured that few practising engineers have made use of the technology.

Mathematics of Weld Mechanics

We begin by defining the structure to be welded as a domain in Euclidean space. Imposing conservation of energy and momentum on that domain pointwise leads to partial differential equations and associated constitutive equations, boundary conditions and initial conditions. Formulating the conservation laws as integral equations over appropriate Sobolev spaces relaxes the restrictions on the partial derivatives. Finally, formulating the integral equations in finite or discrete subspaces leads to the usual finite element formulation, which can be interpreted as projecting the exact solution in the infinite Sobolev spaces onto the finite subspaces. The error in the finite element solution is that part of the exact solution that is orthogonal to the finite subspace. To fix definitions and notations, we briefly state these equations. Detailed proofs and derivations can be found in many books.

Given an open domain in euclidean space, $\Omega \in \Re^3$, with a boundary that is its closure , $\Omega \cup \partial\Omega = \overline{\Omega}$. The boundary is partitioned into sets to which the essential and natural boundary conditions are applied such that $\partial\Omega_e \cup \partial\Omega_n = \partial\Omega$ and $\partial\Omega_e \cap \partial\Omega_n = \emptyset$. Essential boundary conditions (temperature or displacement) are applied to $\partial\Omega_e$ and natural boundary conditions (flux or traction) are applied to $\partial\Omega_n$. The boundary is assumed to be sufficiently smooth, e.g. Lipschitzian.

Partial Differential Equations of Weld Mechanics

Conservation of energy is described by

$$\nabla \bullet \mathbf{j} + \dot{H} + \sigma \bullet \dot{\varepsilon} + Q = 0 \tag{13}$$

$$\mathbf{j} = -\mathbf{k}\, \nabla T \tag{14}$$

the constitutive equation defines the thermal conductivity tensor $\mathbf{k}$. It is symmetric positive definite.

$$\dot{H} = c_p \dot{T} + L_i f_i \tag{15}$$

constitutive equation defines the specific heat and latent heat of the ith transformation. c_p and L_i are strictly positive.

$$q(\mathbf{x}, t) = \mathbf{J} \bullet \mathbf{n}; \; \mathbf{x} \in \partial\Omega_n \tag{16}$$

The internal flux, a vector, is the linear operator that maps the outward normal onto the external flux, a scalar.

$$T(\mathbf{x}, t) = \mathbf{J} \bullet \mathbf{n}; \qquad \mathbf{x} \in \partial\Omega_e \tag{17}$$

This is the essential essential boundary condition.

Conservation of linear momentum is described by

$$\nabla \bullet \sigma + b = m\ddot{x} + c\dot{x} \qquad \sigma \bullet n = \tau \tag{18}$$

Inertia and damping terms are usually neglected in welding, enabling time varying load, rather than full dynamic analyses to be performed.

$$\sigma = \mathbf{D}\varepsilon \tag{19}$$

the constitutive equation defines the tensor $\mathbf{D}$ that maps strain onto stress.

$$\varepsilon = \mathbf{B}\mathbf{u} \tag{20}$$

is the continuity or compatibility equation.

Weak Formulation for Weld Mechanics

We begin by defining the Sobolev spaces of kinematically admissible inhomogeneous and homogeneous displacements.

$$\aleph = \left\{ \mathbf{u} \mid \mathbf{u} \in \left[H^1(\Omega) \right]^3 \mathbf{u} \big|_{\partial\Omega_*} = \bar{\mathbf{u}} \right\} \tag{21}$$

$$\aleph_0 = \left\{ \mathbf{u} \mid \mathbf{u} \in \left[H^1(\Omega) \right]^3 \mathbf{u} \big|_{\partial\Omega_*} = 0 \right\} \tag{22}$$

The spaces of kinematically admissible inhomogeneous and homogeneous displacements are

$$\Sigma = \left\{ \sigma \mid \sigma \in L^2(\Omega), \qquad \sigma = D[Bu], u \in \aleph \right\} \tag{23}$$

$$\Sigma_0 = \left\{ \sigma \mid \sigma \in L^2(\Omega), \qquad \sigma = D[Bu], u \in \aleph_0 \right\} \tag{24}$$

The displacement finite element method leads to the following equation.

$$\int_\Omega v^T B^T DBu \, d\Omega - \int_\Omega v^T b \, d\Omega - \int_{\partial\Omega} v^T \tau \, d\Gamma = 0 \quad \forall \, v \in \aleph_0 \tag{25}$$

For the finite space $u_h \in \aleph_h \in \aleph$, eq. 25 becomes

$$\int_\Omega {v_h}^T B^T DBu_h \, d\Omega - \int_\Omega {v_h}^T b \, d\Omega - \int_{\partial\Omega} {v_h}^T \tau \, d\Gamma = 0 \quad \forall \, v_h \in \aleph_{0h} \in \aleph_0 \tag{26}$$

Equation 26 has a unique solution. Furthermore this solution satisfies eq. 19 and 20 pointwise but only satisfies eq. 18 in the weak sense. Specifically, the nodal forces are in equilibrium at each node.

The spaces of inhomogeneous and homogeneous statically admissible stresses are

$$X = \left\{ \sigma \mid \sigma \in H(\mathrm{div}, \Omega), \, div \, \sigma + f = 0 \text{ in } \Omega, \quad \sigma \bullet n = \tau \text{ on } \partial\Omega_n \right\} \tag{27}$$

$$X_0 = \left\{ \sigma \mid \sigma \in H(\mathrm{div}, \Omega), \, div \, \sigma + f = 0 \text{ in } \Omega, \quad \sigma \bullet n = 0 \text{ on } \partial\Omega_n \right\} \tag{28}$$

The statically admissible stress field field satisfies eq. 18 and its natural boundary condition, pointwise.

A similar formulation can be presented for the energy equation. This formulation for FEM is well established for linear elasticity with small displacements, rotations and strains. For large displacements, rotations and plasticity, visco-plasticity and/or creep, rapid progress has been made in the past decade.

Kinematics, Deformation and Strain

The analysis of large displacements, rotations and strains requires a more rigorous mathematical description than that traditionally used in most analyses. Finite deformation algorithms require specific choices to be made of the discretization model, the deformation description and the stress rate. For our analyses we have chosen an Updated Lagrangian discretization model as most appropriate for incremental problems of this nature. The choice of deformation description and stress rate are not completely arbitrary. Two requirements must be met. First the stress and strain measures used must be conjugate. Second the stress rate must be objective or frame invariant.

Rotation tensors are obtained at the midpoint and end of the step from the polar decomposition of the deformation gradients.

$$ {}^{n+1/2}\mathrm{X} = \frac{\partial^{n+1/2}x}{\partial^{n}x} \qquad (29)$$

$$ {}^{n+1/2}\mathrm{X} = {}^{n+1/2}R\,{}^{n+1/2}U \qquad (30)$$

Because of the unacceptable limitations of the Jaumann stress rate [58,59] the authors have chosen to use the centred strain and Green-Naghdi stress rate as the deformation description and stress rates respectively.

$$ \Delta\varepsilon_{ij}^{Tot'} = \frac{1}{2}\left(\frac{\partial u_i}{\partial^{n+1/2}x_j} + \frac{\partial u_j}{\partial^{n+1/2}x_i}\right) \qquad (31)$$

Since this is in the midpoint orientation it must be rotated to the initial orientation.

$$ \Delta\varepsilon^{Tot} = {}^{n+1/2}R^{T}\,\Delta\varepsilon^{Tot'}\,{}^{n+1/2}R \qquad (32)$$

The constitutive relations are evaluated in the initial orientation and the resulting value of the final stress in the initial orientation is then rotated onto the final orientation.

$$^{n+1}\sigma = {}^{n+1}R\left({}^{n}\sigma + \Delta\sigma\right){}^{n+1}R^{T} \tag{33}$$

Constitutive Equations

In a time step, each point in the domain and its associated neighborhood undergoes a deformation that can be decomposed into a rigid body translation, and the product of a rotation and a stretch. The stretch can be further decomposed into volumetric and deviatoric components. The volumetric deformation can be decomposed into elastic, thermal and phase transformation contributions. The deviatoric part can be decomposed into elastic, plastic, creep, visco-plastic and transformation plasticity contributions. What is the best way to decompose the deformation? This is a critical question and a great deal of progress has been made in the past decade. Traditional treatments use an additive decomposition of the strain rates. The constitutive model for temperature dependent properties becomes

$$\Delta\varepsilon^{Tot} = \Delta\varepsilon^{El} + \Delta\varepsilon^{Pl} + \Delta\varepsilon^{Th} + \Delta\varepsilon^{TrV} + \Delta\varepsilon^{TrP} + \Delta\varepsilon^{Cr} \tag{34}$$

$$\dot{\sigma} = D\dot{\varepsilon}^{El} + \dot{D}\varepsilon^{El} \tag{35}$$

$$\Delta\sigma = D'\left(\Delta\varepsilon^{Tot} - \Delta\varepsilon^{Pl} - \Delta\varepsilon^{Th} - \Delta\varepsilon^{TrV} - \Delta\varepsilon^{TrP} - \Delta\varepsilon^{Cr}\right) + \Delta D^{n}D^{-1\,n}\sigma \tag{36}$$

where

$$D' = {}^{n+1/2}D + \frac{{}^{n+1}D - {}^{n}D}{2} \tag{37}$$

More recent derivations of elasto-plastic problems have begun to use a multiplicative decomposition of the deformation gradient tensor [60].

Some care must be taken when choosing the order of the various strain fields. It has been known for some time that the order of the thermal strain fields and mechanical fields must be matched [9, 61]. The thermal strain comes from the temperature field whereas the mechanical strains are the spatial derivative of the displacements. It is necessary to ensure that the temperature field used to compute the thermal strains is one order lower than that of the displacements. In some cases this is achieved through the use of higher order elements in the stress analysis than in the thermal analysis [9, 61]. In other cases, the use of linear grading elements in the stress analysis has required that, so far as the stress analysis is concerned, only isothermal elements are present. If this requirement is not observed then serious errors occur in the stresses predicted in the fusion zone (FZ) and heat-affected-zone (HAZ) [62]. It is often difficult to determine whether or not this restriction has been observed.

What class of tensor fields can be valid constitutive equations? This is one of the most difficult mathematical issues in mechanics. To our knowledge, all weld mechanics analyses performed to date have been hypo-elastic. They have used either elasto-plastic or visco-elasto-plastic constitutive equations. The elasticity tensor for isotropic solids is a function of the Young's modulus and Poisson's ratio. The yield strength is defined from the effective stress-effective strain relationship obtained from uniaxial tension tests. A von Mises yield function and associated flow rule are used. Both isotropic and kinematic hardening have been used. The coefficients involved are often functions of temperature and effective plastic strain.

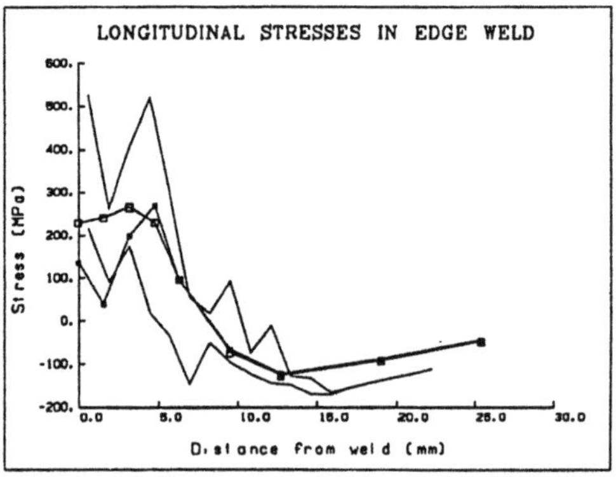

Fig. 9 Residual stress fields computed with consistent (open squares) and inconsistent strain fields (closed squares). Solid lines represent maximum and minimum ranges of measured residual stresses [62].

The inclusion of creep, like transformation plasticity, requires that a stress dependent strain rate be integrated over each increment. This nonlinear, transient response, if not integrated accurately, leads to predictions which can be seriously in error both locally and globally. Integration in a multidimensional strain space is a challenging task. The effective stress function (ESF) algorithm of Kojic and Bathe [63] appears to be stable, accurate and effective method of analyzing elasto-plastic-creep problems. The ESF algorithm reduces the constitutive relations to the iterative solution of single nonlinear equation equation with one variable. For isotropic plasticity the relation is:

$$a^{2\,t+\Delta t}\bar{\sigma}^2 + b^{\tau}\gamma - c^{2\,\tau}\gamma^2 - d^2 = 0 \tag{38}$$

where

$^{t+\Delta t}\bar{\sigma}$ is the effective stress

$^{\tau}\gamma$ is a scalar creep factor

a is a function of the elastic, plastic and creep properties

b is a function of the strain increment and the initial deviatoric stress

c is a function of the time step size and initial effective stress

d is a function of the strain increment

The effective stress function method may also be used in the evaluation of elasto-plastic-creep tangent constitutive matrices.

Alternative constitutive models do exist. Bammann [64] advocates a microstructurally motivated model for the plastic spin. This model has been coupled with an improved continuum damage model [65]. The resulting predictions of ductile failure showed very good agreement with experiment. In the future, improved constitutive and continuum damage models like these will allow better predictions of weld cracking.

Algorithm for Thermal Stress Analysis

Briefly, a FEM step computes a trial displacement. The thermal expansion from the temperature field is usually the main load term. Given this displacement field, the constitutive equation is used to determine the stress at each Gauss point. From this stress and the external loads, a contribution to the residual is computed at each Gauss point. Using this residual a new displacement field is computed. This iterative procedure is repeated until convergence is attained.

Conclusion

It is now possible to compute the transient temperature field in welds with an accuracy that is limited primarily by our knowledge of the input data. The input data include geometry, thermal properties, initial temperature distribution and boundary conditions including the heating effect of the arc. Such computations can now be performed on low cost workstations in a few hours.

The stress in welds can be computed with accuracy comparable to experimental measurements of residual stress. Computing stress in welds is roughly ten times more expensive than computing temperatures. Our knowledge of the stress - strain relationship at high temperatures and the microstructure evolution, are thought to limit the accuracy of the computed stress.

The microstructure evolution in the HAZ of low alloy steels can be computed by integrating ordinary differential equations obtained from modern metal physics. While the first results of such computations are promising, more comparisons of computed microstructures with experiment are needed to assess the accuracy of such computations.

It is expected that creep analyses will become common in the near future. New constitutive models will allow better prediction of multiaxial plasticity. In the longer term, the analyses will focus on fracture and the interaction between the atomic, microstructure and macro-models. Continuum damage models will become more commonly used. The interactions between microstructure, stress, strain and mass diffusion will provide a particularly rich world for research in theoretical and applied mechanics. Another major challenge will be the increased experimental efforts needed to measure more accurately the material properties and those parameters needed to validate the results.

Acknowledgements

The financial support of the National Science and Engineering Research Council (NSERC A2992) is gratefully acknowledged.

References

1. Goldak, J.A., Nguyen, C.S., A Fundamental Difficulty in Notch Toughness Testing Narrow Welds and HAZ , Weld. J., Research Suppl., April 1977, pp. 119s-125s.
2. Holden, T.M., Root, J.H., Holt, R.A., Roy, G., Neutron Diffraction Measurements of the Residual Strain State of a Tubular T-Joint, Proc. 7th Int. Conf. Offshore Mech. and Arctic Eng., Houston, Feb. 1988, pp. 127-131.
3. Rosenthal, D., The Theory of Moving Sources of Heat and Its Application to Metal Treatments, Trans ASME, 1946, Vol. 68, p. 849-865.
4. Okerblom, N. O., The Calculations of Deformations of Welded Metal Structures, London, Her Majesty's Stationery Office, 1958.
5. Vinokurov, V. A., Welding Stresses and Distortion, The British Library Board, 1977.
6. Hibbitt, H.D., Marcal, P. V., A Numerical Thermo-Mechanical Model for the Welding and Subsequent Loading of a Fabricated Structure, Comp. & Struct., Vol. 3, 1975, pp. 1145-1174.
7. Freidman, E., Analysis of Weld Puddle Distortion, Welding Journal Research Suppl., 1978, pp. 161s-166s.
8. Masubuchi, K., Analysis of Welded Structures, Pergamon Press, 1980.
9. Andersson, B.A.B., Thermal Stresses in a Submerged-Arc Welded Joint Considering Phase Transformations, J. Eng. & Tech., Trans. ASME, Vol. 100, 1978, pp. 356-362.
10. Leggatt, R.H., Residual Stresses and Distortion in Multipass Butt-Welded Joints in Type 316 Stainless Steel, Int. Conf. on Residual Stresses, Garmisch-Parten Kirchen, Oct., 1986.
11. Andersson, B., Karlsson, L., Thermal Stresses in Large Welded Plates, J. Therm. Stresses, Vol. 4, 1981, pp. 491-500.
12. Karlsson, L., Thermal Stresses in Welding, Thermal Stresses I, Ed. R. B. Hetnarski, Elsevier, 1986, pp. 300-389.
13. Goldak, J. A., Chakravarti, A. and Bibby, M. J., A New Finite Element Model for Welding Heat Sources, Trans. AIME., Vol. 15B, June, 1984, pp. 299-305.
14. Argyris, J.H., Szimmat, J., Willam, K.J., Computational Aspects of Welding Stress Analysis, Comp. Meth. Appl. Mech. Eng., Vol. 33, 1982, pp. 635-666.
15. Goldak, J., Oddy, A., McDill, M., Bibby, M.J., House, R., and Chakravarti, A.P., Progress in Computing Residual Stress and Strain in Welds, Ed. S. A. David, Conf. Proceedings International Trends in Welding Research, Gatlinburg, Tennesse, U.S.A., May 18-22, 1986, pp 1-6
16. Karlsson, R. I. and Josefson, B. L., Three Dimensional Finite Element Analysis of Temperature and Stresses in Single-Pass Butt Welded Pipe, accepted for publication, J. Pressure Vessel Technology, Trans. ASME, 1987.
17. Lindgren, L., Karlsson, L., Deformations and Stresses in Welding of Shell Structures, Int. J. Num. Meth. Eng., Vol. 25, 1988, pp. 635-655.
18. Tekriwal, P. and Mazumder, J., Finite Element Modeling of the Arc Welding Process, Proceedings of an International Conference on Trends in Welding Technology, Gatlinburg, Tennessee, USA, May 18-22, 1986, ed. S. A. David, pp. 71-80.

19. Ohji, T., Ohkubo, A. and Nishiguchi, K., Mathematical Modeling of the Molten Pool in Arc Welding, IUTAM Symposium, Mechanical Effects of Welding, Ed. L. Karlsson, Lulea, Sweden, June 10-14, 1991.
20. Ueda, Y. and Murakawa, H., Applications of Computer and Numerical Analysis Techniques in Welding Research, Trans. of JWRI, Vol. 13, No. 2, 1984, pp. 337-346.
21. Rybicki, E.F., Stonisifer, R.B., Computation of Residual Stresses due to Multipass Welds in Piping Systems, Trans ASME, J. Press. V. Tech., Vol. 101, May, 1979, pp. 149-154.
22. Leung, C.K., Pick, R.J., Finite Element Analysis of Multi-Pass Welds, WRC Bulletin 356, Aug., 1990, pp. 11-33.
23. Josefson, B.L., Residual Stresses and their Redistribution During Annealing of a Girth-Butt Welded Thin Walled Pipe, Trans ASME, J. Press. Vess. Tech., Vol. 104, 1982, pp. 245-250.
24. Josefson, B.L., Local Annealing of a Multi-Pass Pipe, Computation of Stress Redistributions and Cracking Susceptibilities, 4 Nat. Conf. on Press. Vess. and Piping, Portland, 1983.
25. Malvern, L. E., Introduction to the Mechanics of a Continuous Medium, Prentice-Hall, 1969.
26. Gurtin, M. E., An Introduction to Continuum Mechanics, Academic Press, 1981
27. McDill, M. J., Goldak, J. A., Oddy, A., Bibby, M. J., Isoparametric Quadrilaterals and Hexahedrons for Mesh-Grading Elements, Comm. Applied Nu. Methods, Vol. 3, 1987, pp. 155-163.
28. Goldak, J. A., M. McDill, A. Oddy, R. House, X. Chi, M. J. Bibby, Computational Heat Transfer for Weld Mechanics, Proceedings of an International Conference on Trends in Welding Technology, Gatlinburg, Tennessee, USA, May 18-22, 1986, ed. S. A.David, pp. 15-20.
29. McDill, J.M.J., Oddy, A.S., Goldak, J.A., An Adaptive Mesh-Management Algorithm for 3-D Automatic Finite Element Analysis, Trans CSME, Vol. 15, No. 1, 1991, pp. 57-70.
30. Gu, M., Goldak, J., and. Haaland, K., Mixing Thermal Shell and Brick Elements in FEA of Welds, Proceedings of Offshore Mechanics and Arctic Engineering Conference, Stavanger, Norway, June 23-28, 1991.
31. Mashaie, A. Error Estimates for Finite Element Solutions of Elliptic Boundary Value Problems, Ph.D. thesis, Carleton University, 1990.
32. Christensen, N., Davies, V. de L., Gjermundsen, K., Distribution of Temperatures in Arc Welding, British Welding Journal, Vol. 12, 1965, pp. 54-75.
33. Chandel, R.S., Oddy, A.S., Goldak, J.A., Computer Prediction of Weld Bead Shapes, 2nd Int. Conf. Comp. Tech. in Welding, Abington, June, 1988
34. Shinoda, T., Masumoto, I., Prediction of Weld Geometries for CO_2 Butt-Welded Joints, Mat. Sci, & Tech., Vol. 5, 1989, pp. 293-298.
35. Pardo, E. and Weckman, D. C., Prediction of Weld Pool and Reinforcement Dimensions of GMA Welds Using a Finite Element Model, Met. Trans B., Vol. 20B, 1989, pp. 937-946.
36. Tekriwal, P., Stitt, M., Mazumder, J., Finite Element Modelling of Heat Transfer for Gas Tungsten Arc Welding, Metal Construction, Oct., 1987, pp. 599R-606R.

37. Kou, S., Wang, Y.H., Computer Simulation of Convection in Moving Arc Weld Pools, Met. Trans A, Vol. 17A, 1986, pp. 2271-2277.
38. Matsunawa, A., Yokoya, S., Asako, Y., Convection in Weld Pool and its Effect on Penetration Shape in Stationary Arc Welds, Trans. JWRI, Vol. 16, No. 2, 1987, pp. 1-8.
39. Gu, M., Goldak, J.A., Hughes, E., Steady State Thermal Analysis of Welds with Filler Metal Addition, to be published.
40. E. Hughes, W. Ma, D. Downey, and J. Goldak, Experience with Macro-Micro Modeling of Solidification in Castings, Proceedings of Numerical Simulation of Castings Solidification in Automotive Applications, East Lansing Mi, USA, May 1-2, 1991.
41. Ashby, M.F.and Easterling, K.E., A First Report on Diagrams for Grain Growth in Welds, Acta Met. Vol. 30, 1982, pp. 1969-1978.
42. Ion, J.C., Easterling, K.E. and Ashby, M.F., Acta Met., Vol. 32, 1983, pp. 1949-62.
43. Leblond, J.B., Devaux, J., A New Kinetic Model for Anisothermal Metallurgical Transformations in Steels Including Effect of Austenite Grain Size, Acta Met., Vol. 32, No. 1, 1984, pp. 137-146.
44. Henwood, C., Bibby, M.J., Goldak, J.A. and Watt, D.F., Coupled Transient Heat Transfer-Microstructure Weld Computations, Acta Metal., Vol 36, No. 11, 1988, pp. 3037-3046.
45. Watt, D.F., Coon, L., Bibby, M.J., Goldak, J.A. and Henwood, C., Modelling Microstructural Development in Weld Heat-Affected Zones, Acta Met., Vol 36, No. 11. 1988, pp. 3029-3035.
46. Buchmeyer, B. and H. Cerjak, Mathematical Description of HAZ Behavior of Low-Alloyed Structural Steels, Proc. International Conf. Improved Weldment Control with Special Reference to Computer Technology, Vienna, Austria, July 4-5, 1988, pp. 43-50.
47. Denis, S., Gautier, E., Sjostrom, S., Simon, A., Influence of Stresses on the Kinetics of Pearlitic Transformation During Continuous Cooling, Acta Metall., Vol. 35, No. 7, 1987, pp. 1621-1632.
48. Oddy, A.S., Chandel, R.S., Hardness Prediction for the Repair Welding of 2.25 Cr - 1.0 Mo Pressure Vessels, AWS Annual Meeting, Detroit, 1991.
49. Bhadeshia, H., Svensson, L. E., Gretoft, B., Prediction of Microstructure of the Fusion Zone of Multicomponent Steel Weld Deposits, Proceedings of an International Conference on Trends in Welding Technology, Gatlinburg, Tennessee, USA, May 18-22, 1986, ed. S. A. David, pp. 225-229.
50. Mahin, K.W., MacEwen, S., Winters, W.S., Mason, W., Kanouff, M., Fuchs, E.A., Evaluation of Residual Stress Distributions in a Traveling GTA Weld using Finite Element and Experimental Techniques, 4th Int. Conf. Modeling of Casting and Welding Processes, Palm Coast, April, 1988, pp. 339-350.
51. Greenwood, G.W., Johnson, R.H., The Deformation of Metals Under Small Stresses During Phase Transformations, Proc. Roy. Soc., Vol. 283, 1965, pp. 403-422.
52. Leblond, J.B., Mottet, G., Devauz, J.C., A Theoretical and Numerical Approach to the Plastic Behavior of Steels During Phase Transformations - I. Derivation of General Relations, J. Mech. Phys. Solids, Vol. 34, No. 4, 1986, pp. 395-409.
53. Leblond, J.B., Devaux, J., Devaux, J.C., Mathematical Modelling of Transformation Plasticity in Steel: Case of Ideal-Plastic Phases, Int. J. Plast., in press.

54. Leblond, J.B., Devaux, J., Devaux, J.C., Mathematical Modelling of Transformation Plasticity in Steel:, Int. J. Plast., in press

55. Dubois, D., Devauz, J., Leblond, J.B., Numerical Simulation of a Welding Operation: Calculation of Residual Stresses and Hydrogen Diffusion, Fifth International Conference on Pressure Vessel Technology, San Francisco, Vol. 11, 1984, pp. 1210-1239.

56. Oddy, A.S., Goldak, J. A., McDill, J. M. J., Transformation Effects in the 3D Finite Element Analysis of Welds, Proceedings of the 2nd International Conference on Trends in Welding Research, Gatlinburg, Tennessee, May 15-19, 1989.

57. Oddy, A.S., Three-Dimensional Finite Deformation, Thermal-Elasto-Plastic Finite Element analysis, Ph.D. Thesis, Carleton University, Ottawa, Canada, 1987.

58. Johnson, G.C., Bammann, D.J., A Discussion of Stress Rates in Finite Deformation Problems, Int. J. Solids Struct., Vol. 20, 1984, pp. 757-737.

59. Kojic, M., Bathe, K-J., Studies of Finite Element Procedures - Stress Solution of a Closed Elastic Strain Path with Stretching and Shearing using the Updated Lagrangian Jaumman Formulation, Comp & Struct., Vol. 26., No. 1/2, 1987, pp. 175-179.

60. Weber, G., Anand, L., Finite Deformation Constitutive Relations and a Time Integration Procedure for Isotropic Hyperelastic-Viscoplastic Solids, Comp. Meth. in Appl. Mech. and Eng., Vol. 79, 1990, pp. 173-202.

61. Josefson, B.L., Effects of Transformation Plasticity on Welding Residual-Stress Fields in Thin-Walled Pipes and Thin Plates, Matl Sci. and Tech., Vol. 1, Oct 1985, pp. 904-908.

60. Oddy, A.S., McDill, J.M.J., Goldak, J.A., Consistent Strain Fields in 3D Finite Element Analysis of Welds, J. Press. V. Tech., Trans. ASME, Vol. 112, No. 3, 1990, pp. 309-311.

61. Kojic, M. and Bathe, K. J., The Effective-Stress-Function Algorithm for Thermo-Elasto-Plasticity and Creep, International Journal for Numerical Methods in Engineering, Vol. 24, 1987, pp. 1509-1532.

62. Bammann, D.J.,, Modelling the Large Strain-High Temperature Response of Metals, Proc. of Modeling of Casting and Welding Processes IV, Ed. A. F. Giamei and G. J. Abbaschian, April 17-22, 1988, Pub. Eng. Foundation and TMS/AIME, pp. 329-338.

63. Bammann, D.J., Chiesa, M.L., McDonald, A., Kawahara,W.A., Dike, J.J., Revilli, D., Prediction of Ductile Failure in Metal Structures, ASME Symposium Dallas, Texas, AMD-Vol. 107, Nov. 25-30, 1990, pp. 7-12.

Very High Homologous Temperature Constitutive Models for Semi-Solid and Solid Metals

S. Brown, P. Kumar, and V. Dave

Department of Materials Science and Engineering
Massachusetts Institute of Technology
Cambridge, Massachusetts 02139, U.S.A.

Summary

Although the constitutive behavior of metals is well characterized for homologous temperatures below one-half the melting temperature, very little data exist for very high homologous temperatures and semi-solid or mushy state constitutive behavior. The deformation processes at high homologous temperatures consist of some combination of dislocation glide and recovery-controlled creep, but it is not clear whether one particular process dominates and therefore which constitutive model is most appropriate. This investigation presents constitutive behavior for both very high homologous temperature deformation and semi-solid deformation processes. The experiments assume an internal variable formulation where the flow equation is coupled with the evolution of microstructure via a set of coupled, first order differential equations. The semi-solid model incorporates a single internal variable to capture the effect of solid particle agglomeration. The high homologous temperature behavior includes both steady state and transient data. Data is presented for both lead and lead/tin systems, illustrating the proper experiments to determine the material functions in the models.

Introduction

The simulation of nonlinear, boundary-valued problems through finite element, finite difference, and other computational schemes has revealed significant shortcomings in the constitutive models used to represent mechanical behavior. This is particularly true in the case of molten, semi-solid, and high homologous temperature behavior. Increased computer speed and improved algorithms have reduced the computational difficulties of simulating highly nonlinear problems involving many degrees-of-freedom. Now that these difficulties have become less restrictive, the absence of appropriate constitutive data and models has become a subsequent barrier to accurate simulation.

More sophisticated constitutive models, however, pose additional difficulties. Internal variable models of the form,

$$\frac{d\epsilon_{ij}^{vp}}{dt} = \hat{f}_{ij}(\sigma_{kl}, T, s_1, \ldots, s_m) \tag{1}$$

$$\frac{ds_n}{dt} = \hat{g}_n(\sigma_{kl}, T, s_1, \ldots, s_m), \quad 1 \le n \le m \tag{2}$$

L. Karlsson, L.-E. Lindgren, M. Jonsson (Eds.)
Mechanical Effects of Welding
IUTAM Symposium Luleå/Sweden 1991

where

$$
\begin{array}{lll}
\epsilon_{ij}^{vp} & = & \text{viscoplastic strain,} \\
\hat{f}_{ij}(\ldots) & = & \text{flow function,} \\
\sigma_{kl} & = & \text{state of stress,} \\
T & = & \text{temperature,} \\
s_1, \ldots, s_m & = & m \text{ internal variables, and} \\
\hat{g}_n(\ldots) & = & \text{evolution equation for internal variable } n
\end{array}
$$

require very specific experiments to characterize unambiguously the material functions (or equations) $\hat{f}_{ij}(\ldots)$ and $\hat{g}_n(\ldots)$.

The following sections describe both experiments and models based on investigations of both high homologous temperature and semi-solid constitutive behaviors. The next section describes experiments on high purity lead at temperatures exceeding 0.85 T_m and suggests appropriate constitutive models. The following section presents a constitutive model for semi-solid behavior assuming a single internal variable that represents the agglomeration between solid particles.

High Homologous Temperature Behavior

Microstructural inhomogeneities and elevated temperature test conditions frequently complicate the experimental evaluation of high homologous temperature constitutive behavior. To avoid difficulties with grain boundary melting and high test temperatures, we selected very high purity lead (99.9999%). Compression specimens were machined with a height-to-diameter ratio of 1 to 1. Concentric, shallow end grooves were added to the specimens to hold lubricant. The lubrication was sufficient to give virtually homogeneous deformations for strains exceeding 50 percent true strain. Further details of this work are provided in Brown, Kumar, and Dave [1].

Steady State Response

Figure 1 presents representative constant true strain rate data for the lead at a homologous temperature of 0.95 T_m. The peak in stress is representative of dynamic recrystallization. Figure 2 presents the steady state strain rate dependence on stress for three elevated temperatures. We assume that steady state conditions are reached at a true strain of approximately 0.6. The power law exponent assuming a simple constitutive law of the form

$$\dot{\epsilon}^{vp} = A(T)\sigma^n \tag{3}$$

is approximately $n = 4$. This result was surprising since values of n in this range are normally associated with climb-controlled creep. Dynamic recrystallization is very active at these temperatures, however, so we expected the addition of this microstructural process to change the stress/strain rate dependence. One implication of this result is that at lower strain rates where the peak in stress is less pronounced, a simple power law may represent the constitutive response relatively well.

Rate-Dependence at Constant Structure

Strain rate change experiments on the high purity lead indicate a larger dependence of strain rate on stress. Figure 3 presents a set of strain rate change experiments at a homologous temperature of 0.95 T_m. The stress responses to different new strain rates were compared at identical plastic strain offsets from the point of the strain rate change. Figure 4 presents variation in the constant structure power law exponent determined from the strain rate change data as a function of plastic strain offset. Note that the power law rate dependence decreases from a value of $n = 9$ towards the value of $n = 4$ associated with steady state behavior. A power law dependence of $n = 5$ is reached after a strain rate increment of 0.02, indicating that a power law with a stress exponent in the range of 4 to 5 may be applicable from very small strains up to steady state behavior.

Semi-Solid Behavior

The constitutive behavior of metal semi-solid systems is very complex. These materials frequently exhibit a thixotropic or shear-thinning behavior, where the effective rate-dependence or apparent viscosity decreases as deformation occurs. This behavior is generally assumed to result from the breaking up of particle agglomerates or networks by the continued shear deformation. There is therefore a strong dependence on structure, to the extent that the constitutive response of the semi-solid system may be pseudoplastic in a particular strain and strain rate regime and dilatent in another regime.

A particular model for this behavior was presented by Brown [2] for a semi-solid slurry. The model, based on a procedure proposed by Frankel and Acrivos [4] proposes an apparent viscosity, μ^*, with the functional dependence

$$\mu^* = \mu A \left\{ \frac{(c/c_{max})^{1/3}}{1 - (c/c_{max})^{1/3}} \right\} + \mu D f_s \dot{\gamma}^{(1-n)/n} s \tag{4}$$

where

c	= effective volume fraction solid,
c_{max}	= maximum volume fraction solid,
A	= scaling factor based on assumed spatial particle arrangment,
f_s	= volume fraction solid,
s	= internal variable for agglomeration,
n	= rate dependence of solid particle phase,
D	= temperature-dependent scaling factor, and
$\dot{\gamma}$	= scalar measure of rate-of-deformation tensor.

The evolution of structure in a thixotropic semi-solid slurry behaves in a static hardening, flow softening manner. We therefore assume that the evolution equation for the internal variable is of the form:

$$\frac{ds}{dt} = \hat{H}(1 - s) + s\,\hat{B}\,\dot{\gamma}^m. \tag{5}$$

where

$\hat{H}$	= static hardening function,
$\hat{B}$	= dynamic softening function, and
m	= power law exponent.

We were unable to find a complete set of rheological measurements on a slurry system permitting rigorously deterministic evaluation of the constitutive model. In particular, there is a noticeable absence of strain-rate-jump or stress-jump experiments that would permit decoupling of the flow behavior from the evolution of agglomerate structure. Although incomplete, data obtained by Joly [3] on a semi-solid, tin-15% lead alloy provided sufficient information to evaluate the constants associated with the constitutive model described above under limited conditions of temperature and volume fraction solid.

Only one set of steady state shear rate versus apparent viscosity data was available in the lead tin data. These data are plotted in Figure 5 where the fit of the above constitutive model assuming steady state conditions is included. Nonlinear, least squares minimization subroutines provided by the NAG Library were used to fit the highly nonlinear constitutive model to the test data. Three material parameters, (D, H, B), were determined for the material state indicated in the figure, where the true volume fraction solid was $f_s = 0.45$.

Figure 6 illustrates the dependence of apparent viscosity (μ^*) on the true volume fraction solid (f_s) as measured by Joly for a lead/tin slurry. Notice the rapid increase in viscosity as the volume fraction solid increases, and the strong strain-rate dependence of this increase. Data are included for two steady state strain rates.

Figure 7 illustrates the behavior of the constitutive model for same conditions, using the values of the model parameters determined for the case of $f_s = 0.45$. The values of A and c_{max} in the flow equation were selected to be those used by Frankel and Acrivos. We expect the coefficients of the constitutive model to be temperature dependent. However, there is insufficient data to characterize the temperature dependence of the material functions, particularly that representing agglomeration, H. Nevertheless, the constitutive model represents the steady state behavior of the slurry remarkably well, particularly since the material constants were determined for one set of steady state data at one value of volume fraction solid.

Acknowledgement

Support for this work was provided by the U.S. National Science Foundation (Grant No. DMR-8806901)

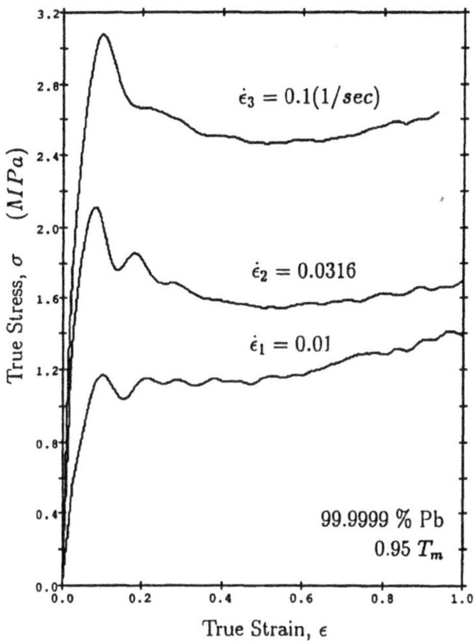

Fig.1. Constant true strain rate experiments at homologous temperature of 0.95.

References

1 S B Brown, P Kumar, and V. Dave, "High Homologous Temperature Constitutive Behavior of FCC Metals" *Constitutive Laws for Engineering Materials*, ASME Press, pp. 871-875, 1990.

2 S B Brown, "An Internal Variable Constitutive Model for Semi-Solid Slurries," *Proceedings, Fifth Conference on Modeling of Casting, Welding, and Advanced Solidification Processes,* Davos, Switzerland, 1990.

3 P A Joly, "Rheological Properties and Structure of a Semi-Solid Tin-Lead Alloy," *Ph.D. Thesis, Department of Materials Science and Engineering, M.I.T.*, 1974.

4 N A Frankel and A Acrivos, "On the Viscosity of a Concentrated Suspension of Solid Spheres," *Chemical Engineering Science," Vol. 22, 1967, pp. 847-853.*

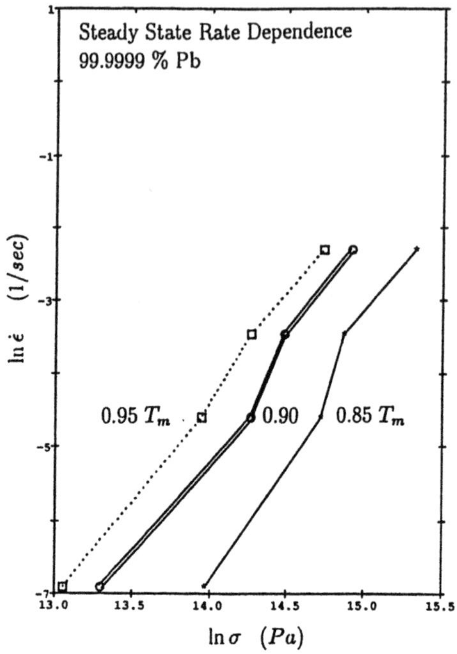

Fig.2. Steady state dependence of strain rate on stress.

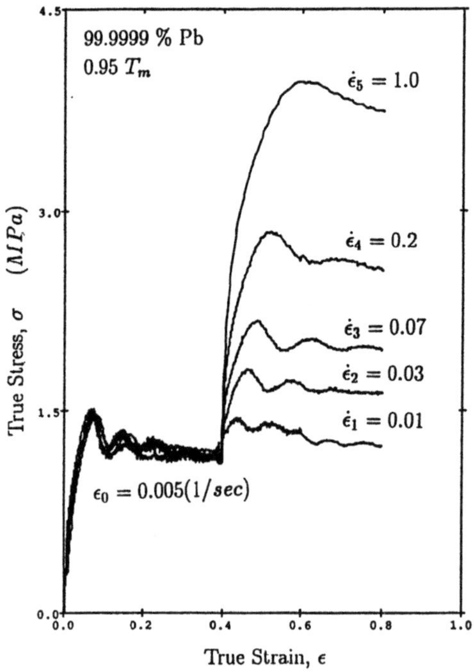

Fig.3. Strain rate change experiments at a homologous temperature of 0.95 T_m.

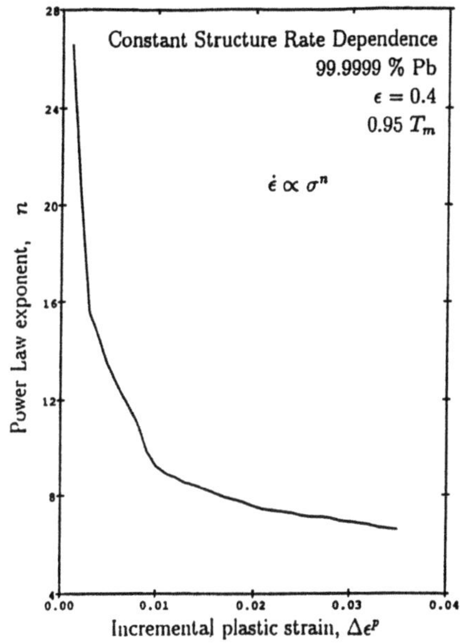

Fig.4. Variation of constant structure power law exponent determined from the strain rate change data as a function of plastic strain offset.

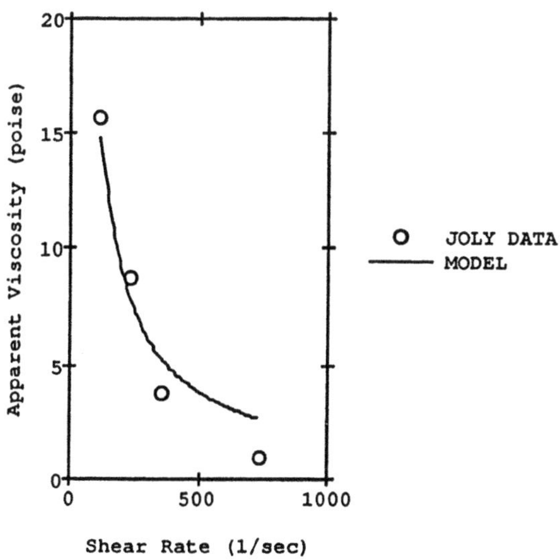

Fig.5. Steady state dependence of apparent viscosity on shear rate. Fraction solid = 0.45, Tin - 15% Lead.

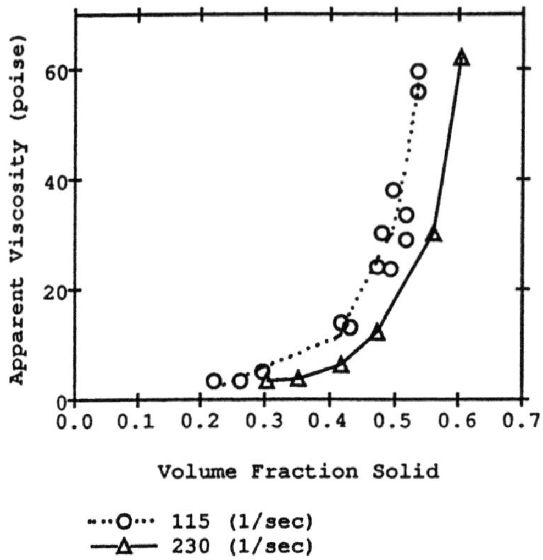

Fig.6. Steady state dependence of apparent viscosity on volume fraction solid. Tin - 15% Lead.

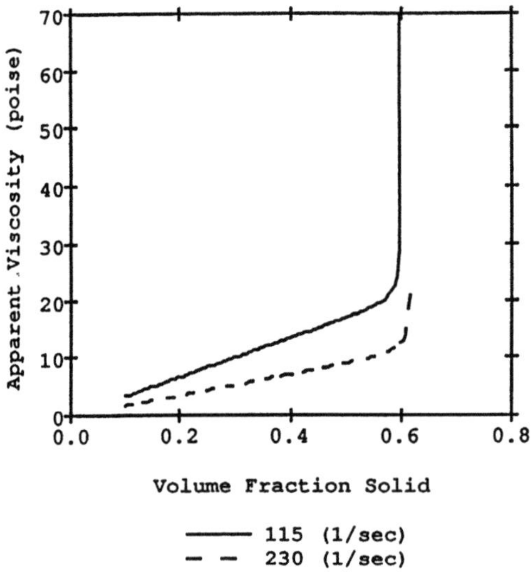

Fig.7. Steady state dependence of apparent viscosity on volume fraction solid. Model prediction.

Transformation Plasticity (TRIP) Under a Triaxial Stress State

F.D.FISCHER

Christian Doppler Laboratory for Micromechanics of Materials
Institute of Mechanics
University for Mining and Metallurgy, Leoben, Austria

Summary

After a short explanation of phase transformations during a welding process a micromechanical model for TRIP is presented. Based on this "natural" concept the TRIP strain tensor -due to a diffusional as well as a displacive transformation of an uniaxially loaded specimen- is derived. The theory is then extended to a triaxial load stress state. The micromechanical approach is also used to develop a kinetic equation for the martensite transformation.

Metallurgical Aspects

Right in front of the welding seam, the fusion zone (FZ) which transfered to the liquid state and the heat affected zone (HAZ) which was nearly totally austenitized during the heating phase of welding can be distinguished, see Fig. 1 taken from Cerjak [1]. During cooling the austenite in the HAZ decomposed by solid-solid transformation to ferrite, perlite, bainite, or martensite depending on the cooling rates. The time-temperature-history (T-T-H) may show different peak temperatures and cooling rates for various distances from the FZ. This is the main reason for an often complete change of the metallurgical structure of the base material in the HAZ. The situation is more complicated in a multipass welding since each new pass produces a certain tempering of the HAZ. Goldak and coauthors demonstrate in [2], [3] a proper calculation of the T-T-H taking into account temperature-dependent material data and latent heats of fusion and transformation. This is done by keeping track of the kinetics based on metal physics research of the last 15 years. A specific, differing method is to use the C-C-T (continuous cooling transformation) diagrams for weldable materials, e.g. as published by Seyffarth [4]. They were generated by extensive experimental procedures reflecting the "real" material behavior for a given T-T-H. Since the T-T-H can very often not be measured in practice, a computer calculation is imperative for applying a C-C-T diagram.

L. Karlsson, L.-E. Lindgren, M. Jonsson (Eds.)
Mechanical Effects of Welding
IUTAM Symposium Luleå/Sweden 1991

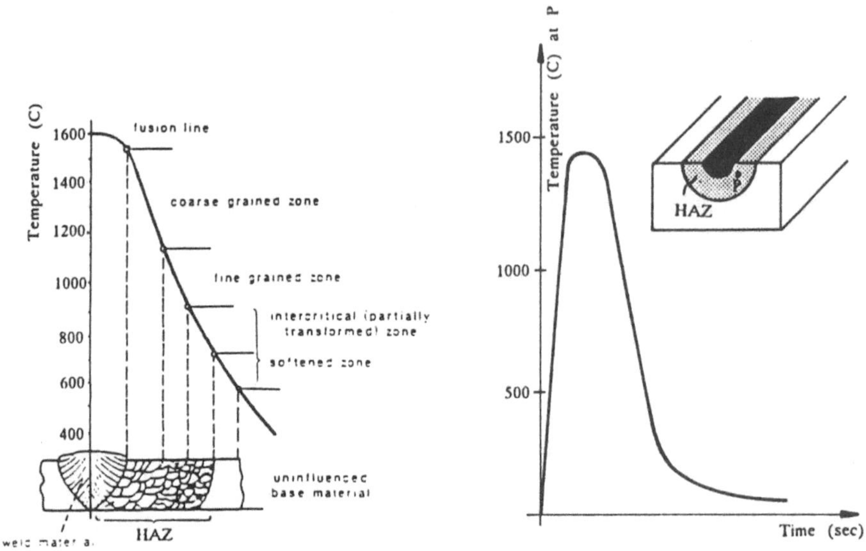

Fig. 1. Temperature distribution in the HAZ

Residual Stress Distribution

In addition to the nonhomogenous distribution of temperature T and material properties of the individual phases, a micro-stress state, developed by the volume (and shape) change of the transforming microregions, is responsible for a residual macro-stress state. The "superposition" of this micro-stress state, varying even within a single grain, with a local external (e.g. load) stress state may result in a significant macroscopic deformation named TRIP. Its influence on residual stresses during quenching was "detected" by Franitza, [5], in 1972. Rammerstorfer et al gave in [6] a simplified consideration by an "artificially" decreased yield stress. This topic was extensively treated by Mitter in his book [7]. Goldak and coauthors included a TRIP term in the local strain tensor for welding simulations [8], [9], [10], thus showing again its important influence on the residual stress distribution. Van Gulick recently discussed aspects of the TRIP implementation in a standard finite element code, see [11].

Modelling of TRIP

In this chapter, emphasis is laid on the modelling of TRIP by distinguishing between a diffusive transformation (e.g. austenite-perlite- transformation) and a displacive transformation (e.g. the austenite- martensite- transformation). In the mechanical-oriented welding literature TRIP is applied **only** for a diffusive-type transformation, see e.g. Goldak and coauthors, [8], [9], [10].

Diffusional Transformation

Treatment of this topic goes back to the famous paper by Greenwood and Johnson, [12]. It was reformulated by Leblond et al [13] and extended by Mitter [7] to the full transformation fraction ξ, $0 \le \xi \le 1$. The product phase is assumed to grow as spherical nodules in an infinite parent phase matrix with a volume change δ, see Fig.2, leading to a plastification of the neighbouring parent phase shell.

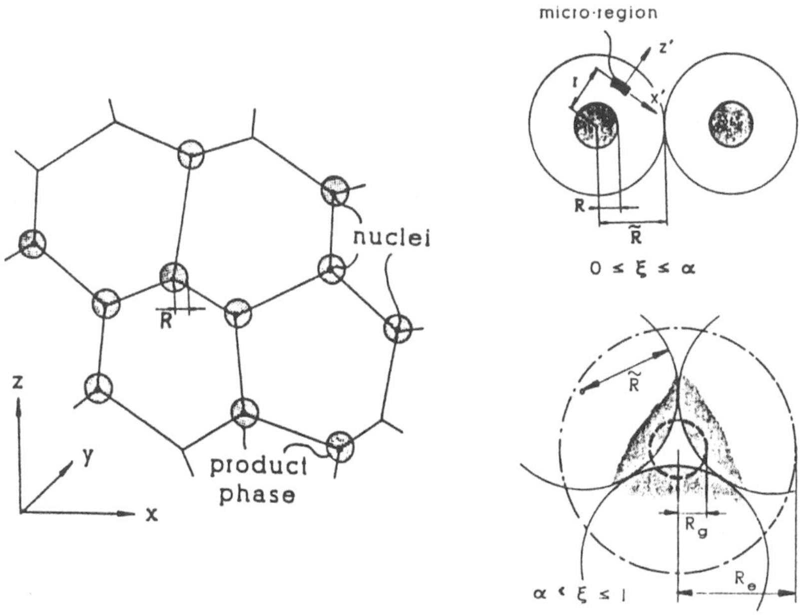

Fig. 2. Micromechanical model for diffusional transformation

Fischer made a rigorous analysis in [14] using the following steps:

* The nodules are arranged in a certain pattern (distance $2\tilde{R}$ of two spheres, for $\xi = \alpha$ the spheres just "touch" each other, R_g is the radius of an equivalent sphere representing the gap for $\xi \ge \alpha$). A small part of the shell is defined as a microregion that may freely deform due to the transformation tensor $\tilde{\varepsilon}_c'$ described in a local coordinate system. The rate $\dot{\tilde{\varepsilon}}_c'$ follows as

$$\dot{\tilde{\underline{\underline{\varepsilon}}}}'_c = \frac{\delta}{3\beta}\dot{\xi}\,\frac{1}{\tilde{\rho}^3}\begin{bmatrix} 1 & 0 & 0 \\ 0 & 1 & 0 \\ 0 & 0 & -2 \end{bmatrix} \begin{array}{l} 0 \le \xi \le \alpha: \beta = \alpha,\ \tilde{\rho} = r/\tilde{R} \\ , \\ \alpha < \xi \le 1: \beta = 1,\ \tilde{\rho} = r(1-\alpha)^{1/3}/R_g \end{array} \qquad (1)$$

* Following a Taylor-Lin assumption, the same global strain tensor $\underline{\underline{\varepsilon}}$ is introduced into each microregion of a homogenously loaded specimen.
* Neglecting the elastic deformations and following standard plasticity, the interaction between a microregion and the neighbouring material is described by

$$\dot{\underline{\underline{\varepsilon}}} - \underline{\underline{R}} \cdot \dot{\tilde{\underline{\underline{\varepsilon}}}}'_c \cdot \underline{\underline{R}}^T = \lambda\, \underline{\underline{s}}\,,\ \underline{\underline{s}} : \underline{\underline{s}} = 2/3\ \sigma_y^2\,, \qquad (2)$$

$\underline{\underline{s}}$ stress deviator, $\underline{\underline{R}}$ orthogonal transformation tensor from the local (x') to the global (x) coordinate system, σ_y yield strength. A component s_{ij} follows now as

$$\frac{s_{ij}}{\sigma_y} = \frac{\dot{\varepsilon}_{ij} - R_{ik}\, \dot{\tilde{\varepsilon}}'_{c,kl}\, R_{jl}}{[3/2(\dot{\underline{\underline{\varepsilon}}} : \dot{\underline{\underline{\varepsilon}}} - 2\underline{\underline{R}}.\dot{\tilde{\underline{\underline{\varepsilon}}}}'_c.\underline{\underline{R}}^T : \dot{\underline{\underline{\varepsilon}}} + \dot{\tilde{\underline{\underline{\varepsilon}}}}'_c : \dot{\tilde{\underline{\underline{\varepsilon}}}}'_c)]^{1/2}} \qquad (3)$$

* An uniaxial load stress σ_L in the global z-direction is applied to the specimen. Global equilibrium enforces

$$\sigma_L = (1-\xi)\ \langle\sigma_z\rangle_O + \xi\ \langle\sigma_z\rangle_N\ ,\ 0 \le \xi \le \alpha \qquad (4)$$

with $\langle\rangle$ being the average symbol. The indices, O, N respectively, refer to the parent and product phases. With $\langle\sigma_x\rangle = \langle\sigma_y\rangle = 0$, the consequence for the deviator $s_{ij} \Rightarrow s_z$ is

$$\sigma_L = 3/2(1-\xi)\ \langle s_z\rangle_O + 3/2\xi\langle s_z\rangle_N. \qquad (5)$$

The average $\langle\rangle$ is calculated due to the Eulerian angles $\underline{\omega}$, $0 \le \phi, \psi \le 2\pi$, $0 \le \vartheta \le \pi$ with a weighting function g as

$$\langle f\rangle = \int_{\underline{\omega}} gf \sin\vartheta\ d\underline{\omega} \,/ \int_{\underline{\omega}} g\sin\vartheta\ d\underline{\omega}\,. \qquad (6)$$

$g=1$ since a uniform distribution can be assumed. $\underline{\underline{\varepsilon}}$ has now the components

$$\varepsilon_{xx} = \varepsilon_{yy} = -\varepsilon/2 \text{ and } \varepsilon_{zz} = \varepsilon,\ \varepsilon_{ij} = 0 \text{ for } i \neq j. \qquad (7)$$

The component $\tilde{\varepsilon}_{c,zz}$ of $\underline{\underline{R}} \cdot \dot{\tilde{\underline{\underline{\varepsilon}}}}'_c \cdot \underline{\underline{R}}^T$ follows as

$$\dot{\tilde{\varepsilon}}_{c,zz} = \delta/3\beta\ \dot{\xi}\ 1/\tilde{\rho}^3\ (1 - 3\cos^2\theta),\ \beta,\ \tilde{\rho} \text{ see (1).} \qquad (8)$$

By considering $\xi \le \alpha$ and taking for $3/2\langle s_z\rangle_N = \sigma_L$ (-since the product phase is assumed to behave elastically-) eqn.(5) reduces to $\sigma_L = 3/2\langle s_z\rangle_O$. Insertion of (3) with σ_{yo} leads to a very complicated nonlinear differential equation for $\varepsilon \doteq \dot{\varepsilon}(\xi, \dot{\xi})$. Only in the case of $\sigma_L/\sigma_{yo} \le \sim 0.5$ can an exact solution be written down for the irreversible TRIP strain ε_{TRIP}

$$\varepsilon_{TRIP} = \frac{10\delta}{6} \frac{\sigma_L}{\sigma_{yo}} [(1+\alpha)\ln(1+\frac{\xi}{\alpha})-\xi], \; 0 \le \xi \le \alpha. \tag{9a}$$

A similar procedure leads to

$$\varepsilon_{TRIP} = \varepsilon_{TRIP}|_{\xi=\alpha} + \frac{5}{6}\delta \frac{\sigma_L}{\sigma_{y_N}} \frac{(\xi^2-\alpha^2)}{(2-\alpha)}, \; \alpha \le \xi \le 1 \tag{9b}$$

and for $\xi = 1$ with an average $\alpha = 0{,}632$ to

$$\varepsilon_{TRIP} = \frac{5}{6}\delta \frac{\sigma_L}{\sigma_{y_o}} (1.0 + 0.439 \frac{\sigma_{yo}}{\sigma_{y_N}}) \; . \tag{9c}$$

This agrees with the famous "Greenwood and Johnson relation", [12], for $\sigma_{yo} \ll \sigma_{yN}$!

* A consistent extension of the algorithm to a triaxial load stress state is possible and under preparation. Arguments such as "reducing" (9c) to an "indical" equation as given by Goldak et al. [8] are a purely heuristic approach.

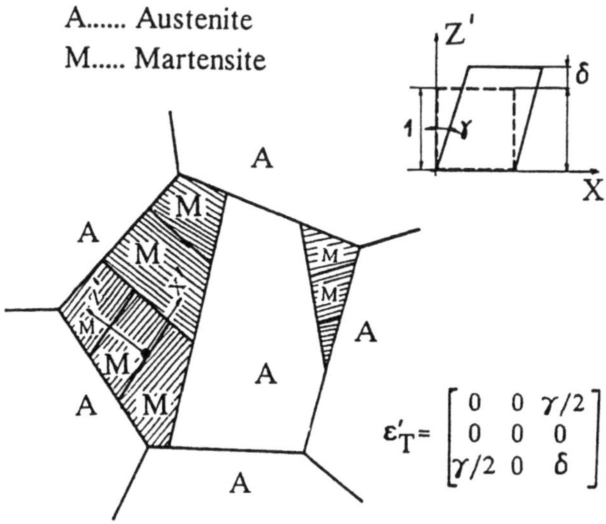

Fig. 3. Micromechanical model for displacive transformation

Displacive transformation

The martensitic transformation is an "instantanous" process, driven by a certain surplus ΔG_C of the chemical free energy, caused by rapid cooling below the martensitic start temperature M_S.

* A microregion, a part of a grain, transforms in a very short time interval $[t_0, t_\xi]$, $t_\xi - t_0 \leq 10^{-6}$ sec, into e.g. a martensite lath, see Fig.3. Crystallographic theory, see e.g. [15], reports a "free" transformation tensor rate as

$$\dot{\tilde{\underline{\varepsilon}}}_c = \dot{h} \begin{bmatrix} 0 & 0 & \gamma/2 \\ 0 & 0 & 0 \\ \gamma/2 & 0 & \delta \end{bmatrix} \quad \begin{array}{l} h = h(t), \\ h(t_0) = 0, \ h(t_\xi) = 1, \end{array} \tag{10}$$

$h(t)$ being a time function defined in $[t_0 \leq t \leq t_\xi]$. δ depends significantly on the Ni and C-content and varies from 0.006 (1.78%C) to 0.04 (30% Ni). The shear component γ is ca. 0.20.

* The next two steps are equivalent to those leading to eqn.(2) and eqn.(3). For steels, both the invariant shear (γ) and the interaction of the microregion are performed by slip (plastification). Instead of σ_{yo} an "average" yield strength σ_y^* must be applied since the microregion transforms from austenite (σ_{yo}) to martensite (σ_{yN}).

* The averaging process needs a proper weighting function $g = g(\xi, \underline{\omega})$. Specifically for $\xi \ll 1$ those martensitic variants are selected which may cause an extreme length change ("Magee"- effect). From

$$\dot{\varepsilon}_{c,zz} = \dot{h}/2 \, [\gamma \sin 2\theta \sin \phi + \delta(1 + \cos 2\theta)] \tag{11}$$

a maximum length change of $1/2(\delta + \sqrt{\gamma^2+\delta^2})$ (~ 12%!) can be calculated if only one variant $\bar{\underline{\omega}}$ ($\phi = \pi/2$, $\theta = 1/2 \operatorname{atan}(\gamma/\delta)$, ψ free) exists, $g = D(\underline{\omega} - \bar{\underline{\omega}})$, D .. Dirac-function.
Progressive transformation introduces an abundance of different variants which are interactive but which may also show some "self"-accomodation of the transformation shear. The influence of ξ in $g(\xi, \underline{\omega})$ is decreasing. For a detailed discussion, see Gautier et al [16].
If a rapid cooling to the martensitic finish temperature M_f takes place, $\xi = 1$, then $\sigma_L = 3/2 \langle s_z \rangle$ follows. A nonlinear differential equation for $\dot{\varepsilon} = \dot{\varepsilon}(h,g)$ can be derived by insertion of (3) with σ_y^* for s_z which reduces to an algebraic equation if the same variation with time, $h(t)$, is supposed for $\underline{\underline{\varepsilon}}'$ and $\tilde{\underline{\underline{\varepsilon}}}_c$. In the case of $\sigma_L/\sigma_y^* \leq \sim 0{,}5$ a Taylor-series expansion leads for a uniform distribution of the variants, $g = 1$, to

$$\dot{\varepsilon}_{TRIP} = \dot{K}^* \sigma_L, \; K^* = \frac{5}{6} \frac{1}{\sigma_y^*} (\delta^2 + \frac{3}{4} \gamma^2)^{1/2} \qquad (12)$$

representing an "extended" "Greenwood and Johnson" relation.

* This consistent concept was then applied to a triaxial load stress state with $\sigma_x=\sigma_1$, $\sigma_y=\sigma_2$, $\sigma_z=\sigma_3$ and $\varepsilon_{xx}=\varepsilon_1$, $\varepsilon_{xy}=\varepsilon_2$, $\varepsilon_{zz}=\varepsilon_3$, $\varepsilon_1+\varepsilon_2+\varepsilon_3=\delta$. A detailed study worked out for g = 1 (but allowing also other types of g), see [17], showed that for even large TRIP strains ε_1, ε_2, ε_3 ($\leq$ 0.08!) a superposition rule can be applied, surprisingly leading to

$$\underline{\underline{\varepsilon}}_{TRIP} = \frac{3}{2} K^* \underline{\underline{s}}_L . \qquad (13)$$

Of course, K^* can not be calculated by (12) for a general distribution g and should be taken from the "classical" TRIP experiment. But it is obviously **not** appropriate to take the "classical" "Greenwod and Johnson" relation including only δ as proposed in [8].

An expansion of (13) to $\xi<1$ by multiplying (13) with $f(\xi)$, $0\leq\xi\leq1$, raises the problem of finding $f(\xi)$. Although some proposals exist, see for a discussion Denis et al [18] and Fischer [14], the problem is unsolved. Increased efforts simulating the transformation process, e.g. by applying finite elements (large deformations, elasto-plasticity) will be necessary.

Transformation Kinetics

The diffusional transformation is an Avrami-type process well described by Johnson and Mehl more than 50 years ago, see Goldak and coauthors [2]. For transformation kinetics of martensite an integrated Avrami-type relation is used going back to Koistinen and Marburger, see e.g. Inoue et al. [19].
The authors of [19] also included the influence of the actual stress state $\underline{\underline{\sigma}}$ (using the invariants J_1, J_2) in the kinetic equations by using a heuristic approach. Patel and Cohen [20], however, had shown how a stress state affects the martensite start temperature M_s ($J_1<0$ decreases M_s!) by a thermodynamical consideration. Tanaka and Fischer prove now in [21] that a local transformation condition does not only depend on the load stress tensor $\underline{\underline{\sigma}}_L$ but also on the local interaction eigenstress tensor $\underline{\underline{\sigma}}_{eq}$ as well as on the stresses $\underline{\underline{\sigma}}_n$ developing during the transformation of the microregion under consideration (here $\underline{\underline{\sigma}}_{eq} + \underline{\underline{\sigma}}_n$ is approximated by equ. (3) with σ_y^*).

The energy balance before and after the complete transformation of a microregion thus takes the form with $\Delta G \leq G_0$, a threshold energy,

$$\Delta G_c - (\underline{\underline{\sigma}}_L + \underline{\underline{\sigma}}_{eq} + \kappa\, \underline{\underline{\sigma}}_n) : (\underline{\underline{R}} \cdot \underline{\underline{\varepsilon}}_c' \cdot \underline{\underline{R}}^T) = \Delta G,\ 0 \leq \kappa \leq 1. \tag{14}$$

Assuming an incremental relation between ξ and the volume average of ΔG,

$$d\xi = -kV\,(1 - \xi)\,\langle d\Delta G \rangle, \tag{15}$$

k, V being material constants, an average of (14) over a mesodomain consisting of an assembly of grains ($\underline{\underline{\sigma}}_L$, T, ξ homogenous) results in

$$\left(\frac{1}{1-\xi} + kV p\delta - \langle \kappa \frac{\partial \underline{\underline{\sigma}}_n}{\partial \xi} : \underline{\underline{\tilde{\varepsilon}}}_c \rangle\right) d\xi = -kV\left(\frac{\partial \Delta G_c}{\partial T}\right) dT + kV \langle \underline{\underline{\tilde{\varepsilon}}}_c \rangle : d\underline{\underline{\sigma}}_L. \tag{16}$$

This is a kinetic equation including only physical terms derived from a consistent micromechanical approach. $\underline{\underline{\sigma}}_{eq}$ is here approximated by $-p\xi\underline{\underline{I}}$ ($\underline{\underline{I}}$ unity tensor). (16) can also be applied to a triaxial load stress state and includes the average of the transformation tensor $\underline{\underline{\tilde{\varepsilon}}}_c = \underline{\underline{R}} \cdot \underline{\underline{\tilde{\varepsilon}}}'_c \cdot \underline{\underline{R}}^T$.

References

1. Cerjak, H.H: The role of welding in component integrity. Bachelet, E. et al. (eds) High temperature materials for power engineering 1990; Proc. Conf. Liege, Sept. 1990, Dordrecht, Boston, London: Kluwer Academic Publishers 1990, Part I, 445 - 472

2. Watt, D.F., Coon, L., Bibby, M., Goldak, J.A., Henwood C.: A reaction kinetics algorithm for modelling microstructural development in weld heat-affected zones, Part (A), Reaction kinetics. Report Dept. Mech. & Aeronautical Engng. Carleton Univ., Ottawa and Dept. Engng. Mat., Univ. Windsor, Canada, 1987

3. Henwood, C., Bibby, M., Goldak, J.A., Watt, D.: Coupled transient heat transfer - Microstructure, Part (B), weld computations, Report. Dept. Mech. & Aeronautical Engng. Carleton Univ., Ottawa and Dept. Engng. Mat., Univ. Windsor, Canada, 1987

4. Seyffarth, P.: ATLAS Schweiß-ZTU-Schaubilder. Bd. 75. Deutscher Verlag für Schweißtechnik, Düsseldorf, 1982

5. Franitza, S.: Zur Berechnung der Wärme- und Umwandlungsspannungen in langen Kreiszylindern. Ph.D. Thesis. T.U. Braunschweig, 1972

6. Rammerstorfer, F.G., Fischer, D.F., Mitter W., Bathe, K.J., Snyder, M.D.: On thermo-elastic-plastic analysis of heat-treatment processes including creep and phase changes. Comp. Struct. 13 (1981) 771-779

7. Mitter W.: Umwandlungsplastizität und ihre Berücksichtigung bei der Berechnung von Eigenspannungen, Materialkundlich-Technische Reihe 7. Berlin, Stuttgart: Gebrüder Bornträger, 1987

8. Oddy, A.S., Goldak, J.A., Mc Dill, J.M.J.: Numerical analysis of transformation plasticity in 3 D finite element analysis of welds. Eur. J. Mech., A/Solids, 9 (1990) 253-263

9. Oddy, A.S., Goldak, J.A., Mc Dill, J.M.J.: Transformation plasticity and residual stresses in single-pass repair welds. PVP-Vol 173. Rybicki, E. etal. (eds) Weld residual stresses and plastic deformation, ASME (1989) 13-18

10. Oddy, A.S., McDill, J.M.J., Goldak, J.A.: Consistent strain fields in 3D finite element analysis of welds. ASME J. Pressure Vessel Technology, 112 (1990) 309-311

11. Van Gulick, L.A: Abaqus welding analysis. H.K.S (eds) ABAQUS user's conf. Proc. Stresa, 1989, 237-250

12. Greenwood, G.W., Johnson, R.H.: The deformation of metals under small stress during phase transformations. Proc. R. Soc. Lond. A 283 (1965) 403-422

13. Leblond, J.B., Devaux J., Devaux J.C.: Mathematical modelling of transformation plasticity in steels, I: case of ideal plastic phases. Int. J. Plasticity 5 (1989) 551-572

14. Fischer, F.D.: A micromechanical model for transformation plasticity in steels. Acta metall. mater. 38 (1990) 1535-1546

15. Wayman, C.H.: Introduction to the crystallography of martensitic transformations. New York, London: The Macmillan Company, 1964

16. Gautier, E., Zhang, X.M., Simon, A.: Role of internal stress state on transformation induced plasticity and transformation mechanism during the progress of stress induced phase transformation. Beck, G. et al. (eds). Proc. ICRS 2, Oct. 1988. London, New York: Elsevier Applied Science, 1989, 777-783

17. Fischer F.D.: Transformation induced plasticity in triaxially loaded steel specimens subjected to martensitic transformation. Paper to be published Eur. J. Mech., A/Solids, 1991

18. Denis S., Gautier, E., Simon, A.: Modelling of the mechanical behaviour of steels during phase transformations: a review. Beck, G. et al (eds). Proc. ICRS 2, Oct. 1988. London, New York: Elsevier Applied Sience, 1989, 393-398.

19. Inoue, T., Wang Z.: Coupling between stress, temperature, and metallic structures during processes involving phase transformations. Mat. Sci. & Technology, 1 (1985) 845-850.

20. Patel, J.R., Cohen, M.: Criterion for the action of applied stress in the martensitic transformation. Acta metall., 1 (1953) 531-538

21. Tanaka, K., Fischer, F.D.: A continuum mechanical approach to the kinetics of martensitic transformation. Paper submitted to Acta metall. mater., 1991

Research on Welding Stresses During Phase Transformation

Jianhua Wang, Chu Chen, Fengming Ding
Shanghai Jiao Tong University,
Shanghai, P.R.China

ABSTRACT

In this paper, the stresses variation during phase transformation and its influence on residual stresses in low-alloy steel are investigated by means of thermal elastic-plastic finite element analysis. Three models were adopted to analyze the stresses during phase transformation. It was found that the residual stresses were lower than expected due to the abrupt variation of expansion coefficient caused by the volume expansion of the metal with its microstructure changes. The more the expansion coefficients changed, the more the residual stresses decreased. Furthermore, the lower the phase transformation temperature, the greater were the effects of phase transformation on residual stresses.

INTRODUCTION

Welding residual stresses have a significant influence on the weld cracking and strength and in-service behavior of welded structures. Recently, large thick-walled pressure vessels have been used more and more widely with the development of petroleum, chemical and nuclear industries. Usually high strength steels are utilized in these welded structures to keep weight to a minimum. Therefore, it is extremely important to evaluate the residual stresses accurately in such welded structures.

Nowadays,much progress has been made in thermal elastic-plastic finite element analysis for transient and residual stresses in welding. However, few of them have introduced the influence of phase transformation.[1,2] For low carbon steels, phase transformation occurs at relatively high temperature when the elasticity of the material becomes negligible. The residual stresses are likely not to be affected by the phase transformation in this case due to low yield stress. In contrast, the phase transformation appears at comparatively low temperature in low-alloy, high strength steels, and high-alloy steels when both the yield stress and elastic modulus are

L. Karlsson, L.-E. Lindgren, M. Jonsson (Eds.)
Mechanical Effects of Welding
IUTAM Symposium Luleå/Sweden 1991

restored. The volume variation due to phase transformation will affect the residual stresses greatly.

In this paper, mechanical properties such as thermal expansion coefficient were measured during the heating and cooling stages for a ship-building low-alloy steel. On the basis of thermal elastic-plastic theory, the stress distributions were analyzed by FEM for the following three cases: (1) a cylindric specimen under the rigid restraint condition with uniform heating and cooling; (2) a cylindric specimen under the rigid restraint condition with simulated welding thermal cycle ; (3) butt welding of thin plate. Verification experiments were also carried out. These agreed with the theoretical analysis very well. The mechanism and relevant trends are discussed for the influence of phase transformation on welding transient and residual stresses.

MEASUREMENT AND ASSUMPTION

Measurement of Expansion Coefficients

A rapid quenching expansion device Lk-02 was used for the test. Its working range covered temperature from -200°C to 1350°C. To simulate the thermal cycle in the HAZ, the austenizing temperature 1330°C was kept for 1 second. Then a series of Expansion-Temperature curves was plotted for various cooling rate. A typical graph obtained in the test is shown in Fig.1(a). The volume changes due to phase transformation. These changes led to deviations on the expansion curve which marked the beginning and finishing points of phase transformation. As the phase transformation occurs at a higher temperature during heating, its effect on residual stresses is very small. So only the expansion variation during phase transformation in cooling is discussed in this paper.

The Cooling Rate-Phase Transformation Temperature graph obtained under various cooling rates is shown in Fig.1(b). It can be seen that the beginning and finishing temperatures decrease gradually with the increase of cooling rate. However, when the cooling rate reaches a certain value (80°C/s in this

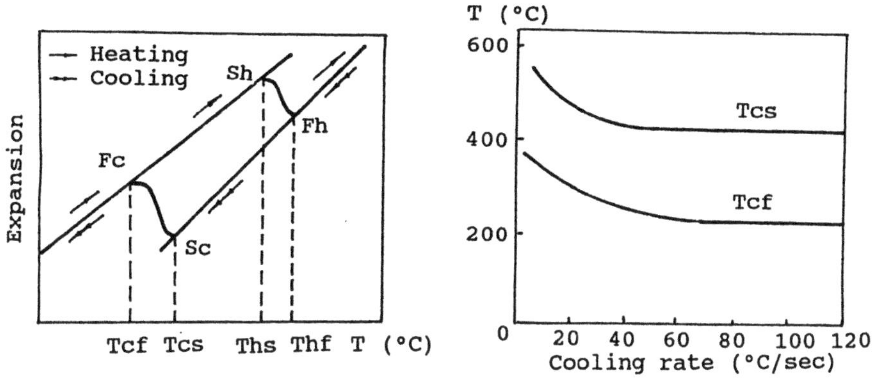

(a) (b)

Figure.1 Relationship between: (a). expansion and temperature or: (b). cooling rate and phase transformation temperature.

figure), the points of phase transformation change little with the increase of cooling rate.

The variations of expansion coefficient during the heating and cooling stages are shown in Fig.2(a) which indicates that changes abruptly from positive to negative during phase transformation in the cooling stage.

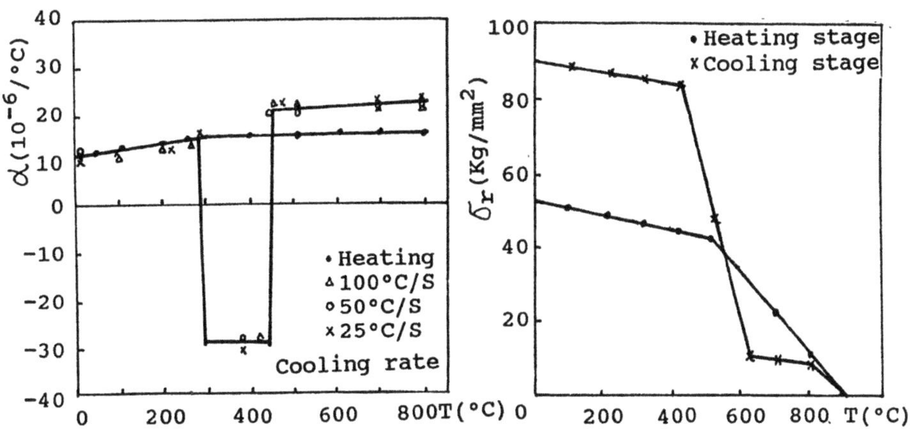

(a). expansion coefficient (b). yield stress

Figure.2 Temperature dependence of (a). or (b). r

Measurement of Yield Strength Data over a Temperature Range

The tests were carried out on the Thermorestor-W welding thermal simulation machine. From Fig.2(b), it can be seen that the yield strength decreased with an increase in temperature during heating and cooling. When the temperature reached a critical value, the yield strength decreased dramatically. Because of very high rate of cooling, the quenching tendency of the materials under phase transformation caused the yield strength at room temperature to increase greatly after passing through the welding thermal cycle.

Relationship Between Elastic Modulus and Temperature

The relationship between elastic modulus and temperature is displayed in Fig.3. The variation of elastic modulus during heating is assumed to be the same as during cooling.

Based on the results of tests described above, the simplified trends of these parameters over a temperature ranges for a phase transformation are of fundamental importance in elastic-plastic stress analysis in welding.

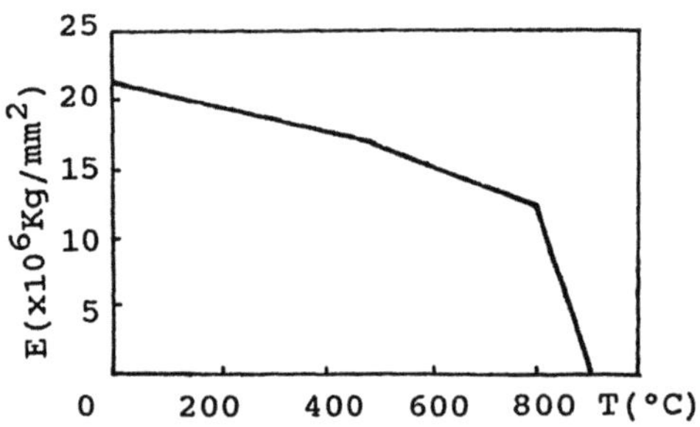

Figure.3 Temperature dependence of Young's modulus.

RESULTS AND DISCUSSION

Rigidly Restraint Model with Uniform Heating & Cooling(Model 1)

The variation of stresses during uniform heating and cooling is displayed in Fig.4(a) for a specimen fixed at the two ends. Curve 1 shows the case in which the variation of expansion coefficient is ignored during phase transformation (no abrupt change). Curves 2, 3, and 4 are the computed results when the expansion coefficients in the phase transformation zone are different. It appears that the following rules are followed regarding the variation of stresses. During the beginning of heating stage, the compressive stresses generated and increased gradually because the two ends of specimen were fixed.

Tensile stresses are produced during cooling as contraction of the specimen is inhibited by the two fixed ends. In the case of curve 1, phase transformation is irrespective, the stresses are tensile, and they reach the tensile yield stress value. Curves 2,3 and 4 show that expansion under transformation causes tensile stresses to relax and then to turn into compressive stresses. After the phase transformation is completed, the expansion coefficient is restored to a positive value and the specimen continues to contract as the temperature decreases, which brings about the gradual decrease of compressive stresses. The compressive stresses turn into tensile stresses

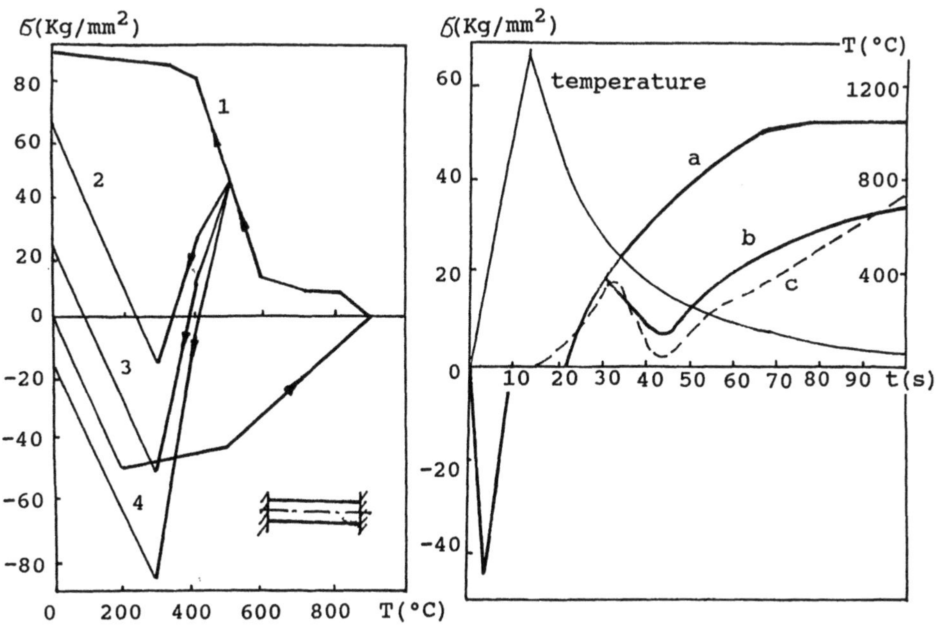

(a). model 1 (b). model 2

a--Phase transformation neglected (FEM)

b--Phase transformation considered (FEM) c--Experiment

Figure.4 Transient and residual stresses of two models.

or remain at a relatively low value when the temperature cools to room temperature. It can be seen that the residual stresses are reduced as a result of the expansion due to phase transformation. The greater the expansion during phase transformation, the more the residual stresses decrease, they even change from tensile to compression.

Rigid Restraint Test with Simulated Welding Thermal Cycle (Model 2)

The specimen is illustrated as Fig.5, with the two ends fixed rigidly, the middle part locally heated and the HAZ simulated with a welding thermal cycle. Figure 4(b) shows the results of finite element calculation and experiments for the variation of temperature and stresses with time during the simulated thermal cycle. The theoretical calculation agrees with the experimental results reasonably when expansion during phase transformation is considered. The trends of stress variation during heating and cooling are similar to those for the rigid restraint model with a uniform temperature field. But in the process of local heating, the expansion due to phase transformation in the phase transformation zone is not large enough for the tensile stresses to turn into compressive stresses because the phase transformation zone is small. So the residual stresses are tensile and are much lower than those for a sample in which the phase transformation is not taken account of. The residual stresses are about 30% lower when phase transformation is considered.

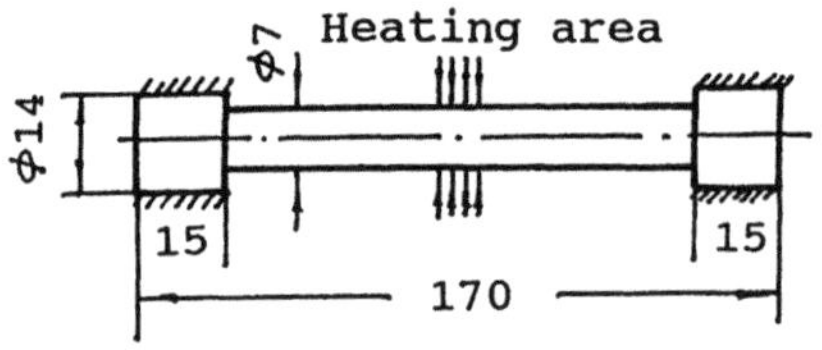

Figure.5 Restraint model 2 with welding thermal simulation

Butt Welding of Thin Plate

the specimen with a weld laid in the center groove is shown in

Fig.6(a). The welding electrode had a diameter of 4mm,a welding current was 170A while the arc voltage of 25V,speed of 0.29cm/s was used. Figure 6(b) displays the results of calculation and measurement of residual stresses along the weld direction.

Tensile residual stresses which reach the value of yield stress exists in the weld and near-weld regions when the expansion due to phase transformation is not taken into account. In contrast to the above, the tensile stresses are much lower in the near-weld zones and maximum stress occurs in the near-weld zone instead of at the center of the weld, as phase transformation being taken into consideration.

The residual stresses in actual welding structures are much more complicated,changing with the thickness of plate,the value of welding heat input and the effect of phase transformation.

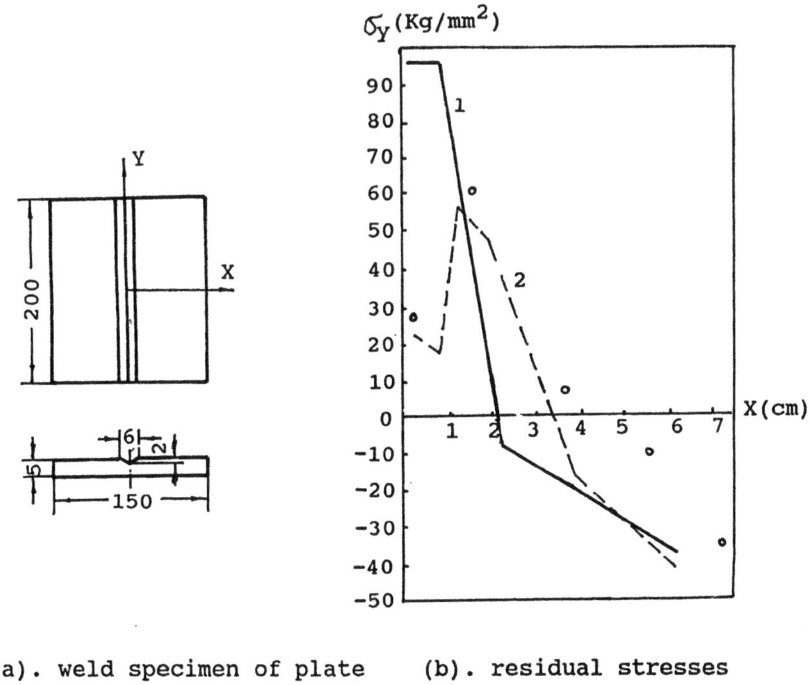

(a). weld specimen of plate (b). residual stresses

1--Phase transformation neglected (FEM)
2--Phase transformation considered (FEM) o--Experiment

Figure.6 Model 3 and its residual stresses

The influence of phase transformation on residual stresses should be considered when the low-alloy and high strength steel or high-alloy steel is used for welding. Using method proposed in this paper, the further investigation will be performed into the influences of phase transformation on residual stresses under various conditions.

CONCLUSIONS

1. Thermal elastic-plastic stress analysis on welding thermal stresses considering the influence of phase transformation has been proved to be effective .

2. The variations in mechanical properties and expansion coefficients during heating and cooling were measured for a ship-building low-alloy steel. The yield stress increased significantly after a welding thermal cycle. In the phase transformation temperature zone during cooling, the expansion coefficients turned from positive to negative.

3. The maximum residual stress decreased as a result of the phase transformation in welding. The more the expansion coefficient changed during phase transformation, the lower the tensile stresses became, which may even change into compressive stresses.

ACKNOWLEDGMENT

This research is supported by National Natural Science Foundation of China, we would like to thank it very much.

REFERENCES

1. C.Chen, J.Wang and H.Q.Yang, The Application of Numerical Analysis in Welding,Shanghai Jiao Tong University,January,1989.

2. Y.Ueda, Y.C.Kim, C.Chen and Y.M.Tang, Mathematical Treatment of Phase Transformation and Analytical Calculation Method of Restraint Stress-Strain, Transaction of JWRI, 1985, Vol.14, No.1, 153-162.

Modelling Phase Transformations for the Calculation of Internal Stresses During Fast Heating and Cooling in Steels

S. Denis, D. Farias, A. Simon

Laboratoire de Science et Génie des Matériaux Métalliques, Ecole des Mines de Nancy, Parc de Saurupt, 54042 NANCY Cedex

Introduction

A model for the prediction of phase transformations during fast heating and cooling can improve the understanding and control of the microstructures and mechanical properties in processes like surface hardening treatments, welding... Such a model could be used more conveniently than the graphical representations (isothermal and continuous heating and cooling transformation diagrams). It is also absolutely necessary in programs predicting the build up of internal stresses during the treatment and the residual stresses at the end of the treatment. In this paper, we present a mathematical model coupling phase transformations and temperature evolutions for a cylindrical specimen during a rapid thermal cycle.

Description of the phase transformation calculation model

Heating

The method used in this paper for calculating phase transformations during continuous heating from isothermal data is based on a rule of additivity. It has been used by several authors (1-3), we recall briefly the principle of the method. The temperature-time curve is discretized in a series of isothermal steps. On each step the volume fraction of new phase formed is calculated by using isothermal transformation kinetics.

The isothermal transformation kinetics is modelled according to the law developed by Johnson-Mehl (4) and by Avrami (5) :

$$y_k = y_{max\,k}\,(1 - \exp(-b_k t^{n_k}))$$

where y_k is the volume fraction of constituent k transformed into austenite (k = 1 pearlite, k = 2 ferrite) and bk and nk are temperature dependent parameters. At each temperature, the coefficients n_k and b_k are calculated by using two points corresponding to a given percentage of phase formed (10 pct and 90 pct for example) obtained from the isothermal kinetics of transformation or from the Isothermal Heating Diagram. $y_{max\,k}$ is the maximum volume fraction of austenite that can be formed. The method for quantifying the phase transformations

L. Karlsson, L.-E. Lindgren, M. Jonsson (Eds.)
Mechanical Effects of Welding
IUTAM Symposium Luleå/Sweden 1991

is the same as the one used previously on cooling [3], it is based on the concept of fictitions time.

The I.T heating diagram used in our model for a hypoeutectoïd carbon steel with a ferrite-pearlite microstructure is presented on figure 1.

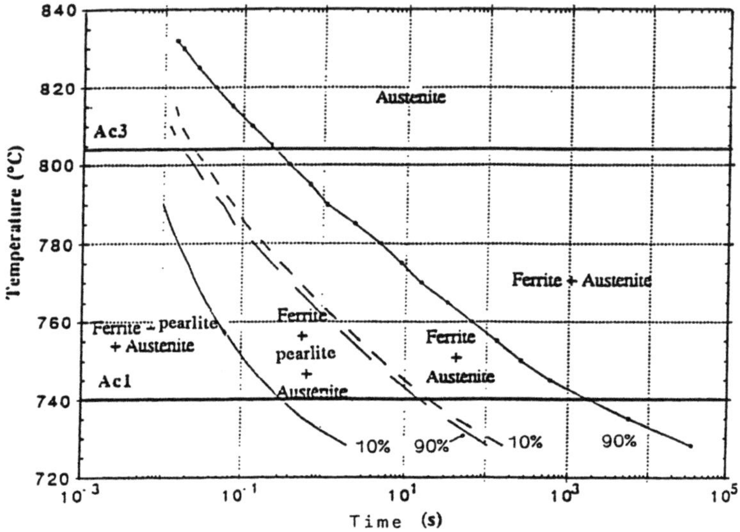

Fig. 1. - Isothermal transformation heating diagram for a XC42 carbon steel

The different steps of the transformation from ferrite-pearlite microstructures into austenite, the mechanisms and the kinetics of these transformations have been widely treated and discussed in the literature [6-11]. From the IT diagram, we can see that austenite formation is considered to occur in two steps : firstly the pearlite dissolution and then the transformation of ferrite. For the first step, the nucleation of austenite occurs instantaneously (above temperature AC_1) so that no incubation period is considered. The diffusion distances are short and the transformation is rapid. For the second step the growth of austenite into the ferrite regions is slower : it is controlled by carbon diffusion. At temperature between AC_1 and AC_3, at completion of the transformation some ferrite remains. The maximum amount of austenite that is formed is calculated from the Fe-C equilibrium diagram. Above temperature AC_3, the completion of the transformation corresponds to a full austenitic structure. At each temperature, the growth of austenite is modelled by two Johnson-Mehl-Avrami laws, one for the dissolution of pearlite and the other for the transformation of ferrite [12].

- Evolution of the carbon content in austenite

During rapid heating austenite is not homogeneous in composition. This inhomogeneity will have an effect on the kinetics of transformation during cooling. Thus it is important to know the carbon content in austenite at the end of the heating process. Analytical and numerical treatments of austenite homogenization from one or two phases have been proposed [13-16].

They are based on the solution of the diffusion equation. In the present model, we have chosen to use experimental carbon distributions. From the experimental study on a XC42 carbon steel [17-18] an evolution law of the carbon content in austenite originating from the pearlite and ferrite regions as a function of the temperature difference $\Delta T = T_{aus} - AC_1$ has been obtained (figure 2) (T_{aus} is the austenitization temperature ; AC_1, the beginning temperature of the transformation is here heating rate dependent).

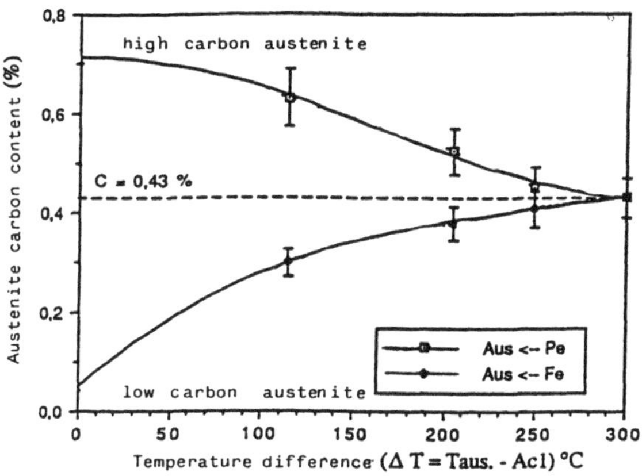

Fig. 2. - Variation of carbon content in austenite as a function of heating conditions. The heating rate ranged from 590 to 800°C/s.

When the temperature difference between AC_1 and T_{aus} is low, the pearlite becomes austenite with 0.7 % carbon and ferrite becomes austenite with a low carbon content (~ 0.02 %). When the austenitization goes on (with time and temperature), carbon diffuses towards the low concentration regions until the mean carbon content of steel is reached (here 0,43 % C).

- Calculation of austenite grain growth

Several authors [19-21] have studied the effect of time and temperature on the grain growth of austenite during rapid heating. According to the rule of additivity (thermal cycle divided into isothermal time steps), austenite grain growth is described by the following relationship [21] :

$$G^a - G_o^a = ko \sum_i \Delta ti \exp\left(-\frac{Q}{RTi}\right)$$

where G is the austenite grain size during heating, G_o the austenite grain size at the beginning of full austenitization, Δti length of time step i, Ti temperature of step i. a, ko, Q are constants and R is the gas constant.

Cooling

This model has been put together with the existing model for the calculation of transformations during continuous cooling based on a rule of additivity. We give here only the principal features of the model [3,22].

Incubation and growth periods are treated separately for diffusion-controlled transformations. The incubation period is characterized according to Scheil's method.

The phase growth is modelled according to the law developed by Johnson-Mehl and Avrami. It applies when the austenite transforms into proeutectoïd constituent (ferrite/cementite), pearlite or bainite. On the other hand, for martensitic transformation the progress of transformation is calculated using the relation established by Koistinen and Marburger.

A hardness calculation is associated with this phase transformation calculation [3,22].

Specific aspects related to rapid heating

The specific aspects of rapid heating (inhomogeneity of austenite and grain size of austenite) have needed new developments of the existing model for calculating phase transformation during cooling. For diffusion dependent transformation, the increase of the austenite grain size leads to a slowing down of the transformation. This effect is modelled by a shifting in the time scale of the IT cooling diagram. When austenite is inhomogeneous, we consider a spreading out in time of its isothermal transformation kinetics on cooling. The modelling of these effects is described in detail in reference [12]. For the martensitic transformation, when austenite is inhomogeneous, the model takes into account a variation of Ms temperature with carbon content. The martensitic start temperatures are respectively $Ms_{\gamma p}$ and $Ms_{\gamma F}$ for the austenite originating from the pearlite regions (high carbon austenite : γp) and for the austenite originating from the ferrite regions (low carbon austenite : γF):

$$Ms_{\gamma p} = Ms_O + h\,\Delta C_{\gamma p}$$
$$Ms_{\gamma F} = Ms_O + h\,\Delta C_{\gamma F}$$

Ms_O is the martensite start temperature for the homogeneous austenite. $\Delta C_{\gamma p}$ ($\Delta C_{\gamma F}$) is the difference between the carbon content of γp (γF) and the mean carbon content of the steel - h is a constant. Thus, the two austenites will have a different progress of martensitic transformation with temperature. In addition, the hardness of martensite is calculated as a function of its carbon content. When austenite is homogeneous, Ms temperature is taken as a function of austenite grain size [12].

This phase transformation calculation model has been coupled with the calculation of the temperature distribution in a cylinder during rapid heating and cooling. The temperature distribution is calculated by solving the heat conduction equation. This equation contains a term describing the rate of energy released by the phase transformation. The details of the thermal calculation model have been given previously [3,17].

Application of the computational model

In order to work out the model, an experimental characterization of phase transformations during rapid heating and cooling has been performed on a XC42 carbon steel (study of the kinetics of transformations, microstructural analysis, hardness measurements... [17,18].

The complete set of input data has been described previously [12]. Firstly, the model has been applied to dilatometric specimens (without radial temperature gradients). We have shown that our model, with the different concepts taken into account, correctly represents the state of austenite at the end of the heating and the following martensitic transformation during cooling [12]. In this paper, we shall only present how the model works on cylindrical specimens with high thermal gradients during heating and cooling. For this purpose we have carried out a numerical simulation of the rapid heating and cooling of a cylinder with 16 mm in diameter. A constant heat flux density is imposed at the surface for heating and for cooling.

Figure 3 shows the temperature evolutions at different locations on the radius of the cylinder. The maximum temperature that is reached at the surface is 1100°C. This temperature lies below AC_1 temperature ($AC_1 \sim 755°C$) in an area between r = 0 and r = 3,5 mm.

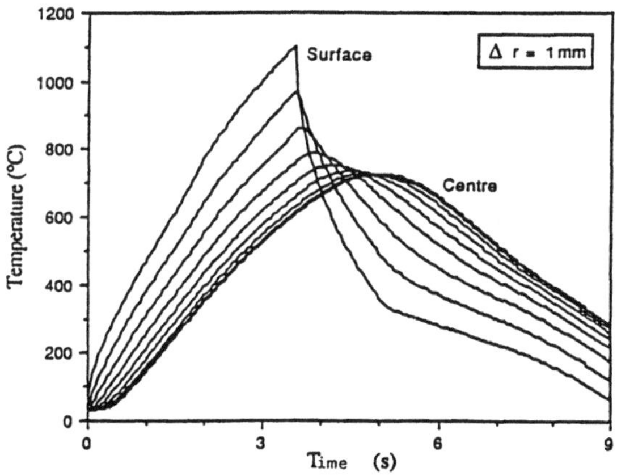

Fig. 3. - Temperature evolutions at different locations on the radius of the cylinder

In order to analyse the state of the austenitization along the radius of the cylinder, we show the volume fraction of austenite in figure 4, the carbon content and the grain size of austenite in figure 5.

Five zones can be distinguished along the radius :

1 : no transformation during heating

2 : only the pearlite is transformed into austenite with a high carbon content

3 : all the pearlite is transformed and the ferrite becomes austenite with a low carbon content

4 : the austenitization process has reached completion and the homogeneisation of austenite

occurs; 5 : austenite is homogeneous and the grain size increases from 6 µm to 15 µm at the surface.

Of course, these different austenitization states will lead to different transformation kinetics during cooling. Figure 6 shows the radial distribution of microstructures at the end of cooling.

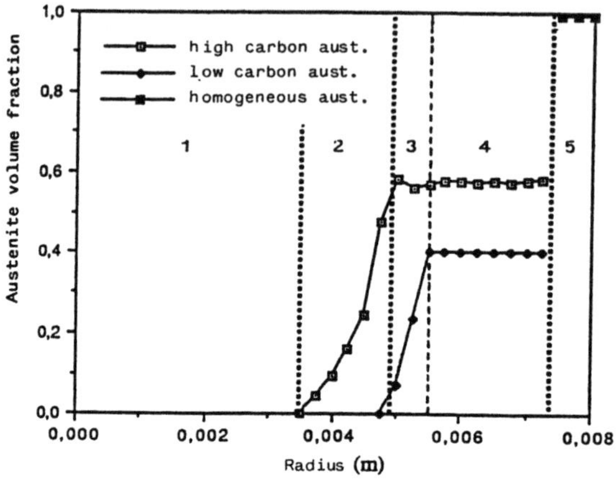

Fig. 4. - Volume fraction of austenite along the radius

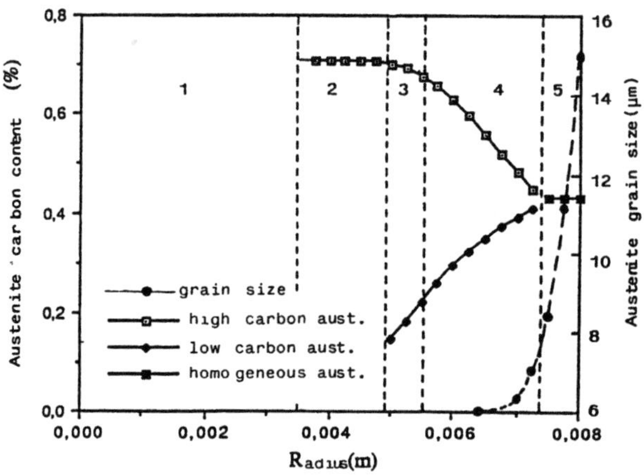

Fig. 5. - Carbon content and grain size of austenite

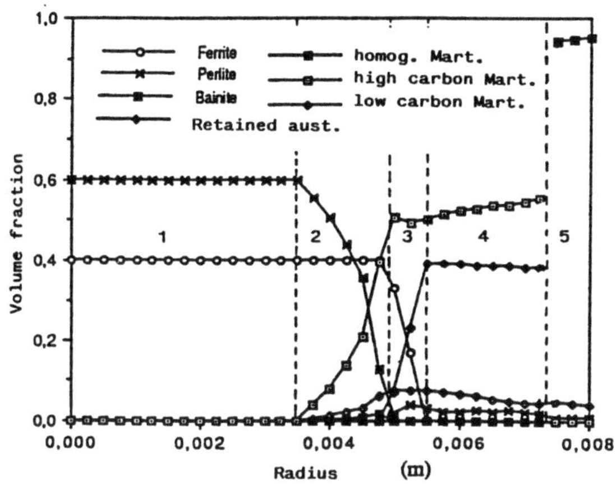

Fig. 6. - Radial distribution of microstructures

As expected in zone 1, we find the original microstructure (60 % pearlite, 40 % ferrite). In zone 2, the volume fraction of high carbon martensite increases at the expense of pearlite. As a consequence of the low Ms temperature (Ms ~ 180°C) an amount of retained austenite is obtained. There remains 40 % ferrite. In zone 3, the amount of high carbon martensite remains nearly constant and the amount of low carbon martensite increases. Zone 4 is made of a mixture of high and low carbon martensites. The amount of high carbon martensite increases slightly and the amount of low carbon martensite decreases slightly. These evolutions are to be related to the variations in Ms temperatures with carbon content. Near the surface (zone 5), the amount of homogeneous martensite reaches 95 %. There is 5 % retained austenite. These microstructure distributions lead to a radial hardness distribution which is characteristic of a surface hardening treatment. Hardness is high (760 HV) and nearly constant down to a depth of 2 mm and then decreases progressively until the original hardness of the steel (240 HV) is reached at 4.5 mm in depth.

In order to test the capability of our model to predict the microstructural evolutions during heating and cooling we have applied it to the induction hardening treatment of a cylinder. Figure 7 shows the comparison between the measured and the calculated hardness profiles along the radius of the cylinder. The calculation represents correctly the hardness evolution excepted in the central zone of the cylinder where the calculated hardness is higher than the experimental one. This difference has to be attributed to a discrepancy between the calculated and experimental temperature evolutions in the centre of the cylinder [12].

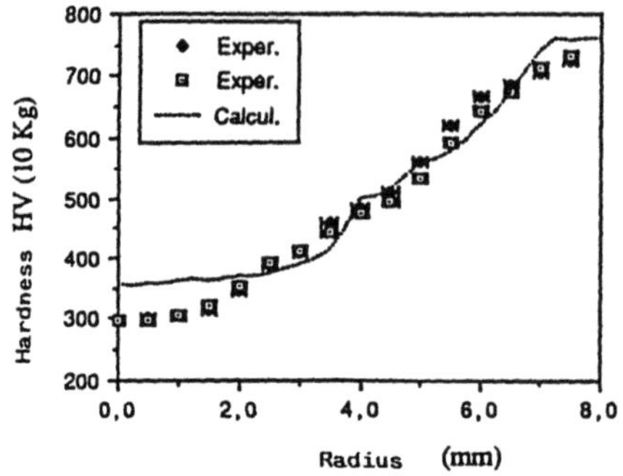

Fig. 7. - Measured and calculated hardness profiles in the case of an induction hardening treatment

Conclusion

The model for calculating phase transformations during rapid heating and cooling that has been developed describes the state of austenite at the end of the heating process and the transformations during cooling.

This model, associated with thermal calculation programs already enables the prediction of microstructure and hardness distribution in workpieces subject to a surface heat treatment. For cylindrical pieces, this model is coupled with a calculation that predicts the development of internal stresses and the distribution of residual stresses too. This type of temperature phase transformation-stress calculation model will be a valuable guide for controlling and optimizing heat treatments.

References

1. B. HILDENWALL : Prediction of the residual stresses created during quenching : especially the quench response in carburized steels. Dissertation n°39, Linköping University (Sweden), 1979.
2. M. MELANDER : A computational and experimental investigation of induction and laser hardening. Dissertation N° 124, Linköping University (Sweden), 1985.
3. F. FERNANDES, S. DENIS, A. SIMON : Prévision dc l'évolution thermique et structurale des aciers au cours de leur refroidissement continu. Mém. et études scient. Revue Métallurgie, 1986, 83, 355.
4. W.A. JOHNSON and R.F. MEHL : Reaction kinetics in processes of nucleation and growth. Trans. AIME, 1939, 135, 416.
5. M. AVRAMI : Kinetics of phase change. Part I : General Theory. J. Chem. Phys. 1939, 7, 1103. Part II : Transformation time relations for random distribution of nucleï. J. Chem. Phys. 1940, 8, 212. Part III : Granulation, phase change and microstructure. J. Chem. Phys. 1941, 9, 177.
6. G.R. SPEICH and A. SZIRMAE : Formation of austenite from ferrite and ferrite-carbide aggregates. Trans. AIME, 1969, 245, 1063.
7. B. KARLSSON : Transformation kinetics for the formation of austenite from a ferritic-pearlitic structure. Z. Metalkde, 1972, 63, 160.

8. A.P. SUROVTSEV and V.V. YAROVOI : Features of polymorphic transformation kinetics during heating for low carbon steels - Met. Sci. Heat Treat. 1984, 9, 649.
9. A.P. SUROVTSEV and V.V. YAROVOI, V.E. SUKHANOV et A.J. PROKLOVA Kinetics of polymorphic transformation in low-carbon steels - Met. Sci. Heat Treat. 1986, 2, 104.
10. V.M. ZALKIN : Some controversial problems of the kinetics of the transformation of pearlite into austenite during heating of steel - Met. Sci. Heat Treat.1986, 1-2, 96.
11. G.R. SPEICH, V.A. DEMAREST et R.L. MILLER : Formation of austenite during intercritical annealing of dual-phase steels. Metall. Trans. 1981, 12 A, 1419.
12. D. FARIAS, S. DENIS, A. SIMON : Modélisation des transformations de phases des aciers en cycles thermiques rapides. Proc. Ecole de Printemps CNRS-EPFL "Lasers de puissance et Traitement des matériaux" 27-31 mai 1991 (à paraître).
13. B. KARLSSON et L.E. LARSSON : Homogenization by two-phase diffusion. Mater. Sci. Eng. 1975, 20, 165.
14. M.G. ASHBY et K.E. EASTERLING : The transformation hardening of steel surfaces by laser beams I - Hypoeutectoïd steels. Acta Metall. 1984, 32, 1935.
15. W. LI : Laser transformation hardening of steel surfaces. Doctorat Thesis, University of Luleä (Sweden) 1984.
16. H.U. FRITSCH, H.W. BERGMANN : Influence of the carbon diffusion during laser transformation hardening. Numerical simulation and experimental verification.European Scientific Laser Workshop on Mathematical Simulation (organized by H.W. BERGMANN) Lisabon 1989. Sprechsaal Publishing Group - D 8630 COBURG, 31.
17. D. FARIAS, S. DENIS, A. SIMON : Les transformations de phases en cycle thermique rapide et leur modélisation - Cas d'un acier XC42. Traitement Thermique, 1990, 237, 63.
18. D. FARIAS : Traitement thermique laser de l'acier XC42 et modélisation des transformations de phases en cycles thermiques rapides au chauffage et au refroidissement. Thèse de Doctorat I.N.P.L. Nancy 1991 (à venir).
19. J. WYSZKOWSKI : Grain growth of austenite on rapid heating. Iron and Steel, 1970, 44, 77.
20. R.A. GRANGE : The rapid heat treatment of steel. Metall. Trans. 1971, 2 A, 65.
21. H. IKAWA, S. SHIN, H. OSHIGE et Y. MEKUCHI : Austenite grain growth of steel during thermal cycles. Trans. JWS, 1977, 8, 46.
22. F.M.B. FERNANDES : Modélisation et calcul de l'évolution de la température et de la microstructure au cours du refroidissement continu des aciers. Thèse de Doctorat de l'Institut National Polytechnique de Lorraine, Nancy 1985.

The Analytical Method of Bainite Transformation Evaluation in Microalloyed Steels in the Process of Rapid Cooling

C. ORLOWSKI

Mechanical Technology Faculty,
Technical University of Gdansk, POLAND

Summary

The investigation aims at the elaboration of an easy, operative and statistically precise estimation method of bainite quantity in underwater weldment HAZ. The scope of investigation has been limited to the representative of microalloyed steels of C-Mn type, low alloy steel used in the Polish shipbuilding industry. The accepted cooling time interval is $t_{800-500}$, for the underwater welding from 1.5 to 10s. It has also been assumed that mainly bainite-martensite structure appear in HAZ in these conditions.The statistical method of transformation analysis has been carried out in the following stages:

a) model investigation of the kinetics of the structural transformation process; the construction of the physical model, its mathematical description and computer simulation;

b) qualitative analysis of the factors influencing structural transformation process based on experiments;

c) dependence of bainite quantity and the factors influencing the transformation process based on the model investigation for the thermal cycles in underwater conditions;

d) statistical relationship between the quantity of bainite, steel composition, the linear energy of welding and hardness.

The Aim of Investigation

In the process of underwater welding of higher strength steel weldments, using a wet method, there appear martensite structures which are disadvantageous from the point of view of structure strength and cause numerous cracks.

L. Karlsson, L.-E. Lindgren, M. Jonsson (Eds.)
Mechanical Effects of Welding
IUTAM Symposium Luleå/Sweden 1991

The use of the local dry cavity method of welding reduces the amount of martensite and limits the risk of cracking. The processes occur in HAZ due to the reduction of cooling rate, thanks to the use of special welding head. Still, as experimental data shows, high cooling rates (although lower than during wet welding) cause the HAZ and the grain size shrinking, (as compared to the welding in the air). Thus the investigations into the kinetics of transformations in underwater welding require the usage of other methods than the ones traditionally used during in the air welding.

Physical Model of Bainite Transformation

The assessment of austenite before transformation (local carbon concentration and austenite grain size) was used as the basis for the physical model determining bainite quantity, its distribution and grain size in HAZ overheated zone after welding. As matalographic analyses (in underwater weldments) showed only the presence of bainite and martensite, the assessment of bainite data made the structural analysis of the material after welding possible.

The physical model has been based on the following factors:

a) austenite assessments after transformation during the process of heating by austenite nucleation and its homogenization processes;

b) modeling of bainite nucleation and cementite precipitation considering:

   - martensite type nucleation for bainite;
   - precipitation of cementite from austenite in upper bainite formation;
   - precipitation of cementite from ferrite in lower bainite formation. (Fig 1.).

The II Fic's law was used for the analysis of austenite homogenization:

$$\partial u/\partial t = D1 * (\partial^2 u/\partial x^2 + \partial^2 u/\partial y^2) \qquad (1)$$

where:

u – carbon concentration in the point, which is the function of coordinates and time t;

$\partial u/\partial t$ – change of carbon concentration in time;

D1 – carbon diffusion coefficient in austenite.

As the process of carbon diffusion takes place in the austenite structure, diffusion coefficient value in austenite was taken for D1.

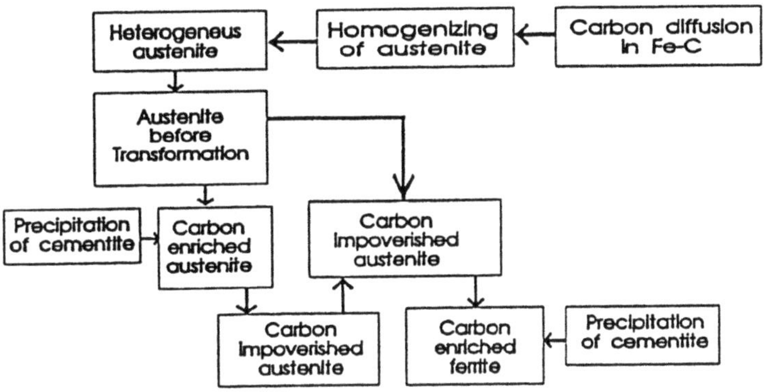

Fig.1. Scheme of physical model.

The value of diffusion coefficient D1 was assumed as for the homogeneous area, disregarding the changes of its values during carbon diffusion across the the grain boundaries. This assumption was connected with the fact that the influence of a grain size on the carbon diffusion process in austenite was not taken into account.

The significant dependence of diffusion coefficient on temperature was accepted for calculations. Because the sample was heated and cooled uniformly, one may assume that the process of diffusion takes place with the same probability in any sufficiently large area. Hence for the determination of carbon concentration the sample area of a rectangular shape has been assumed:

$$S = \{(x, y): \quad a \leq x \leq b, \quad c \leq y \leq d\} \tag{2}$$

It has been assumed that on the area boundaries the diffusion carbon exchange does not take place which determines the following boundary conditions:

$$\partial u/\partial x = 0, \quad x = a, x = b \qquad \partial u/\partial y = 0 \quad y = c, y = d \tag{3}$$

For an univocal solution of the equation (1) with the boundary conditions (3) the determination of an initial condition is indispensable. It has been assumed that the process of diffusion during the heating of a sample from the ambient temperature to the temperature marking the end of austenite is insignificant, thus carbon concentration hasn't changed in this time.

Modeling of Bainite Nucleation and Cementite Precipitation

The following equation has been used for the analysis of bainite nucleation:

$$P(ls) = (-0.0009) * (T - 873) + 0.02 \ [\%] \tag{4}$$

which is the dependence of the temperature on the local carbon concentration in austenite before transformation where:

T – Temperature [K];

P(ls) – carbon fracture in the structure.

The dependence in (4) is accurate in the range of carbon concentration up to 0.8% (max. carbon concentration in the tested austenite).

The quantity of upper and lower bainite has been analysed using equation (5)

$$\begin{aligned} R(I,J) &= k10 * \Delta t * (D1)^{0.5} \ [\%] \\ S(I,J) &= K11 * (G4 * \Delta t)^{0.5} \ [\%] \end{aligned} \tag{5}$$

where:

k10,k11 – kinetic coefficients;

D1,G4 – carbon diffusion in austenite,in ferrite.

The Analysis of the Factors Influencing The Process of Transformation.

The model investigation presented above shows the process of the transformation of bainite from austenite. Its mathematic description allows the qualitative analysis of the model in any thermo-structural conditions. Thus it may be used for the analysis of the transformation in the investigations of welding conditions in the air and underwater. Yet one must start with defining these conditions for a given type of welding first.The presented above specificity of underwater welding is characteristic for its very narrow HAZ and conse- quently an overheated zone. Hence the thermal gradient in this zone is very narrow as well. The vicinity of a very active weld influences carbon diffusion and fine unrecrystallized zone of material limits diffusion and blocks the grain growth. The conditions influence the basic parameters of the model in the following way:

a) Thermal diversification caused by thermal gradient decides upon the process of diffusion with varying probability in a chosen area. Thus the process of diffusion has the highest vicinity of a weld and lower speed in the recrystalization zone. As compared to air welding the process is slower. That is why the quantity of upper bainite is smaller and that of lower bainite is higher. The effective amount of bainite is smaller that in the air;

b) The restriction of grain growth is caused by: an increase of thermal gradient (which results in the change of diffusion and boundary conditions), an increase of compressive stresses;

c) The decrease of an austenite grain growth size slows down the process of bainite formation which increases HAZ hardness;

d) The change of chemical composition influences carbon diffusion and austenite grain growth.

The analysis of the presented factors points to the necessity of quantitative evaluation of the influence of both the thermal diversity and the grain size as well upon bainite formation. One may thus consider the following factors influencing transformation in the statistic investigation:
- chemical composition of steel;
- linear energy of welding;
- grain size of former austenite in HAZ overheated zone.

The Experimental Investigation of Bainite Transformation Kinetics

The experiment was performed in underwater conditions, using the local dry cavity method for steel 15GA and 15G2ANb. The chemical composition of both steel types is presented in table 1.

Table 1.

	Chemical composition of tested steel						
	C	Mn	Si	P	S	Al	Nb
15GA	0.18	1.19	0.35	0.021	0.021	0.055	
15G2ANb	0.11	1.26	0.37	0.016	0.016	0.055	0.044

Butt joints, of 12 mm thickness were welded in the horizontal position at the depth of 20 m with the following parameters: I=280A, V=45V, v=0.13m/min. Joints were made on the copper backing bar using wire SpG2S1NCu diameter 1.6 mm. The results were there statistically described, which resulted in the regression equation:

$$B = -4.66 * x + 5.64 * y + 17705.8 * z - 5.09 \tag{6}$$

where:

B – amount of the formed bainite [%], x-carbon equivalent [%]

y – time of exposition to temperatures $t_{800-500}$ [s]

z – grain size [$*10^{-5m}$]

The equation (6) has the correlation coefficient R=0.99 and statistical parameters: F-ratio = 149.4 and P-value=0.0592. The image analysis method has been used for the identification and evaluation of bainite structures with Micro-Videomat (Opton) image analyzer. The measurement of grain size has the accuracy of $2.5*10^{-6}$ m and that for the amount of structures up to 5%.

The Analysis of the Results

The function dependence represented by the equation (6) made possible the construction of the simple function dependence of bainite quantity on carbon equivalent, $t_{800-500}$ and grain size. The analysis of model investigations enables the substitution of linear energy for $t_{800-500}$ and overheating hardness replaces the grain size in equation (6). The regression analysis has been done for there parameters. The equation (7) presents the correlation dependence (r=0.99) between bainite quantity(B [%]), carbon equivalent(x [%]), hardness (z1 [HV10]) and the linear energy(y1 [KJ/cm]) (F-ratio=68.56, P-value=0.08)

$$B = 150.1 * x + 0.79 * y1 - 0.05 * z1 - 52.85 \qquad (7)$$

Fig.2 presents the graphical values equivalent to the amount of bainite obtained using regression equations for underwater conditions and on the basis of the elaborated equation.

Conclusions

1. The model investigations of bainite transformation kinetics in underwater welding pointed to the necessity of considering the grain size of the former austenite, the chemical composition of steel and thermal cycle parameter in the statistical investigations.
2. The experiments carried out in underwater conditions confirmed the results of model investigation. The regression equation (6) between the amount of bainite and carbon equivalent, $t_{800-500}$, the grain size of former austenite has the correlation coefficient R=0.99.
3. Having considered the results of the physical model, dependence (7) was suggested for practical application. The parameter of the thermal cycle $t_{800-500}$ substitutes linear energy of welding and the maximal HAZ hardness replaces the grain size of the former austenite.

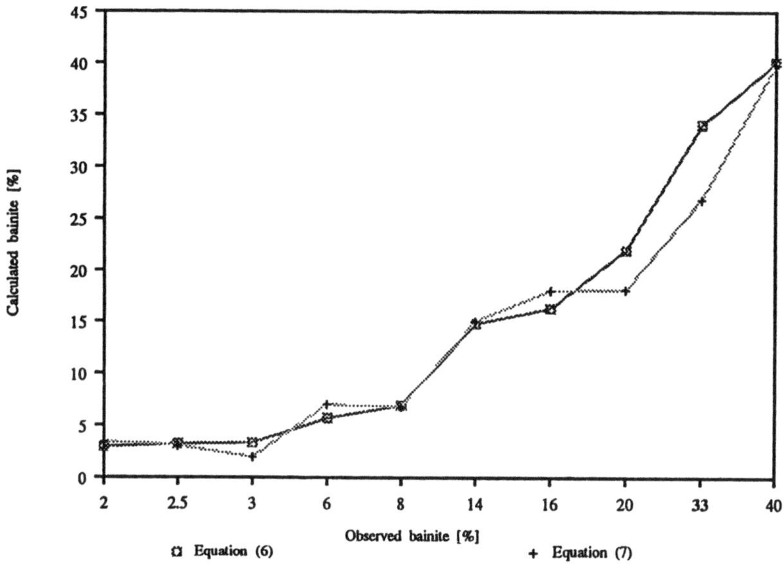

Fig.2. Results of experiments

References

1. Lesinski K.J., Orlowski C., Numerical Method of Austenite Transformation in Micro - Alloyed Steels Applied to Steel Weldability Problems, Phase Transformations '97, Cambridge 1987.

2. Orlowski C. Experimental and Analytical Investigations of the Structural Changes Kinetics in Underwater Conditions. Doctoral Thesis, Gdansk, 1989.

3. Seyffarth P., Kuscher G,. Schweiss ZTU Schaubilder VEB Verlag Technik Berlin, 1982.

4. Hurkans A,. Duit G.A., Hoogendoorn Th.M., Accelerated Cooling and the Transformation of Steel, Transformation of Steel Conference Proceedings "Accelerated Cooling of Steel", Pittsburgh, 1985.

5. John J., C., Easterling K., E., Ashby M., F., A Second Report of Diagrams of Microstructure and Hardnees for Heat Affected Zones in Welds, Technical Report, Lulea University, Sweden 1984.

Thermal Problems of Welding

Jacek RONDA [1]
Oscar MAHRENHOLTZ

Technische Universität Hamburg-Harburg, Meerestechnik II,
D 2100 Hamburg 90, Eissendorfer Str. 42, FRG.

Summary

The present work is aimed at the discussion of the essential problems associated with modelling underwater welding of inelastic structures. A thermomechanical model of underwater welding is proposed assuming rate dependent material reaction, full coupling of thermal and mechanical effects, latent heat accompaning the solid-liquid phase change and environmental water boiling effects. An incremental variational formulation of boundary value problems involving finite strain thermo- elastic-viscoplastic deformation is discussed. A finite element approximation of the variational problem is constructed, and the algorithm for solving the discrete problem is based on ABAQUS and TF-3D. The paper closes with the evaluation of transient and residual stresses and thermal dilatations for a simple butt joint.

Introduction

In underwater welding there occurs heating, when a weld metal is deposited, and fast cooling, when an arc shield displaces the following arc. The model of heat exchange between a weld joint and surrounding water is one of most important subjects of analysis and is proposed assuming the following heat fluxes:

- heat source produced by electric arc,
- internal heat flux resulting from the dissipation of inelastic deformation energy,
- latent heat generated by the solid-liquid phase transformation,
- surface outfluxes.

The phase transformation entails the study of the "thermal free boundary" problem because of the latent heat generation or absorbtion, which occurs on an unknown singular surface S_a [14, 15] defined by the temperature θ_{PH}. The description of the behaviour of the welded material is given in terms of thermo-elastic-viscoplastic theory [10] assuming temperature dependent material parameters following [2, 6].

[1] at present at the University of Cape Town, Department of Applied Mathematics, Private Bag, Rondebosch, 7700 South Africa.

L. Karlsson, L.-E. Lindgren, M. Jonsson (Eds.)
Mechanical Effects of Welding
IUTAM Symposium Luleå/Sweden 1991

Transient or residual stresses, dilatations and temperatures are established by the balance of momentum and the balance law of internal energy, which in integral form encompasses the balance relations held on the singular surface [15]. These relations are often called the Rankine-Hugoniot conditions. The variational form of the incremental boundary value problem is derived following [5, 17] in which the plastic strain rate, equivalent plastic strain rate and temperature rate are unknowns. This problem is solved by using the Galerkin method which is the basis for the finite element method. The thermomechanical problem of a butt welding of plates is a model of an underwater welding process. Numerous examples for this model are solved to show numerical effects of constitutive modelling of thermo-elasto-plastic material behavour. Contour lines of temperature and residual stresses are compared considering or neglecting latent heat, viscous effects and thermomechanical coupling. Examples are evaluated by ABAQUS [1] and TF-3D [11, 12].

Thermomechanical problem formulation

The coupled thermomechanical welding problem is defined by the balance of internal energy, balance of momentum, and thermal and mechanical boundary conditions sufficient to approximate surface and internal phenomena occuring during welding of metals.

Balance of internal energy is expressed by

$$\rho_0 c_V \dot{\theta} + \text{div } q = \text{tr } \{\underline{\tau} : \dot{\underline{\varepsilon}}\} + \rho_0 R \tag{1}$$

where: θ = temperature; c_V = thermal capacity of constant volume; ρ_0 = initial mass density; q = heat flux per unit area; $\underline{\tau} = (\rho_0/\rho)\underline{\sigma}$ is the Kirchhoff stress tensor; $\underline{\sigma}$ is the Cauchy stress and $\dot{\underline{\varepsilon}}$ is the strain rate in the actual configuration; R is energy radiation; $\underline{\tau} : \dot{\underline{\varepsilon}}$ is the scalar product of $\underline{\tau}$ and $\dot{\underline{\varepsilon}}$. The additive strain rate decomposition of the form

$$\dot{\underline{\varepsilon}} = \dot{\underline{\varepsilon}}^{\text{el}} + \dot{\underline{\varepsilon}}^{\text{th}} + \dot{\underline{\varepsilon}}^{\text{VP}} \tag{2}$$

is assumed, where $\dot{\underline{\varepsilon}}^{\text{el}}$, $\dot{\underline{\varepsilon}}^{\text{th}}$ and $\dot{\underline{\varepsilon}}^{\text{VP}}$ are the elastic, thermal and visco-plastic parts of the total strain rate $\dot{\underline{\varepsilon}}$.

A description of the behaviour of welded materials is given in terms of thermo-elasto-viscoplasticity theory [10]. The elastic and thermal parts of strain rate are defined by the following constitutive equations:

$$\dot{\varepsilon}_{ij}^{\text{el}} = \frac{1}{2\mu}\left(\overset{\triangledown}{\tau}_{ij} - \frac{\kappa}{2\mu + 3\kappa}\overset{\triangledown}{\tau}_{kk}\,\delta_{ij}\right), \qquad \dot{\varepsilon}_{ij}^{\text{th}} = \frac{1}{3}\alpha\dot{\theta}\delta_{ij}, \tag{3}$$

where: α = dilation (i.e. thermal expansion); μ = shear modulus; κ = Lame's coefficient; stress rate is given by $\overset{\triangledown}{\underline{\tau}} = (\rho_0/\rho)(\dot{\underline{\sigma}} + \underline{\sigma}\,\text{tr}\,\underline{L} - \underline{\omega}\cdot\underline{\sigma} + \underline{\sigma}\cdot\underline{\omega})$; $\dot{\theta}$ = rate of temperature; $\underline{L}$ = velocity gradient, $\underline{L} = \dot{\underline{\varepsilon}} + \underline{\omega}$; $\underline{\omega}$ = spin, $\underline{\omega} = \frac{1}{2}(\underline{L} - \underline{L}^T)$; $\dot{\underline{\varepsilon}} = \frac{1}{2}(\underline{L} + \underline{L}^T)$. The constitutive equation for the inelastic strain is defined [10] by

$$\dot{\varepsilon}_{ij}^{\mathrm{VP}} = \gamma \langle \Phi(f) \rangle s_{ij} \tag{4}$$

where: γ = fluidity parameter; s_{ij} = deviatoric stress, $s_{ij} = (\rho_0/\rho)(\sigma_{ij} - \frac{1}{3}\sigma_{kk}\delta_{ij})$; f = Mises yield function used to characterise the material, $f(\underline{\tau}, \underline{\tau}_0) = \overline{\tau} - \tau_0(\overline{\varepsilon}^{\mathrm{VP}}, \theta)$; $\overline{\tau}$ = Mises equivalent stress defined in terms of Cauchy's deviatoric stress, $\overline{\tau} = (\rho_0/\rho)\sqrt{\frac{3}{2}s_{ij}s_{ij}}$; $\tau_0(\overline{\varepsilon}^{\mathrm{VP}}, \theta)$ = yield limit; $\overline{\varepsilon}^{\mathrm{VP}}$ = equivalent strain; $\overline{\varepsilon}^{\mathrm{VP}} = \int_0^t \dot{\overline{\varepsilon}}^{\mathrm{VP}} dt = \int_0^T \sqrt{\frac{2}{3}\dot{\varepsilon}_{ij}^{\mathrm{VP}}\dot{\varepsilon}_{ij}^{\mathrm{VP}}}\, dt$; $\langle \Phi \rangle$ is taken as nonzero for positive values of the yield function f. Actual forms of the fluidity function γ and the overstress function Φ are:

$$\gamma = A\exp(-Q/RT) \tag{5}$$

$$\Phi = [\sinh(\beta\overline{\tau})]^n \tag{6}$$

where A, B, Q, n, R are material parameters [8]. Expressing $\overset{\nabla}{\tau}$ in terms of elastic strain rate $\dot{\underline{\varepsilon}}^{\mathrm{el}}$ by inverting the first relation of 3 and substituting it in 1 together with 2 and the second part of 3 gives

$$\rho_0 c_V \dot{\theta} - \lambda_{,k}\theta_{,k} - \lambda\theta_{,kk} - \dot{\theta}\left[-\gamma_{kl}\dot{\varepsilon}_{kl}^{el} + \left(2\frac{\partial\mu}{\partial\theta} + \frac{\partial\kappa}{\partial\theta}\right)\varepsilon_{kl}^{el}\dot{\varepsilon}_{kl}^{el}\right] = \rho_0 R + \xi\overline{\tau}\dot{\overline{\varepsilon}}^{\mathrm{VP}} \tag{7}$$

where: grad $q = \lambda_{,k}\theta_{,k} + \lambda\theta_{,kk}$; $\xi = f/J$, $\gamma_{kl} = (3\kappa + 2\mu)\alpha\delta_{kl}$; f = fraction of plastic work that turns into heat; J = mechanical equivalent of heat; λ = the thermal conductivity; $\underline{\tau} : \dot{\underline{\varepsilon}}^{\mathrm{VP}} = \overline{\tau}\dot{\overline{\varepsilon}}^{\mathrm{VP}}$.

The dissipation of inelastic energy $\xi\overline{\tau}\dot{\overline{\varepsilon}}^{\mathrm{VP}}$ is treated as an internal heat flux. The balance law on a singular surface $S_a \subset \partial V$, where either solid-liquid or solid-solid phase transformations take place, is expressed as the Rankine-Hugoniot condition [14, 15] for a heat flux $\underline{q}$ transferred through the surface, i.e.

$$\lambda_{kl}\theta_{,l}n_k = Q_{LH} - \underline{q}_{ext} \cdot \underline{n} \tag{8}$$

where: Q_{LH} = latent heat defined by the constitutive equation with respect to a Helmholtz free energy and a fusion temperature; q_{ext} = external heat flux, which has to be found by solving the problem of thermal contact between a weld and an environment. The surface S_a is not necessarily a real physical surface and therefore it may be a thin region [15] in the case of phase transformations occuring in alloys. These two balance laws 7 and 8 could be combined, only expressing 7 in the functional form. The stationary condition for such a functional is

$$\int_V \rho c_V \dot{\theta}\,\delta\theta\, d\Omega - \int_V \left[-\gamma_{kl}\dot{\varepsilon}_{kl}^{el} + \left(2\frac{\partial\mu}{\partial\theta} + \frac{\partial\kappa}{\partial\theta}\right)\varepsilon_{kl}\dot{\varepsilon}_{kl}^{el}\right]\dot{\theta}\,\delta\theta\, d\Omega - \int_V \lambda_{kl}\theta_{,k}\,\delta\theta_{,k}\, d\Omega$$

$$+ \int_{\partial V/S_a} \lambda_{kl}\theta_{,l}\,\delta\theta\, n_k\, dS + \int_{S_a} (Q_{LH} - \underline{q}_{ext} \cdot \underline{n})\delta\theta\, dS = \rho_0 \int_V R\,\delta\theta\, d\Omega + \xi \int_V \overline{\tau}\dot{\overline{\varepsilon}}^{\mathrm{VP}}\,\delta\theta\, dV \tag{9}$$

where V is a volume of a body, i.e., a weld joint.

Heat transfer between the weld joint and the surrounding water is modelled considering convection, radiation and boiling phenomena. Therefore the underwater welding problem is solved with thermal boundary conditions determined in respect to the surface temperature range and radiation shape factors suitable for sides of a joint. The heat flux on the upper surface is defined by

$$q = (h_{cu} + h_{ru} + h_{bu})(\theta_E - \theta_S) - Q^* \delta \tag{10}$$

where: h_{cu} = convection heat transfer coefficient [9]; h_{ru} = radiation heat transfer coefficient; h_{bu} = heat transfer coefficent [4] in the stable film boiling region on a horizontal surface; θ_E = environmental temperature; θ_S = boundary surface temperature; Q^* = heat flow supplied to weld joint; $\delta = 1$ for the heating period of a weld cycle; $\delta = 0$ for the cooling period of a weld cycle. These heat transfer coefficients are given by the following expressions:

$$h_{cu} = 0.14 \left(\frac{\rho c_W g \beta k_W^2}{\nu_W} \right)^{1/3} (\theta_W - \theta_\infty)^{1/3}, \tag{11}$$

$$h_{ru} = \frac{\sigma \varepsilon (\theta_W^4 - \theta_{sat}^4)}{\theta_W - \theta_{sat}}, \tag{12}$$

$$h_{bu} = 0.425 \left\{ \frac{k_{vf}^3 \rho_{vf} (\rho_L - \rho_v) g (h_{fg} + 0.4\, c_{pv} \Delta\theta_x)}{\mu_v \Delta\theta_x \sqrt{g_c \sigma_t \,/\, g(\rho_L - \rho_v)}} \right\}^{1/4}, \tag{13}$$

where: $\Delta\theta_x = \theta_w - \theta_s$; θ_w = mean wall temperature; θ_∞ = free stream temperature; θ_{sat} = saturation temperature; β = volume coefficient of expansion for water; g = acceleration of gravity; c_W = specific heat of water; c_{pv} = specific heat of vapour at constant pressure; ρ = density of water; ρ_L = density of saturated liquid; ρ_v = density of saturated vapour; ρ_{vf} = vapour density evaluated at film temperature; L = characteristic length of a weld joint; ν_W = kinetic viscosity of water; k_W = thermal conductivity of water; k_{vf} = thermal conductivity of vapour at film temperature; σ_t = surface tension of liquid-vapour interface; ε = emissivity of the surface; h_{fg} = enthalpy of vapourization; μ_v = viscosity of vapour; g_c = conversion factor $g_c = 1\,\mathrm{kg\,m/N\,s^2}$; σ = Stefan-Boltzman constant.

The heat flux on the bottom and lateral surfaces is defined by

$$q = (h_{cl} + h_{rl})(\theta_E - \theta_S) + q_{bl} \tag{14}$$

The heat transfer coefficients are given by 13 for h_{rl} and the following empirical relation [9]

$$h_{cl} = 0.27 \left(\frac{\rho c_W g \beta k_W^3}{L \nu_W} \right)^{1/4} (\theta_w - \theta_\infty)^{1/4}. \tag{15}$$

The heat flux for nucleate pool boiling q_{bl} is defined by [13]

$$q_{bl} = \left(\frac{C_l \Delta\theta_x}{h_{fg}\, Pr_l\, C_{sf}}\right)^3 \frac{\mu_l h_{fg}}{\sqrt{g_c \sigma_t \,/\, g(\rho_l - \rho_v)}}, \tag{16}$$

where: C_l = specific heat of saturated liquid; Pr_l = Prandtl number of saturated liquid; C_{sf} = constant, determined from experimental data.

Equation 9 must be solved simultaneously with the equation expressing balance of momentum, i.e.,

$$\int_V \overset{\nabla}{\tau}_{ki}\, \delta\dot{\varepsilon}_{kj}\, dV = \int_{S_t} \dot{f}_i \delta v_i\, ds + \int_V \dot{b}_i \delta v_i\, dV, \tag{17}$$

where: $\dot{f}_i$ = rate of surface traction defined on the surface S_t; $\dot{b}_i$ = rate of body force per unit actual volume; v_i = velocity of displacement. The stress rate $\overset{\nabla}{\tau}_{ki}$ may be replaced in 16 by the use of the elastic constitutive equation

$$\overset{\nabla}{\tau}_{ij} = C^E_{ijkl}\left[\dot{\varepsilon}_{kl} - \dot{\varepsilon}^{VP}_{kl} - \dot{\varepsilon}^{th}_{kl}\right], \tag{18}$$

where the fourth order elastic tensor C^E_{ijkl} is expressed by Lame and Poisson constants to assure the consistency with the form 3. Eulerian strain rate components at the actual configuration are given in relation to a velocity displacement by

$$\dot{\varepsilon}_{ij} = \frac{1}{2}\left(\frac{\partial v_i}{\partial x_j} + \frac{\partial v_j}{\partial x_i}\right). \tag{19}$$

Finite Element Approximation of Thermomechanical Problem

Introducing linear time approximations and finite element interpolation, the matrix form of the internal energy equilibrium equation for the element assemblage is

$$(\underline{K}^t_\lambda + \underline{K}^t_{b(u+l)} + \underline{K}^t_\gamma + \underline{K}^t_{\mu\kappa})\Delta_{(i)}\underline{\theta} + {}_{(i-1)}\underline{C}^{t+\Delta t}\,{}_{(i)}\dot{\underline{\theta}}^{t+\Delta t} = {}_{(i-1)}\underline{Q}^{t+\Delta t}_K, \tag{20}$$

where $\underline{K}^t_{b(u+l)}$, $\underline{K}^t_\lambda$, $\underline{K}^t_\gamma$, $\underline{K}^t_{\mu\kappa}$ are stiffness matrices corresponding to boundary conditions defined on the upper and other surfaces, and to thermo-elastic constitutive equation 3; $\Delta_{(i)}\underline{\theta}$ = vector of nodal temperature increments at i-th iteration of $(t + \Delta t)$ time step; ${}_{(i)}\dot{\underline{\theta}}^{(t+\Delta t)}$ = vector of nodal temperature rates at i-th iteration of $(t + \Delta t)$ time step; ${}_{(i-1)}\underline{C}^{t+\Delta t}$ corresponds to the term of 9 with ρc_V; the right hand side heat flux ${}_{i-1}\underline{Q}^{t+\Delta t}_K$, which includes inelastic dissipation, is defined in [11, 12].

The finite element equation corresponding to the balance of momentum 16 can be written in the form [17]

$$\underline{K}^t \cdot \delta\underline{v} = \underline{\mathcal{R}}^{t+\Delta t} - {}^{IN}\underline{F}^t \tag{21}$$

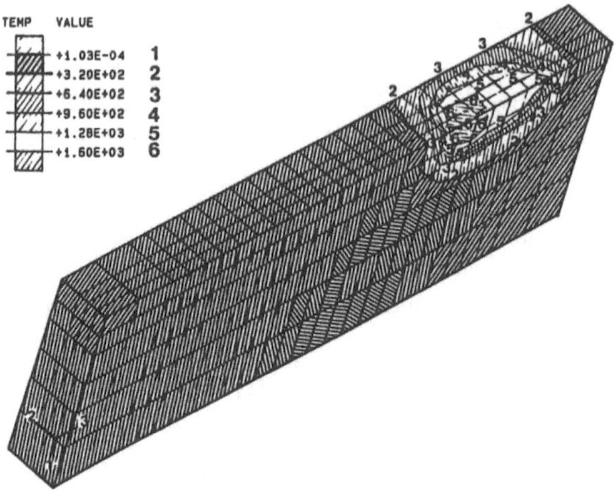

Figure 1: Temperature at the start of cooling period.

where: $\underline{K}^t$ = incremental stiffness matrix arising from the constitutive equations 3 and 4; $\delta\underline{v}$ = vector of nodal displacement variations; $\underline{\mathcal{R}}^{t+\Delta t}$ = vector of the externally applied loads; ${}^{\mathrm{IN}}\underline{F}^t$ = vector of nodal point forces equivalent to the element stresses.

Two different programs, based on the finite element approximations, are used to solve the thermomechanical problem in the heating and cooling periods respectively. Thus, the final temperature at the end of the heating period is evaluated by the use of TF-3D [12] and the residual stresses at the end of the fast cooling period are calculated by ABAQUS [1] for the defined initial temperature distribution. The coupled system of finite element equations 20 and 21 is solved simultaneously by ABAQUS in the case of the full coupling of thermal and mechanical effects.

Results

The model used in the analysis of transient and residual stresses in the butt welded plates is a symmetric plate with a lengthwise single V- groove. Due to symmetry one plate of $120 \times 40 \times 10$ is considered. The shape of the V-groove root is standard, i.e., the angle is 45° and the depth is 8.5 mm. The molten pool is roughly assumed to be a moving ellipsoid, whose projection onto the welding surface is 5 mm long and 4 mm wide. The power of a welding arc transmitted to a weld is 3085 J/s which results because of welding current $A = 190$ A, voltage $V = 28$V and efficiency $\eta = 0.58$. The electric arc covers the distance $l_W = 40$ mm with the welding speeed $v = 4$ mm/s.

The finite element mesh is composed of 624 eight-noded brick isoparametric elements in four layers and 945 nodes. Material parameters such as heat capacity, heat conductivity,

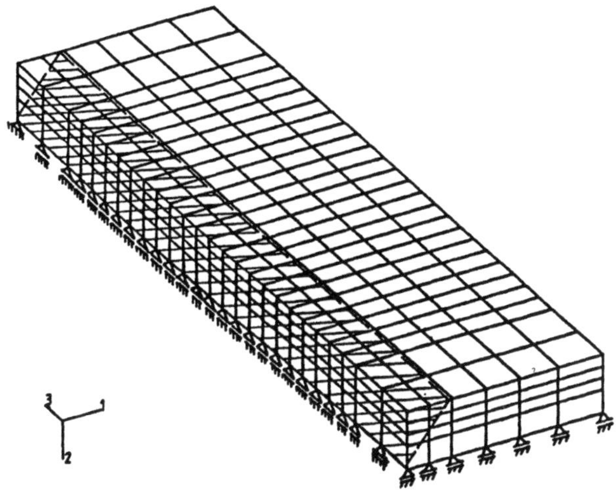

Figure 2: Support of the plate.

yield limit and Young's modulus are temperature dependent and follow the material data appropriate for HT36-steel [2, 6]. The Young's modulus for the range of temperature 0°C to 1600°C changes from 2.11×10^5 MPa to 10 MPa. The plastic yield limit goes through the region 362.9 MPa at 20°C to 1 MPa at 1600°C. The Poisson's ratio is included with a lower limit of 0.255 at 20°C and an upper limit of 0.485 at 1600°C. The phase change is determined by the temperature $\theta_{PH} = 1410$°C and its tolerance $\Delta\theta_{PH} = 50$°C, and the portion of latent heat energy $Q_{LH} = 271.6$ kJ/kg. The data accepted for the viscoplastic constitutive equations 5 and 6, recommended by [8] are the following: $A = 1.2 \times 10^{12}\,s^{-1}$, $\beta = 0.0012\,mm^2/N$; $Q = 351$ kJ/mol; $n = 3.5$; $R = 8.31$ J/mol · K.

The water pressure is assumed to be $1.013 \times 10^6\,N/m^2$ which represents welding conditions at a depth of 10 m. The underwater welding problem has been solved accounting for the following three constitutive models of welded plates:

- thermo-elasto-plastic material with one-sided thermomechanical coupling,
- thermo-elasto-plastic material with full coupling of thermomechanical effects,
- thermo-elasto-viscoplastic material with full coupling of thermomechanical effects.

In all cases the isotropic work hardening is considered. This paper presents the results obtained for the cooling period of welding, which starts after $t = 10.0$ s of the heating period. The initial temperature for the cooling, distributed on the surface of welding and the contact surface of welded plates, is shown in fig. 1. The maximum temperature is less than 1600°C. The cooling period is terminated after 30 s, when the temperature of the weld joint reaches 15°C. The welded plate is supported on the bottom such that the top left hand corner (see fig. 2) is fixed and vertical movement on the other nodes is assumed zero.

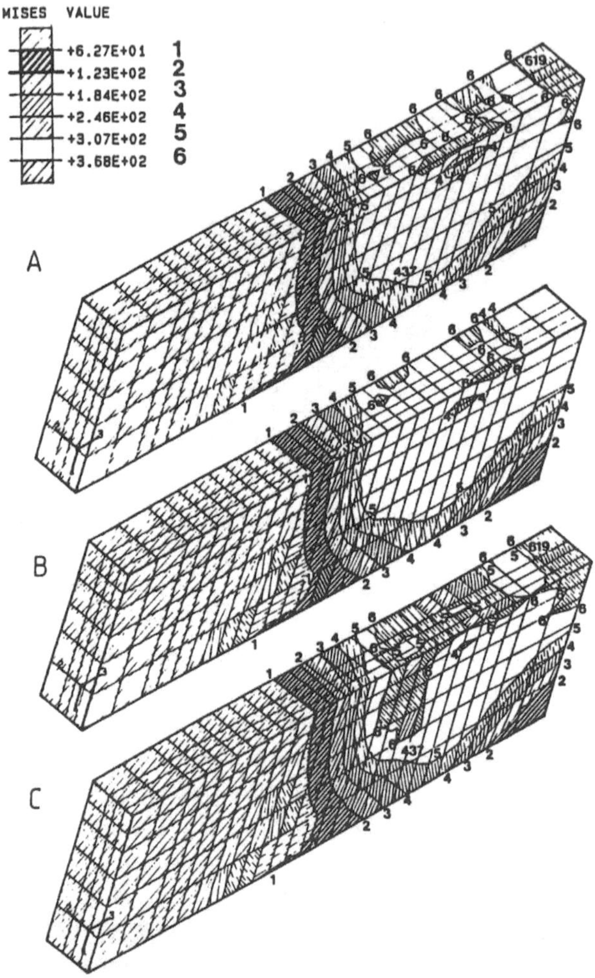

Figure 3: The second invariant of residual stresses: (A) plasticity with one-sided coupling; (B) plasticity with full coupling; (C) viscoplasticity with full coupling.

CONSTITUTIVE THEORY	MAXIMUM OF THE SECOND INVARIANT OF RESIDUAL STRESSES		
	element no.	$\bar{\tau}$ value MPa	r_f %
plasticity with one-sided coupling	621	$\bar{\tau}_{os} = 362.5$	0
plasticity with full coupling	437	359.7	−0.77
	621	357.1	−1.51
viscoplasticity with full coupling	619	384.4	6.04
	621	380.2	4.89

Figure 4: The relative factor r_f is defined by $r_f = (\bar{\tau} - \bar{\tau}_{os}) \;/\; \bar{\tau}_{os} \times 100\%$, where $\bar{\tau}_{os}$ is the maximum value of the second invariant of stress at node 621.

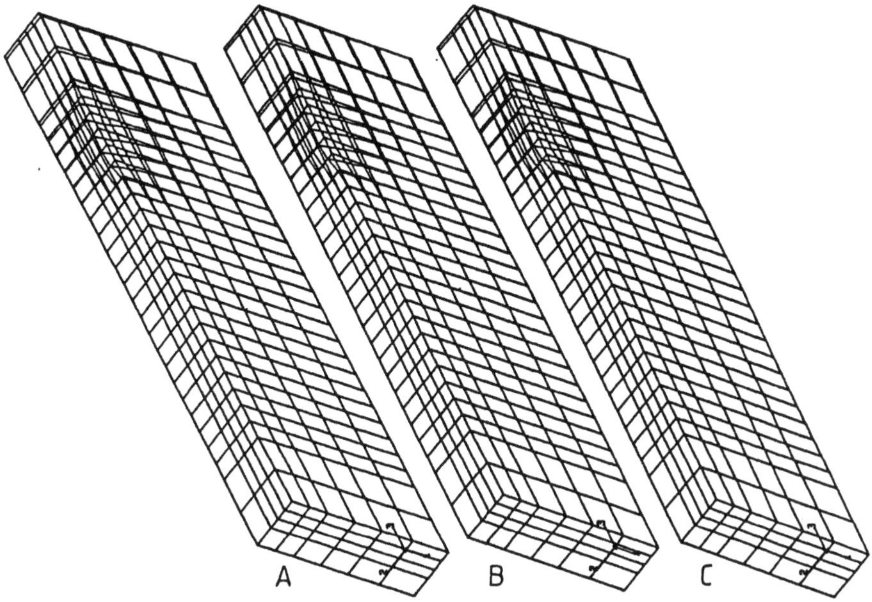

Figure 5: Undeformed and deformed networks for welded plate: (A) plasticity with one-sided coupling; (B) plasticity with full coupling; (C) viscoplasticity with full coupling

Residual stresses generated after cooling are evaluated by **ABAQUS** for each of the three constitutive models. These stresses, represented by the second invariant of the Kirchhoff stress $\overline{\tau}$, are shown in fig. 3. One may conclude, from this figure, that the greater mean stresses occur for the solution with rate dependent effects, and the smaller mean stresses can be noted for the plastic theory with the full thermal coupling. Morever, the residual stresses are compared in respect to the maximum of $\overline{\tau}$ appropriate for the plastic solution with one-sided thermal coupling. Such comparisons are given in fig. 4. The thermal stress maximums are of the same range, but their locations are completely different. The element 437 is placed at the upper layer of the plate, but the elements 619 and 621 are situated on the bottom layer and near the front lateral surface. The permanent displacements conjugated with residual stress are presented in fig. 5 where deformed networks are shown against the background of the virgin plate.

Acknowledgements

J. Ronda would like to thank Prof. J.B. Martin of CERECAM for the use of computing facilities, Graeme Oliver for data preparation and programming, and Kevin Colville for the excellent text preparation.

References

[1] ABAQUS, FE Program: Hibbit, Karlsson & Sorensen, Inc.

[2] B.A.B. Anderson: Thermal stresses in a submerged arc-welded joint considering phase transformations. Trans. ASME, vol. 100 (1978) 356–362.

[3] K.J. Bathe: Finite Element Procedures. Prentice-Hall, Englewood Clifs, 1982.

[4] P. Berenson: Transition boiling heat transfer from a horizontal surface. AICHE Paper 18, ASME-AICHE Heat Transfer Conference, Buffalo, August (1960)

[5] T.J.R. Hughes and R.L. Taylor: Unconditionally stable algorithms for quasi-static elasto/visco-plastic finite element analysis. Computers & Structures 8 (1979) 169–173.

[6] L. Karlsson: Thermal stresses in welding, in R.B. Hetnarski (ed.) Thermal stresses I. Elsevier Publ. (1986) 300–389.

[7] W.H. McAdams: Heat transmission. McGraw-Hill, New York, (1954)

[8] M. Kleiber and Z. Mróz: Computer aided inelastic analysis for design and instruction, in: D.R.J Owen, E. Hinton and E. Onate (eds): Computational Plasticity, Models, Software and Applications. Pineridge Press (1988) 161–179.

[9] R. Krengel: Anpassung von Warmwaltz-Stichplanen an Veranderungen der Waltzguttemperatur. Institut für Werkstoffunformung, T.V. Clausthal (1985)

[10] P. Perzyna: Fundemental problems in viscoelasticity. Advances in Applied Mechanics 9 (1966) 243–377.

[11] J. Ronda and O. Mahrenholtz: Thermomechanical simulation of underwater welding processes. To appear in Ing. Archive 1991.

[12] J. Ronda and A.J. Galicki: TF-3D Finite Element Program, User's Guide. für die Teilbearbeitung des Teilprojektes A4 im SFB 264, Technische Universität Hamburg-Harburg, MT II, Harburg. (1989)

[13] W.M. Rohsenow: A method of correlating heat transfer data for surface boiling liquids. Trans. ASME 74 (1952) 969–982.

[14] W.O. Williams: Axioms for work and energy in general continua II, surfaces of discontinuity. Arch. Rat. Mech. Anal. 49 (1972) 225–240.

[15] K. Wilmanski: Thermodynamic foundations of thermoelasticity, in G. Lebon and P. Perzyna (eds): Recent developments in thermomechanics of solids. Springer-Verlag (1980) 1–94.

[16] K. Wilmanski: Propagation of the interface in the stress-induced austemite-martensite transformation. Ingenieur Archiv 53 (1983) 291–301.

[17] L.M. Taylor and E.B. Becker: Some computational aspects of large deformation, rate-dependent plasticity problems. Comp. Meth. Appl. Mech. Engng. 41 (1983) 251–277.

Three-Dimensional Simulation of a Laser Surface Treatment Through Steady State Computation in the Heat Source's Comoving Frame

J.M. BERGHEAU(*), D. PONT(*) and J.B. LEBLOND(**)

(*) *FRAMASOFT+CSI, 10 rue Juliette Récamier, 69398 Lyon Cedex 03, France*
(**) *Laboratoire de Modélisation en Mécanique, Université Paris VI, tour 66, 4 place Jussieu, 75005 Paris, France*

ABSTRACT

Superficial laser treatments are currently used to generate hard metallurgical structures and compressive residual stresses on the surface of the component treated. This paper presents a three-dimensional numerical simulation of such a treatment, based on a computation of the thermal, metallurgical and mechanical steady state in the laser's comoving frame. This simulation notably shows that very high tensile stresses are generated below the external compression zone; this result yields important consequences upon fatigue crack initiation and propagation.

INTRODUCTION

Laser surface treatments are becoming a standard process in the metallurgical industry. The benefits expected from such treatments are a hard, martensitic structure and also compressive residual stresses, which should improve the material behaviour with respect to the initiation and propagation of fatigue cracks.

Numerical simulations offer a promising means of determining the metallurgical structure and the residual stresses inasmuch as direct measurements are costly and obviously cannot provide the quantities of interest at all points of the structure. Such thermal, metallurgical and mechanical simulations have become more and more common in the last decade in the context of welding and quenching procedures. However numerical simulations of laser treatments have remained limited to thermal and metallurgical computations; see *e.g.* Lunéville [1]. The aim of this paper is to present a complete, thermal, metallurgical and mechanical simulation of such a treatment, using the SYSWELD software [2].

This calculation is fully three-dimensional. Very few such computations are reported in the literature (see, however, Oddy *et al.* [3]), because they are extremely costly and time-consuming. In order to circumvent this difficulty, the present simulation uses a trick which consists in directly evaluating the stationary state in the laser's comoving frame without calculating the intermediary transient states; this greatly shortens the calculation by reducing the number of time-steps to only one. This idea was originally proposed by Nguyen and Rahimian [4], who used it to study a moving crack in an elastoplastic medium; it was later applied by Dang Van *et al.* [5] and Maitournam [6] to the evaluation of stresses induced in rails by the roll of wheels.

L. Karlsson, L.-E. Lindgren, M. Jonsson (Eds.)
Mechanical Effects of Welding
IUTAM Symposium Luleå/Sweden 1991

The results of the computation are compared with measurements performed by the *Institut de Soudure* using X-ray diffraction and the incremental hole drilling method. The effect of taking the third direction into account is also assessed through comparison with two-dimensional simulations performed under both plane strain and generalized plane strain assumptions.

GENERAL FEATURES OF THE SIMULATION

The simulation is performed in two steps: first, thermal and metallurgical calculation; second, mechanical calculation. The second step uses the results of the first one but feedbacks are disregarded. Both of them involve a calculation of the stationary state followed by a transient computation aimed at simulating the laser beam's leaving the plate.

The thermal simulation accounts for the dependence of the heat capacity and the conductivity upon temperature (the calculation is non-linear). Losses due to free convection and radiation are introduced through the use of a temperature-dependent heat transfer coefficient. Though the SYSWELD software permits to take the effect of transformation latent heats into account in a fully rigorous manner (the thermal and metallurgical calculations are then fully coupled), this effect is simply schematized here by means of a peak of the heat capacity near 700 °C. The metallurgical calculation is based on a kinetic model described in [7], which applies to martensitic as well as diffusion-controlled transformations.

The second step is basically a thermoelastoplastic calculation, but is somewhat more complex due to the influence of transformations, which arises from several phenomena: first, the dependence of the various mechanical charateristics upon the proportions of the phases; second, the transformation-induced volume changes; third, the so-called "transformation plasticity" effect. The model used to describe the latter phenomenon is based on both theoretical and numerical studies described in [8] and [9].

With regard to the calculation of stationary states in a moving frame, the first step uses special, discontinuous test functions originally proposed by Hughes and Brooks [10] in order to avoid spurious oscillations often observed for high Péclet numbers. In the second step, the only difference with respect to a classical transient calculation lies in the plastic stress correction: values of mechanical quantities at a given Gauss point at the beginning of the time-step considered are replaced by values at a "preceding" Gauss point in the direction of the heat source movement. This necessitates the definition of sequences of Gauss points, which is done by a pre-processor. This also puts some restrictions on envisageable meshes: namely, Gauss points must form straight lines parallel to the heat source velocity.

FIRST SIMULATION

We consider a plate made of 35NCD16 steel (AFNOR standard) with dimensions 60×50×25 mm; due to symmetry about the mid-plane, only one half of the structure is represented in the finite-element mesh (*Figure 1*). The trace of the beam on the treated surface is circular; the energy per unit surface is constant over a disk of radius 4.6 mm, then decreases linearly down to zero over a 5.3 mm distance. In a first simulation, the beam power is assumed to be 3100 W. This power is absorbed by the metal *via* an absorption coefficient which is a strongly increasing function of temperature.

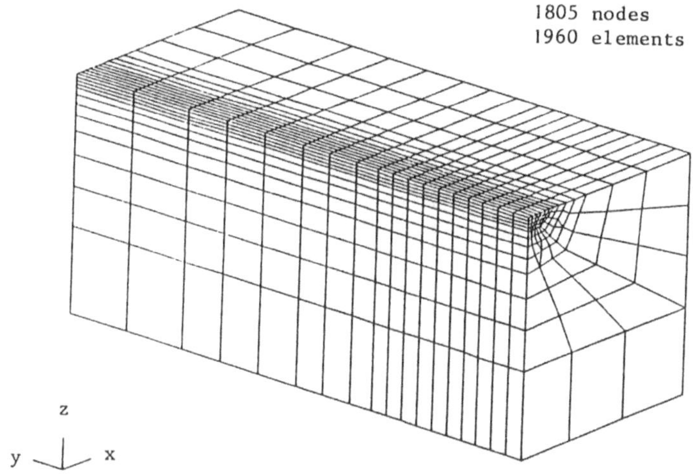

Figure 1: Finite element mesh

The cooling rate is very high: about 360°C/s at 700°C. The metallurgical structure is consequently completely martensitic in the heat affected zone (HAZ); *Figure 2* shows the computed martensite distribution. The depth of the "AC_1 line" is about 1.3 mm.

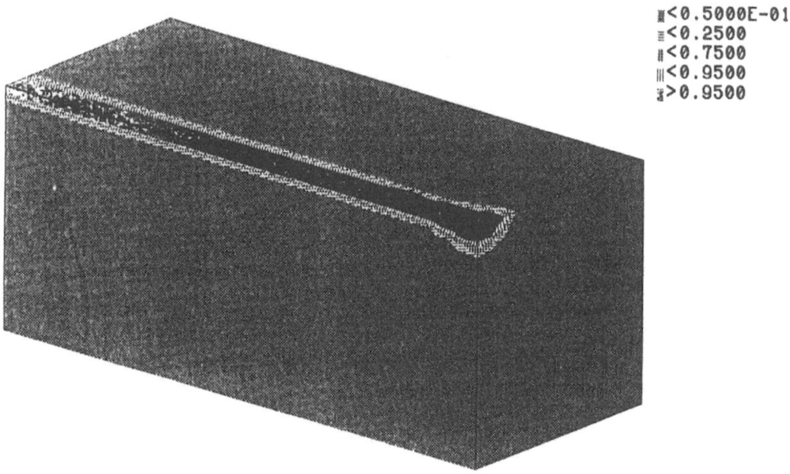

Figure 2: Final martensite distribution

Figure 3 shows the distribution of the final transverse (σ_{xx}) stress. It appears that this stress is highly compressive in the HAZ (due to the large volume increase induced by the martensitic transformation), but highly tensile just below it; the minimum and maximum stresses amount to about - 800 and + 800 MPa respectively. The longitudinal (σ_{yy}) stress is not shown but its behaviour is the same except for a global shift toward higher values. (On the other hand, the through-the-thickness stress σ_{zz} is much lower in absolute value). This bears important consequences upon the subsequent behaviour of the component with respect to fatigue crack initiation and propagation. Indeed, while the laser treatment tends to inhibit crack initiation in the HAZ, it favors it at somewhat larger depths; furthermore cracks should propagate toward the interior of the plate rather than toward the external surface, which means that they should remain undetected if only surface observations are made. It should also be observed that *Figure 3* exhibits very high stress gradients, which casts some doubts upon the ability of experimental procedures to capture such stress fields; this point will be further discussed below.

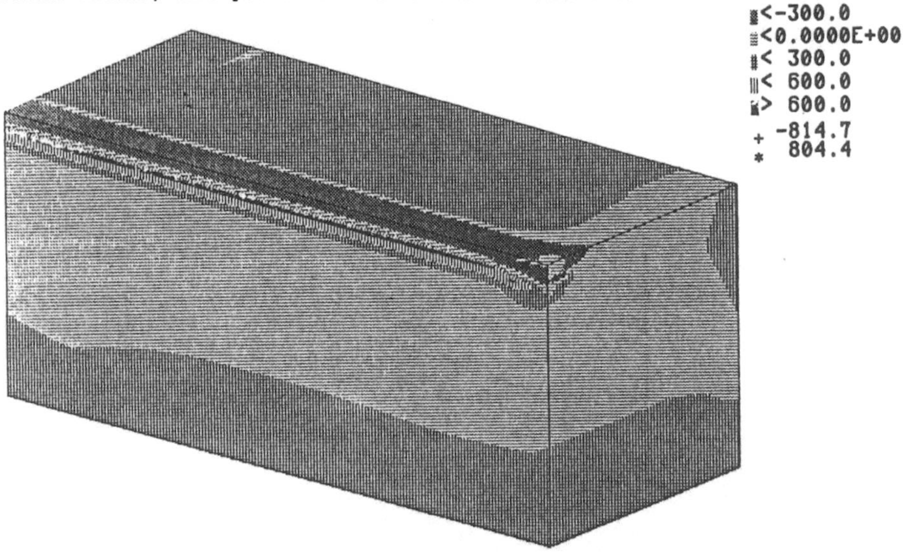

Figure 3: Distribution of the residual transverse stress

Figures 4 to 7 show the variations of the residual transverse and longitudinal stresses halfway along the plate, on the symmetry plane (from the upper to the lower surface) and on the upper surface (from the symmetry plane to the edge); points plotted include those obtained from the three-dimensional simulation, from X-ray diffraction experiments conducted by the *Institut de Soudure*, and from two-dimensional (plane strain or generalized plane strain) simulations. (In the latter simulation, the planes orthogonal to the laser velocity can have a translatory motion along the y-axis *and* rotate about the x-axis; on the other hand, because of the symmetry with respect to the yz-plane, rotations about the z-axis are prohibited). It appears that the results of the three simulations are quite close for the transverse stress, but not so for the longitudinal stress in the HAZ, for which they increase in the order (3D) - (generalized plane strain) - (plane strain), in obvious correspondence with the increasing stiffness of the structure in the y-direction. On the other hand, the comparison with experimental results is somewhat disappointing: in the HAZ, none of the simulations yield a really satisfactory convergence for the transverse stress, and those giving the best agreement for the longitudinal stress are two-dimensional. One possible interpretation would be that the experiments tend to smooth out brutal stress variations; indeed the experimental minimum and maximum stresses are strikingly lower in absolute value than the corresponding computed quantities. This interpretation is also supported by the fact that in about half the experiments reported here, cracks (which of course tend to relax the stresses) were observed below the HAZ.

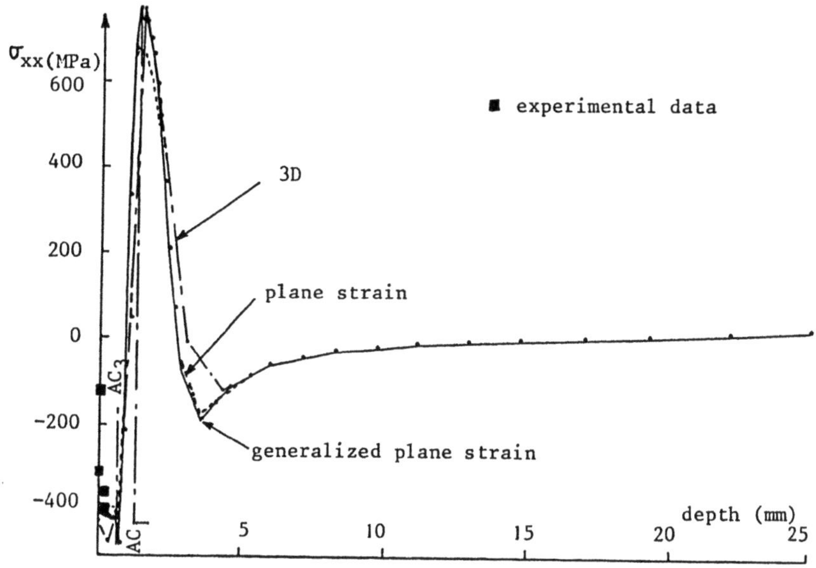

Figure 4: Variations of the residual transverse stress on the plane of symmetry (laser beam power: 3100 W)

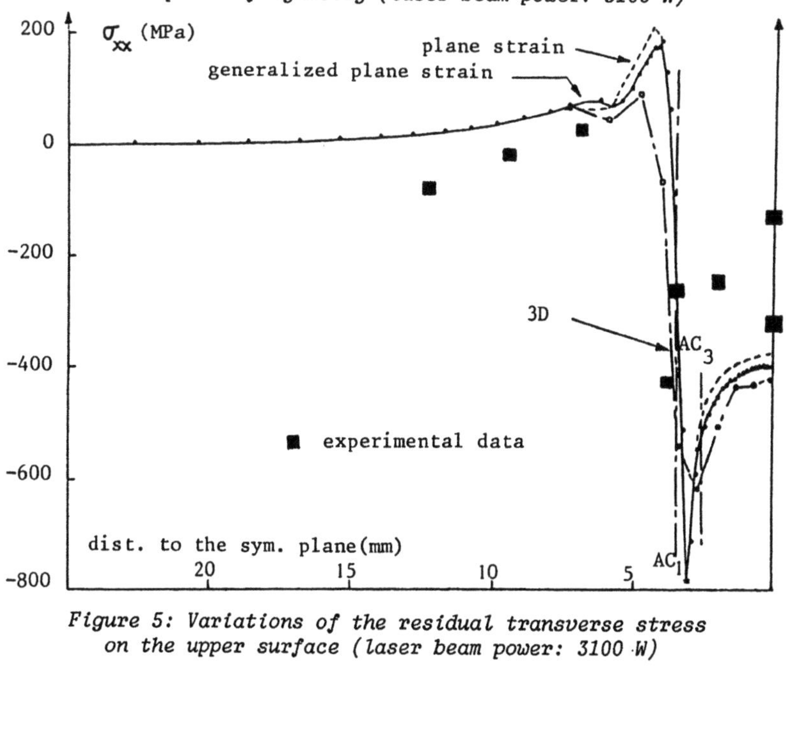

Figure 5: Variations of the residual transverse stress on the upper surface (laser beam power: 3100 W)

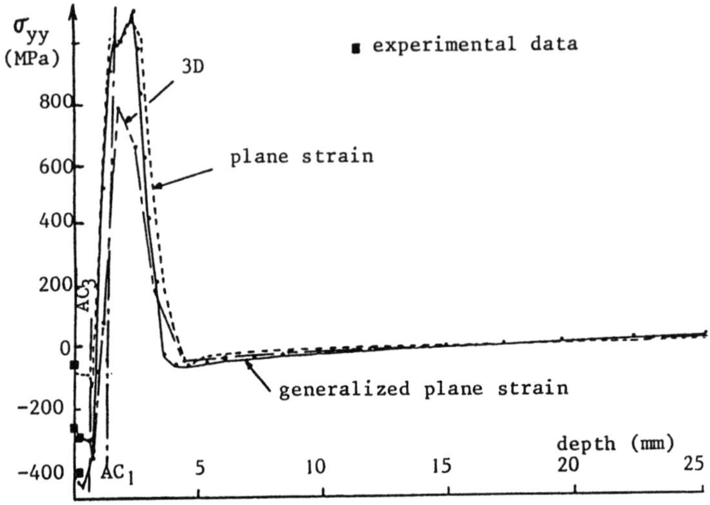

Figure 6: Variations of the residual longitudinal stress on the plane of symmetry (laser beam power: 3100 W)

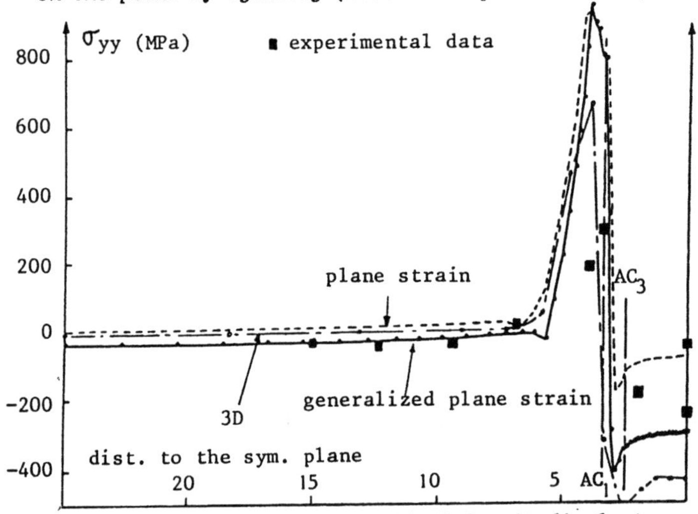

Figure 7: Variations of the residual longitudinal stress on the upper surface (laser beam power: 3100 W)

SECOND SIMULATION

We now study the case of a slightly higher beam power (3150 W instead of 3100); since the absorption coefficient strongly increases with temperature, this results in a notably higher heat input and a larger HAZ. The aim of this new simulation is to assess the influence of the HAZ size upon the residual stresses, and also to perform new comparisons with other experiments carried out for this beam power by the incremental hole drilling method.

Since in the previous case, two-dimensional simulations provided the best fit with the experiments, a two-dimensional computation is carried out (under a generalized plane strain assumption). The depth of the "AC_1 line" is notably greater than before (2 mm instead of 1.3), as expected; however *Figures 8 and 9*, which show the variations of the residual transverse and longitudinal stresses halfway along the plate on the symmetry plane, exhibit little differences with respect to the case of the lower beam power: hence the HAZ size appears to have a relatively minor influence upon residual stresses, at least in the range studied. On the other hand, the convergence of calculated and measured results is much better than before; in particular, the experiments reproduce quite closely the stress peaks in compression and tension obtained numerically. Furthermore, though no three-dimensional computation was performed for this beam power, it may be inferred from the previous simulations that such a computation would result in values of the longitudinal stresses in the HAZ lower than those obtained here by about 130 MPa; the agreement with experiments would then become even better. All this again supports the idea that the somewhat disappointing convergence observed for the lower beam power should be attributed primarily to the experiments, and not the simulations.

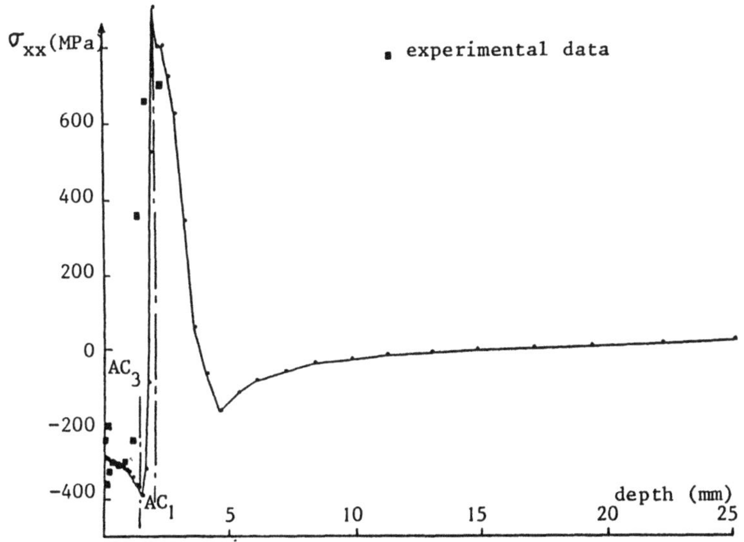

Figure 8: Variations of the residual transverse stress on the plane of symmetry (laser beam power: 3150 W)

Finally, one should mention that a similar study was carried out very recently by Deperrois, as part of his thesis work [11]. Though the mesh used was rather crude, the results obtained qualitatively confirmed the present ones. Also, a systematic study of the influence of the beam power and velocity was undertaken, and the agreement of the results with experiments was encouraging; in particular, the computations confirmed the experimental observation that surface stresses can become tensile for very large penetration depths. The crudeness of the mesh unfortunately made it impossible to numerically optimize the process.

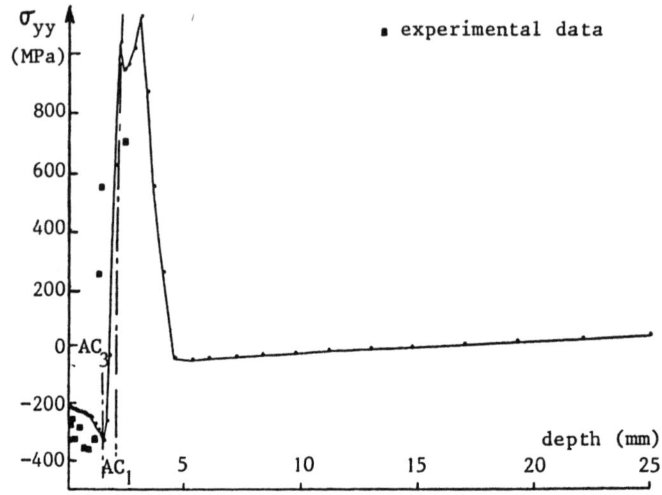

Figure 9: Variations of the residual longitudinal stress on the plane of symmetry (laser beam power: 3150 W)

REFERENCES

1. Lunéville, E. *Simulation et contrôle de la trempe superficielle par laser*. Thèse de Doctorat de l'Université Paris VI, 1988.

2. SYSWELD User's Manual, FRAMASOFT Report n° SM/NT.88/1010 (1988).

3. Oddy, A.S., Goldak, J.A., and McDill, J.M.J. *Numerical analysis of transformation plasticity in 3D finite element analysis of welds*. **Eur. J. Mech. A/Solids, 9**, 253 (1990).

4. Nguyen, Q.S., and Rahimian, M. *Mouvement permanent d'une fissure en milieu élastoplastique*. **J. Méc. Appl., 5**, 95 (1981).

5. Dang Van, K., Inglebert, G., and Proix, J.M. *Sur un nouvel algorithme de calcul des structures élastoplastiques en régime stationnaire.* Conference presented at the 3rd **"Colloque sur les tendances actuelles en calcul des structures"**, Bastia, France (1985).

6. Maitournam, H. *Résolution numérique des problèmes élastoplastiques stationnaires*. Thèse de Doctorat de l'ENPC, 1989.

7. Leblond, J.B., and Devaux, J. *A new kinetic model for anisothermal metallurgical transformations in steels, including effect of austenite grain size*. **Acta Met., 32**, 137 (1984).

8. Leblond, J.B., Mottet, G., and Devaux, J. *A theoretical and numerical approach to the plastic behaviour of steels during phase transformations - I: Derivation of general relations; II: Study of classical plasticity for ideal-plastic phases*. **J. Mech. Phys. Solids, 34**, 395 (1986).

9. Leblond, J.B., Devaux, J., and Devaux, J.C. *Mathematical modelling of transformation plasticity in steels - I: Case of ideal-plastic phases; II: Coupling with strain hardening phenomena*. **Int. J. Plast., 5**, 551 (1989).

10. Hugues, T.J.R., and Brooks, A. *A theoretical framework for Petrov-Galerkin methods with discontinuous weighting functions: application to the streamline-upwind procedure*, in **Finite Elements in Fluids**, edited by Gallagher, Norrie, Oden and Zienkiewicz, Wiley, New York (1982).

11. Deperrois, A. Private communication (1991).

Temperature Fields and Stress States in Welded Tubes of Rectangular Cross Section

R.Parkitny, A.Pawlak and W.Piekarska
The Institute of Mechanics and Machine Designe,
Technical University of Częstochowa, 42-200 Częstochowa, Poland

Summary

The model for calculating the temperature field in welded elements is presented. The algorithm is founded on the method of elementary heat balances. Melting of electrode rod away and melting of parent material and then solidification of molten pool are considered. The calculated temperature courses at characteristic points in welded tubes of rectangular cross-section are presented. On the basis of calculated temperature field the uniaxial state of stress was determined. The residual stresses in welded tubes are shown and the analysis of stresses in tubes under the load was carried out.

Introduction

Welding is an important process applied in the joining of elements. It is connected with the use of concentrated powerful sources of heat producing melting of elements in the region of joint. Each point of the welded element is influenced by the thermal cycle, so the process of welding is accompanied by a changing in time temperature field. The nonuniform temperature field results in generation of residual stresses that remain in the material after welding is completed and produces material changes in the heat affected zone (HAZ). The material heterogenity and residual stresses have an essential influence on the ability of welded joints to carry loads and should be taken into account in the process of designing. It applies particulary to the carrying elements of machine and devices and such is the tube of rectangular cross-section being considered. Temperature distribution in welded elements depends on the type of welding method, its parameters, geometry of elements and their thermal properties. The complex phenomena of heat transfer in welding result in the fact that numerous simplifying assump-

L. Karlsson, L.-E. Lindgren, M. Jonsson (Eds.)
Mechanical Effects of Welding
IUTAM Symposium Luleå/Sweden 1991

tions are made in calculation. This concerns mainly the analitical models [3,4,10]. More accurate solutions are obtained for models founded on the finite element method [1,2,6,8].The methods applied and solutions obtained have been widely described and discussed by L.Karlsson [9].However, they do not include the phenomenon of material transport from the electrode to the joint and the phenomena of generation and then solidification of a spot of molten metal.

In this model the characteristic phenomena accompanying welding are thoroughly considered. Therefore, one considers the electrode rod filler metal that melts away and fills the gap between the elements being joined. Heat is supplied to the joint from the burning arc and from the spot of molten metal. The division into finite elements can be performed for any arbitrary geometry of the joint. By considering the welded tube as a straight prismatic bar being in a varying temperature field, the uniaxial state of actual and residual stresses generated by welding has been calculated on the basis of the theory of small elasto-plastic deformations.

Algorithm of temperature calculation

The region Ω of a plate being welded is divided into three dimensional elements in the form of triangular prisms. Total area of triangles is equal to the cross-section area of the element being welded in the plane x-y and the sum of heights of elements of the same bases is equal to the length of the plate. For each element the heat balance in successive intervals of time is made, as below:

$$c\,\rho\,dV\,\frac{\partial T}{\partial t}\,dt = \sum dQ_i - dQ_0 + dQ_Z \qquad (1)$$

where:

- T - temperature,
- dV - volume of element,
- dt - interval of time,
- dQ_i - heat supplied by element "i" (i-th element),
- dQ_o - heat transferred to environment,
- dQ_Z - heat supplied by arc,
- c - heat capacity of parent material,
- ρ - density of parent material.

Heat dQ_i supplied by element "i" is determined on the basis of Fourier equation:

$$q = -\lambda \frac{\partial T}{\partial n} \qquad (2)$$

where:

λ - thermal conductivity.

Heat dQ_o transferred to surrounding is determined on the basis of Newton equation:

$$q = \alpha \left(T - T_0 \right) \qquad (3)$$

where:

α - heat transfer coefficient,

T_0 - surrounding temperature.

In welding , to the elements being in the region of burning arc heat dQ_Z is supplied:

$$dQ_Z = \eta \, q_Z \qquad (4)$$

where:

q_Z - effective output of heat source,

η - $\eta(r^2)$ - heat source concentration coefficient.

By considering heat balance (1) for all elements, the following system of equations is obtained:

$$[\,A\,]\,[\,T^{K+1}] = [\,B\,(T^K)\,] \qquad (5)$$

where:

T^K - temperature vector for time $t = t_K$,

T^{K+1} - temperature vector for time $t_{K+1} = t_K + dt$.

If the temperature field T_0 at the beginning of welding is known, on the basis of equation (5), one will get the temperature fields for successive times. In order to obtain stable solutions, the Cranck-Nicholson differential method was used. In the presented algorithm the peculiarity resulting from the MAG welding method has been taken into account. It is assumed that there is an intensive exchange of heat between the element borders and gas stream in the region encircled through the gas stream.

Temperature field in welded tubes

On the basis of the presented model a programme in PASCAL has been written for IBM-PC class computers. This programme has then been applied in calculating the temperature field in a welded tube of rectangular cross-section shown in Fig. 1. Tubes

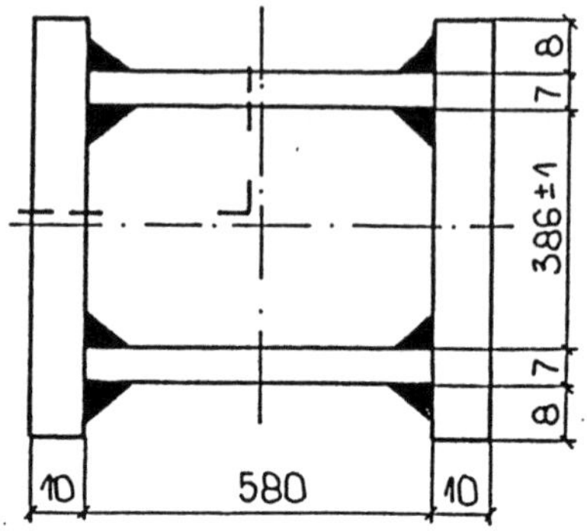

Fig.1. The cross-section of the welded tube

of this profile are used in the gibbets of telescopic car cranes. They are made of low-alloy steel of the following chemical composition :C - 0.14, Mn - 0.5, Si - 0.3, Cu - 0.43, Cr - 0.36 Ni - 0.075, Al - 0.075, Mo - 0.42, V - 0.05, B - 0.002.

The tube has been welded in the MAG process of the following parameters:

$$U = 26 \text{ [V]} \quad I = 260 \text{ [A]} \quad V = 30 \text{ [m/h]}$$

Because of the symmetry of making welded joints, only one quater, that is shown in Fig. 1, has been considered.

The following values of material constants have been used:

$\rho = 7800$ [kg/m^5] - density,

$L = L^* = 27.10^4$[J/kg] - specific heat of melting and solidification,

$T_m = T_l = 1773$ [K] - melting and solidification temperat.

In Table 1 the values of heat conductivity, specific heat and coefficient of heat transfer to surrounding used in calculation, are presented.

In Fig.2.a-c the temperature field isotherms during the making of outside joint (filet weld 1) for times 1.79, 2.2 and 5.0 s are presented. In the first case (Fig.2.a) it is seen that the process of welding has just started; the weld is not filled yet

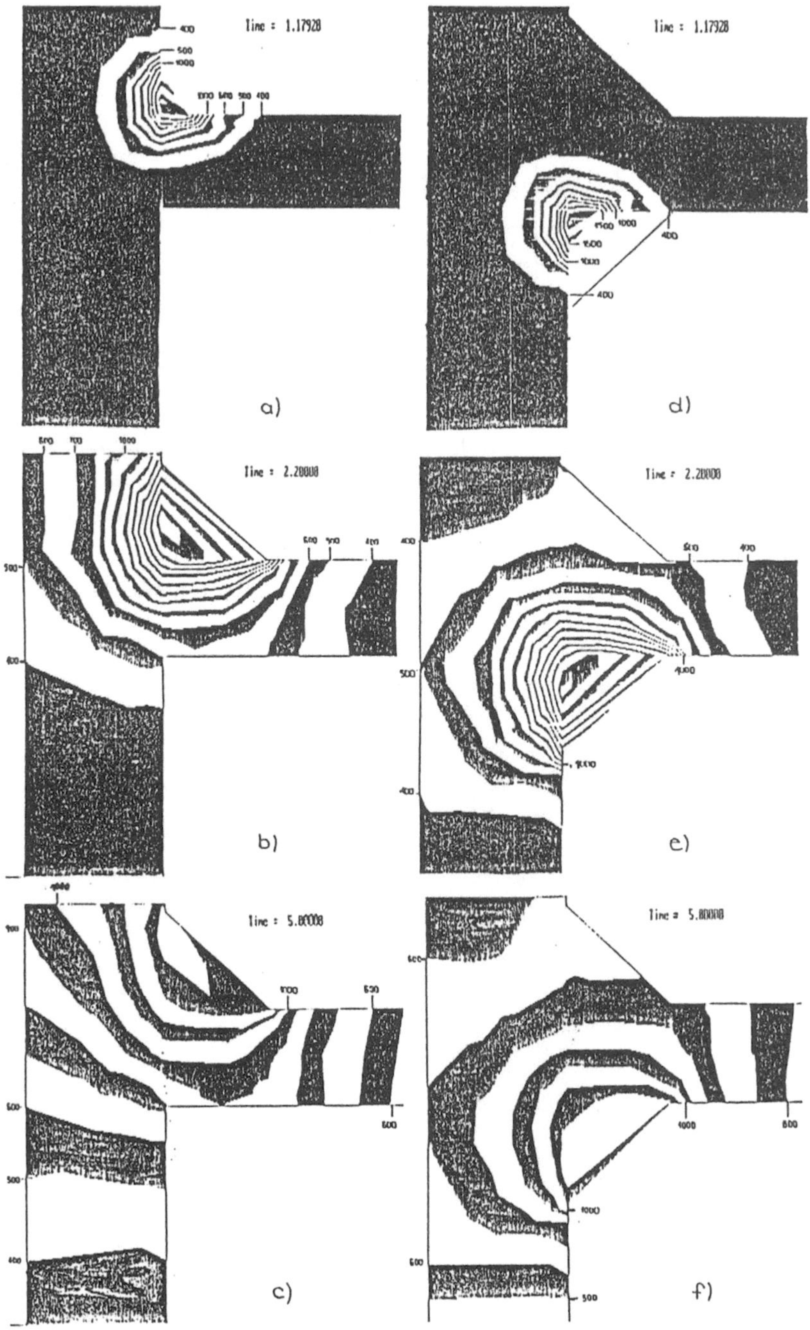

Fig.2. The isotherms of temperature field during making joint 1 (2.a-c) and joint 2 (2.d-f) at selected times

Table 1.

Temperature versus thermal conductivity, specific heat and coefficient of heat transfer to surrounding.

T [K]	273	473	673	873	1073	1273	1473	2073
λ [J/W m]	64	56	48	39	36	35	36.4	37
C [J/kg]	480	500	540	580	620	650	660	660
α [W/m K]	10	20	30	50	80	120	200	500

and there is an extensive area of not heated material of temperature equal to the surrounding temperature. In the second case (Fig.2.b) the weld is already well shaped and the cooling process starts ; an area of liquid filler metal and melted base material (molten pool) is seen. The third case concerns the period in which cooling of material happens; the weld is already congealed and the area of weld neighbourhood is heated.

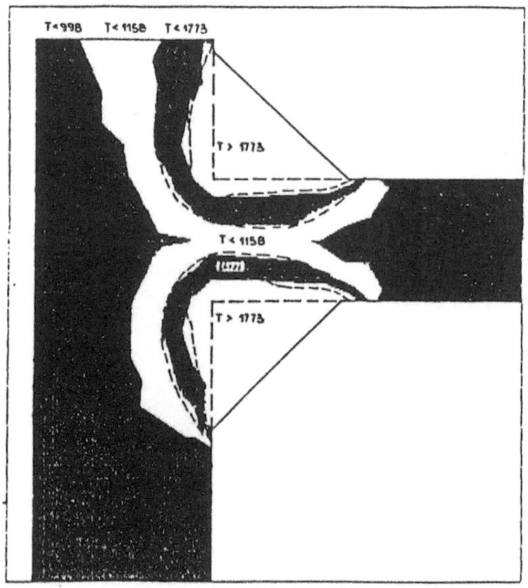

Fig.3. The characteristic areas of welded joints obtained experimentally and on the base of calculation

Analogously, in Fig.2.d-f, the temperature field isotherms of the joint 2 (filet weld 2) making process are shown. In Fig. 2.d at time 1.79s , that is just after the welding process is started. In Fig.2.e the isotherms after the weld is completed are presented. In the last figure (Fig.2.f) the isotherms for time 5s are shown.

Each point of the welded element is influenced by subsequent thermal cycles. In the corner considered, only the thermal cycles connected with making the weld 1 and 2 are essential. Based on the calculated temperature field, the area of weld (including the fusion zone) has been determined and also the area of transformation zone , the heat affected zone HAZ and the area of not changed parent metal. These areas are shown in Fig.3. For the sake of comparison, in the same figure the areas obtained experimentally, from the microsection of the joint, are marked with an interrupted line. The slightly difference in the upper corner in the area of HAZ of joint 1 may results from the fact that in welded sample its corner was longer: 15 and not 8 mm, as in considered tubes, taken for the calculation. The agreement achieved is considered to be good and the calculated temperature field has been used to determine of the instantaneous and residual stresses in welded tubes.

Stresses

By considering the welded tube as a straight prismatic bar being in a varying temperature field the uniaxial state of actual and residual stresses generated by welding has been calculated on the basis of the theory of small elasto-plastic deformations The Young's modulus E, varying with temperature, was determined from the stress-strain curves which were obtained on the basis of own tensile tests. The tests also allowed to determine the elastic limit, for considered low-alloy-steel of chemical composition mentioned above, as a function of temperature. The obtained values are shown in Table 2.

Table 2.

Mechanical properties of considered low-alloy-steel as a function of temperature.

T [K]	293	473	573	673	773	873	973
E [10^{-5}MPa]	2.03	1.92	1.81	1.68	1.59	1.4	0.51
Y [MPa]	830	770	730	616	594	400	84

The dilatometric tests were carried out and the diagram : rate of cooling($\upsilon_{800/500}$) - temperature(T) - transformation(η), was prepared, where $\upsilon_{800/500}$ is a medium rate of cooling for temperature interval (T-273) $\in$ <800,500>, whereas η is a unit volume fraction of the created phase. The schemat of experimetically obtained diagram used in the mechanical model of stresses calculation is presented in Fig.4. Material of the element being welded in the heat affected zone after heating (for full transformation $T_{max}>Ac_3$ and $T_{max}\in <Ac_1,Ac_3>$for incomplete transformation) contained austenite in the rate η^H_A ($\eta^H_A = 1$ for $T_{max} > Ac_3$ and $\eta^H_A < 1$ for $T_{max}\in <Ac_3,Ac_1>$) while the remaining part was bainite ($\eta^H_B=1-\eta^H_A$),which as a product after cooling is a mixture of bainite η^H_B and martensite η^H_M arising from austenite, with mutual share dependent on cooling rate $\upsilon_{800/500}$. The volume fraction of austenite η^H_A while heating was determined by Avrami's equation [5] :

$$\eta^H_A(t) = 1-\exp(-b\ t^n) \ ; \quad t\in<t_s,t_f> \qquad (6)$$

where :

$$n = \frac{\ln\left(\dfrac{\ln(1-\eta^H_{Af})}{\ln(1-\eta^H_{As})}\right)}{\ln\left(\dfrac{t_s}{t_f}\right)} \quad , \quad b = -\frac{\ln(1-\eta^H_{As})}{(t_s)^n}$$

and :

$$Ac_1 = T(t_s) \quad , \quad Ac_3 = T(t_f)$$

$$\eta^H_{As} = 0.01 \quad , \quad \eta^H_{Af} = 0.99$$

The volume fraction of bainite and martensite while cooling was determined by the analogous formula :

$$\eta^c = 1 - \exp(-b\ t^n) \qquad (7)$$

where :

$$t = \frac{T - T_s}{\upsilon_{800/500}} + t_s \quad \text{and} \quad T_s = T(t_s) \quad ,$$

$$T_f = T(t_f) \quad , \quad \eta^c_s = 0.01 \quad , \quad \eta^c_f = 0.99$$

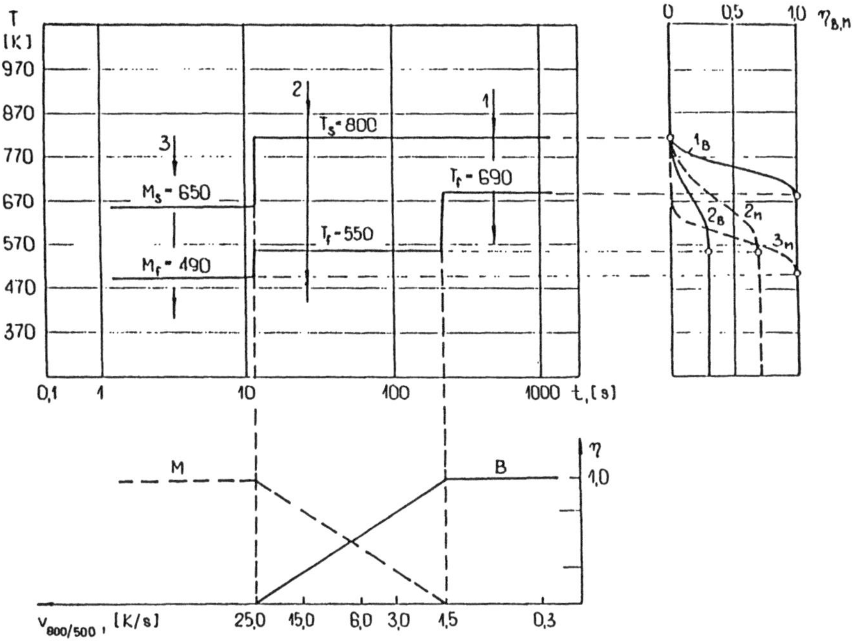

Fig.4. The rate of cooling - temperature - transformation schematic diagram.

For cooling rate $\upsilon_{800/500}$ from interval <1.5,25> [K/s] we have to do with a mixture of bainite and martensite. The volume fractionof these structures is suggested as a linear function against cooling rate $\upsilon_{800/500}$ as shown in Fig.4. Isotropic strains caused by temperature and by phase transformations were determined considering the calculated thermal cycles as well as complete and incomplete austenitization. The strains ε^T were :

$$\varepsilon^T = \begin{cases} \alpha_1(T-T_o) & \text{for } T \in \langle T_o, Ac_1 \rangle \\ \alpha_1(Ac_1-T_o) + \eta_B^H\alpha_1(T-Ac_1) + \eta_A^H(\alpha_A(T-Ac_1)-\gamma_A) & \\ & \text{for } T \in \langle Ac_1, T_S \rangle \\ \alpha_1(Ac_1-T_o) + \eta_B^H\alpha_1(T-Ac_1) + \eta_A^H(\alpha_A(T_S-Ac_1)-\gamma_A) + & \\ + \eta_A^H(\eta_B^c(\gamma_B-\alpha_B(T-T_S)) + \eta_M^c(\gamma_M-\alpha_M(T-T_S)) + & \\ + \eta_A^H(1-\eta_B^c-\eta_M^c)\alpha_A(T-T_S) & \text{for } T \in \langle T_S, T_o \rangle \end{cases} \quad (8)$$

$$\eta_B^c = \eta^c \frac{25 - \upsilon_{800/500}}{25 - 1.5} , \quad \eta_M^c = \eta^c \frac{\upsilon_{800/500} - 1.5}{25 - 1.5}$$

where: α_1 - a coefficient of linear thermal expansion of a structure of parent material or a structure achieved in the former thermal cycles of welding.

$\alpha_A, \alpha_B, \alpha_M$ - coefficients of linear thermal expansion of austenite, bainite and martensite, respectively.

$\gamma_A, \gamma_B, \gamma_M$ - transformation strains for transformation bainite into austenite, austenite into bainite and austenite into martensite, respectively.

Algorithm for stresses calculation is based upon the formulation of non-isothermal plasticity theory [5,7]. The mechanical properties used in calculation are shown in Tabel 2.

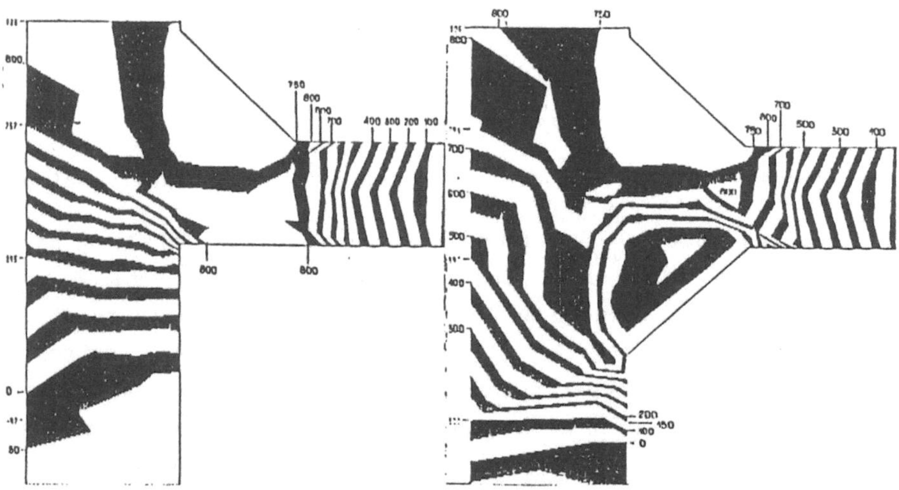

Fig.5. The state of axial residual stresses in the welded tube
a). after filet weld 1 is completed (left)
b). after filet weld 2 is completed (right)

In Fig.5 the state of residual stresses in a quater of the tube being considerd is presented. Corner of the tube in the area near the weld has a significant tensile stress, quite varying. Maximal stresses of the value of 700-800 MPa were localized at the exterior part of the corner in the area of outer weld. These stresses are characterized by a high gradient.

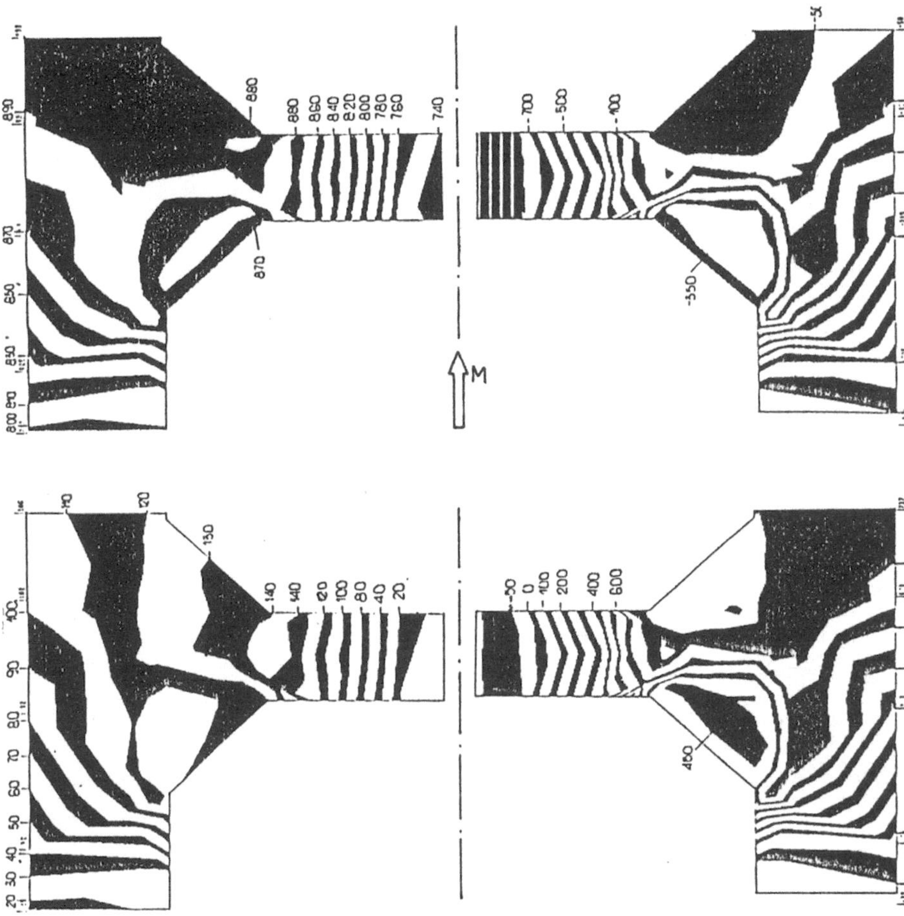

Fig.6. The states of stresses in welded tube under the load (up) and after the load is taken off (down)

Final remarks

Welded box-tubes are important load carrying elements of construction machines. Designing of such elements should take into consideration residual stresses. Calculation of stresses subjected to external loading including internal stresses confirm this fact. For example, if we load a tube with moment M=2.576 MNm it will cause elastic states for tube in natural state but it will cause elastic-plastic states for welded one. After the load in welded tube is taken off appear residual sresses different from stresses created by welding (compare Fig.6.).

References

1. B.A.Andersson and L.Karlsson, Thermal Stresses in Large Butt Welded Plates, J.Thermal Stresses 4, 497-500 (1981)

2. J.H.Argyris, J,Szimmat and K.J.Willam, Computtional Aspects of Welding Stress Analysis, Comp. Methods Appl.Mech. Eng.33, 635-666 (1982)

3. H.S.Carlsow and J.C.Jaeger, Conduction of Heat in Solids, 2nd Edn. (Oxford Univ. Press, Oxford 1973)

4. N.Christensen, de L.Davies and K.Gjermundsen, Distribution of Temperatures in Arc Welding, Br Welding J.12, 54-75 (1965)

5. A.J.Fletcher, Thermal Stress and Strain Generation in Heat Treatment, Elsevier Science Publishers Ltd., (1989)

6. E.Friedman, Thermomechanical Analysis of the Welding Process Using the Finite Elements Method, J.Pressure Vessel Technol. 97, 206-213 (1975)

7. T.Inoue, B.Raniecki , Determination of Thermal-Hardening Stresses in Steels by use of Thermoplasticity Theory. Jour. of the Mechanics and Physics of Solids, vol. 26, No 3, 187-212, (1978)

8. M.Johnsson, L.Karlsson and L.E.Lindgren, Thermal Stresses and Plate Motion in Butt Welding, 3rd Int.Conf. on Numer. Methods in Thermal Problems (Seattle, WA), (Pineridge Press, 1983)

9. L.Karlsson, Thermal Strssses in Welding, Thermal Stresses I, Edited by R.B.Hetnarski, Elsevier Science Publishers B.V. (1986)

10. D.Rosenthal, Mathematical Theory of Heat Distribution During Welding and Cutting, Welding J. 20, 220-234 (1941)

The Characteristics of the Source of Welding Residual Stress (Inherent Strain) and its Application to Measurement and Prediction

Y. UEDA

Welding Research Institute,
Osaka University, Osaka

M. G. YUAN

Production Engineering Center,
Daikin Industries, Ltd., Osaka

Abstract

Welding residual stress in a joint is produced as a result of thermal elasto-plastic behavior during welding. The source of residual stress is incompatible strain, called inherent strain in this report, which is composed of shrinkage strain of the weld metal, plastic strain in the vicinity of welded zone, etc. Based on the characteristics of inherent strain, measurement and prediction of three dimensional residual stress have become possible using the inherent strain as a parameter. Here, the characteristics of inherent strain is described by a simple example and the predicting and measuring methods of welding residual stress are outlined.

Introduction

The welding residual stress is inevitably produced in a welded structure. For accurate evaluation of the safety of the structure, accurate prediction of the welding residual stress is indispensable. Prediction methods of welding residual stress may be classified into the following three: (1) thermal elasto-plastic analysis(theoretical analysis), (2) measurement on similar welded joints, and (3) simple formulae based on accumulated experimental data and theoretical information.

From the theoretical viewpoint, thermal elasto-plastic analysis [1] is the most general method to predict welding residual stress taking account of many influential factors. By this method, complex calculation can be performed simulating the entire mechanical behavior during welding and the welding residual stress is obtained at the end of the calculation. There should exist the source to cause the residual stress.

L. Karlsson, L.-E. Lindgren, M. Jonsson (Eds.)
Mechanical Effects of Welding
IUTAM Symposium Luleå/Sweden 1991

Without tracing the entire process of elasto-plastic behavior, there exist simple predicting methods [2-4] including simple formulae which predict directly welding residual stress, and a few [2,3] which are concerned with the source of residual stress. However, no papers have proposed methods to determine the source, except a series of papers [5-9] published by one of the authors. In these papers, the characteristics of the source of residual stress was clarified, a method of determining the source was presented and several measuring methods of three-dimensional residual stress in a body were developed using the source as a parameter. Later, a new aspect of the characteristics was discovered, and a prediction method of welding residual stress [10-12] has been developed. In this method the source is also used as a parameter.

In this paper, the characteristics of the source of welding residual stress will be clarified and the predicting and measuring methods of welding residual stress developed by using the source as a parameter will be introduced.

Characteristics of the Source of Residual Stress

Generally, the stress contained in a self-balanced body is called residual stress. In the body, the source of residual stress should exist, which is incompatible strain at room temperature and the stress produced by incompatible strain is called inherent stress. In this connection, the source of residual stress is called here inherent strain. In the case of welded joints, the inherent strain may be composed of many components, such as shrinkage strain in the weld metal, plastic strain in the vicinity of the weld line.

When residual stresses are dealt at a fixed temperature (such as room temperature in most cases), inherent strain may be regarded as plastic strain. The characteristics of inherent strain will be described using such example as a circular plate of mild steel in plane stress state. The circular plate of R radius is heated up from 0 ℃ under the following four different conditions, assuming that the heating time is very short and no heat transfer occurs:

(1) the entire plate is heated uniformly to T ℃ ;

(2) a small area inside of R_H radius is heated to 100℃ and cooled down to 0 ℃ ;

(3) a small area inside of R_H radius is heated up to 300 ℃ and cooled down to 0 ℃ ;

(4) Increasing the radius R, a small area inside of R_H radius is heated up to 300℃ and cooled down to 0 ℃ .

In case (1), thermal strain is generated in the entire plate, but no thermal stress is produced, since the thermal strain is compatible strain. In case (2), as thermal strain is confined by the surrounding area of unheated zone, thermal stress is produced, since thermal strain is incompatible strain. As the behavior is regarded as being elastic in this heating process, no residual stress is generated after the plate is cooled down to 0℃ , since the incompatible strain disappears at 0 ℃ . In case (3), the behavior is elasto-plastic at the heating stage because large thermal compressive stress is produced to reach the yield stress, and a certain amount of plastic strain is generated above this stress according to the maximum temperature. These thermal strain and plastic strains are incompatible strains. Once the plate is cooled down to 0 ℃ , the thermal strain disappears while the plastic strain remains as incompatible one. Consequently, the plastic strain causes residual stress. In case (4), as the size of plate is increased, the degree of restraint against thermal strain generated in the central small area is also increased and it approaches rapidly to that for an infinite plate. Consequently, the magnitude of plastic strain generated at the heating stage converges to a certain value(Fig.1(a)). However, the residual stress distribution varies according to the size of plate(Fig.1(b)).

The pattern of plastic strain distribution is simple and a simple relationship between the magnitude of plastic strain and radius R should exist. For prediction of the residual stress in the circular plate, the thermal elasto-plastic analysis may be applied. On the other hand, it can be calculated by elastic analysis, imposing the inherent strain to a stress-free plate of the same size. The inherent strain should have been obtained separately, either by experiments or theoretical analysis. In the case where the size of a square plate is large, its residual stress due to the heating condition of case (4) may be calculated by imposing this inherent strain. This is the basic idea of the new prediction method of welding residual stress.

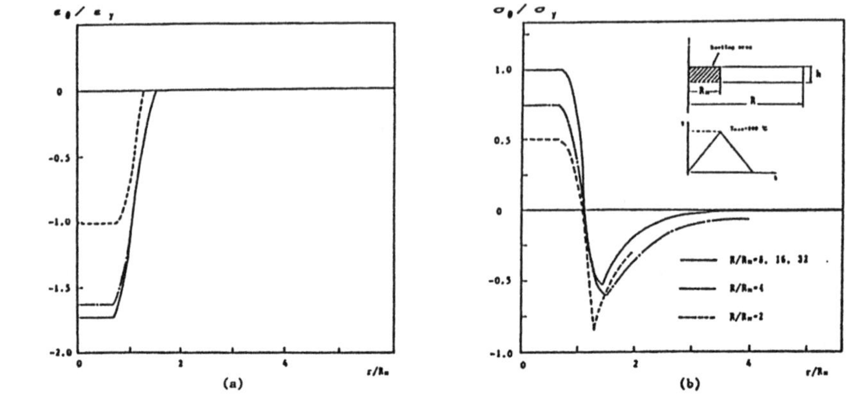

Fig.1 Distributions of circumferential inherent strain and residual stress with respect to the ratio of disk radius to heating area radius
(a) Circumferential inherent strain
(b) Circumferential residual stress

Next, consider cutting a circular plate containing the residual stress into two parts. By the cutting, redistribution of the residual stress should occur due to release of the stress acting on the cutting surfaces. If the cutting procedure is well chosen, the behavior of the plate during the cutting should be elastic and inherent strain does not change. The new residual stress distribution in each part can be computed by applying the released stress on the cutting surfaces in the opposite direction. In contrast with this, if the unchanged inherent strain is imposed in each part of the plate, the entire residual stress distribution can be computed. Inversely, the residual stress in each part of the plate may be measured and the inherent strain may be computed by solving conversely the relationship between inherent strain and residual stress.

The inherent strain in the entire plate is composed of the inherent strains in the two separate parts. If the entire strain is imposed to the original stress-free circular plate, the original residual stress is reproduced. In a case of a welded joint, the joint may be separated into such pieces as being thin plates, etc. whose residual stresses can be easily observed. From the observed stresses, the inherent strain distribution in each piece may be obtained and the entire distribution of welding residual stresses can be computed by applying the entire distribution of the inherent strain to the stress-free welded joint. This is a basic procedure of measurement of welding residual stress using inherent strain as a parameter.

In this above, the characteristics of inherent strain was described using a simple example and the new approaches for prediction and measurement of welding residual stress were outlined. Although details of the basic approaches are described in the original papers, these will be demonstrated using actual welded joints in the following sections.

Relationship between Inherent Strain and Welding Residual Stress[5-9]

The inherent strain $\{\varepsilon^*\}$ in a welded joint is produced only in the limited region near the welds. However, welding residual stress $\{\sigma\}$ (or elastic strain $\{\varepsilon\}$) due to the inherent strain is produced over the whole plate. This relation is expressed in the following equations.

$$\{\varepsilon\} = [H^*]\{\varepsilon^*\} \quad (1)$$

$$\{\sigma\} = [D]\{\varepsilon\} \quad (2)$$

where,

$[D]$: Stress-strain matrix,

$[H^*]$: Elastic response matrix.

The finite element method is used for practical problems. The number of finite elements in which the inherent strain is produced is assumed to be m. The number of finite elements in the whole joint is n (> m).

In order to determine the inherent strains, Eq.(1) is necessary to be solved conversely giving known strains of $\{\varepsilon\}$ more than m. When elastic strain $\{\varepsilon\}$ is determined by measurement, it may contain various errors. According to the condition which minimize the sum of the square of residual, the most probable value of inherent strain $\{\varepsilon^*\}$ is calculated from the following equation.

$$\{\varepsilon^*\} = ([H^*]^T [H^*])^{-1} [H^*]^T \{\varepsilon\} \quad (3)$$

Substituting $\{\varepsilon^*\}$ into Eq.(2), the most probable value of welding residual stress $\{\sigma\}$ at arbitrary positions of the joint can be obtained.

$$\{\sigma\} = [D][H^*]\{\varepsilon^*\} \tag{2'}$$

Predicting Method of Welding Residual Stress[10-12]

In this chapter, the characteristics of the source of welding residual stress generated in a butt joint of plates is studied. It will be shown that the source distributes in a simple form along the weld line and is almost independent of the size of the joint for the specified material and welding condition if it is sufficiently large. Concerning the general theory of predicting welding residual stress using inherent strain as a parameter, the details is described in the original reference [5-9]. In the new predicting method, the elastic analysis is only necessary on a stress-free plate imposing the above mentioned source of residual stress, which should have been known either experimentally or theoretically.

Inherent Strain Distribution in Butt Welded Joint

In this study, a butt welded joint of thin plate subjected to a moving heat source is adopted as the analysis model(Fig.2), and the length, width and thickness of the welded plate are denoted by 2L, 2B and h(=6mm: constant), respectively. The CO_2 gas shielded arc welding is applied in which electric current I=180(A), voltage V=20(V), welding speed v=30(m/h), and the heat efficiency η =75 % is assumed. This heat-input corresponds to Q=900(J/mm) for the instantaneous plane heat source.

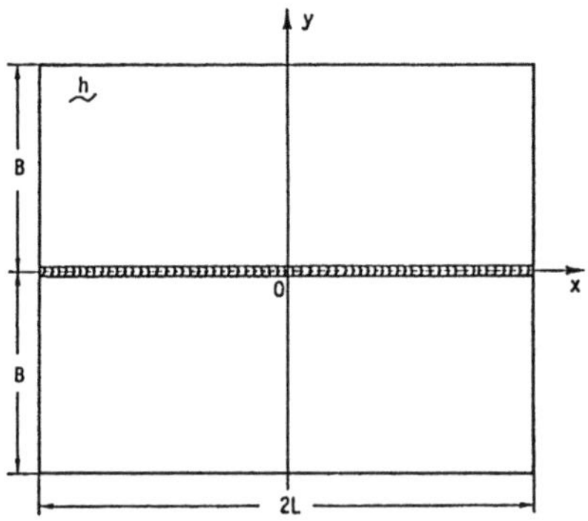

Fig.2 Model of butt welded joint for analysis

The thermal elasto-plastic analysis by the finite element method is performed using the temperature dependent physical and mechanical properties of the material shown in reference [10-13] and assuming that the material is mild steel and the model is in plane stress state.

The first analysis is performed on model MO(2L=1200, 2B=1000) in cooperation with thermal conduction analysis. Using the obtained elastic strain for $\{\varepsilon\}$ of Eq.(3), the distribution of inherent strain $\{\varepsilon^*\}=\{\varepsilon_x^*, \varepsilon_y^*, \varepsilon_{xy}^*\}^T$ is computed and shown by solid lines in Fig.3. The shearing component of inherent strain, ε_{xy}^*, is only produced near the end of the plate.

Neglecting ε_{xy}^*, for simplicity, the inherent strain is calculated as $\{\varepsilon^*\} = \{\varepsilon_x^*, \varepsilon_y^*, 0\}^T$. This result is also represented by a dotted line in Fig.3.

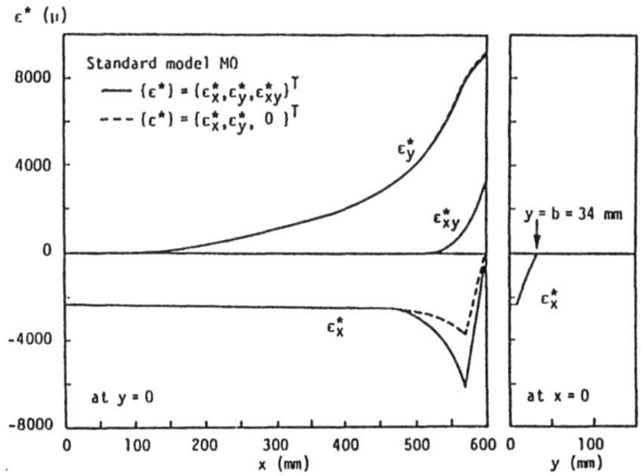

Fig.3 Inherent strain distribution of ε_x^*, ε_y^* and ε_{xy}^*

The entire distributions of $\{\varepsilon^*\}$ are illustrated in Fig.4. It is found that ε_x^* exists in a narrow width spanning the weld line up to y=34 mm. Furthermore, the distribution of ε_x^* along the direction of welding is the same at each cross-section except near the ends of the plate and the distribution of ε_y^* exists only in the vicinities of the two ends and is in a parabola along the weld line.

In order to confirm the accuracy of reproduction of welding residual stress by the inherent strain $\{\varepsilon^*\} =\{\varepsilon_x^*, \varepsilon_y^*, 0\}^T$, welding residual stress is computed by the elastic analysis

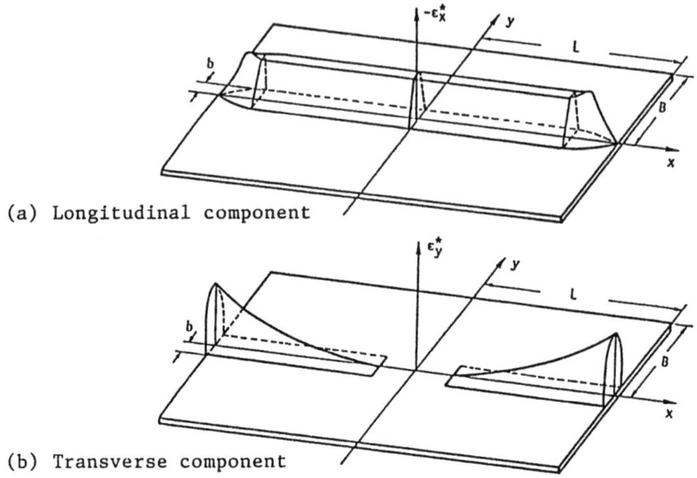

Fig.4 Schematic representations of inherent strain distributions

imposing to model M0 in stress-free state. This welding residual stress is almost coincident with that of the thermal elasto-plastic analysis.

Under the welding condition mentioned before, the thermal elasto-plastic analysis is performed on the five models and the inherent strain in each model is computed. The inherent strains are represented in Figs.5(a) and (b).

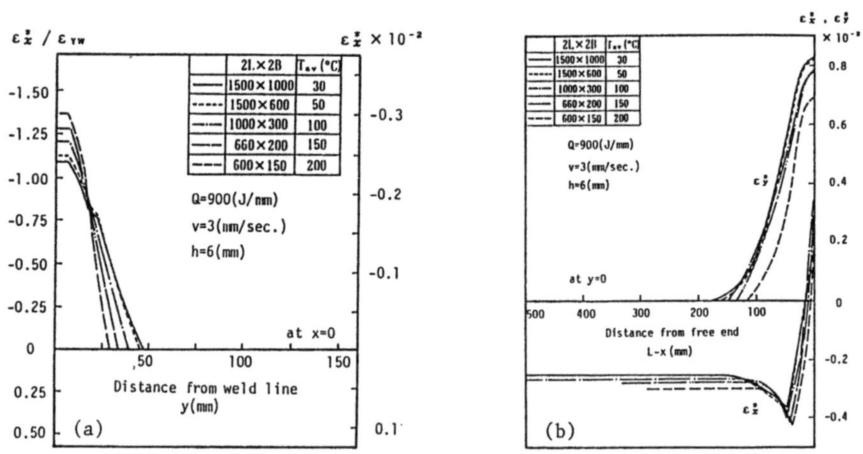

Fig.5 Inherent strain distributions

(a) ε_x^*

(b) ε_x^* and ε_y^*

In the case where the width B of the plate is sufficiently large (B≥ 300mm) to the specified heat input, the distribution of the inherent strain ε_x^* is almost independent of the dimensions of the plate. On the other hand, Fig.5(b) shows that the magnitude and distribution of the inherent strains(including ε_x^* and ε_y^*) near the free ends of the five models are nearly the same in spite of different sizes of the plates.

Figure 6 shows the distribution of residual stress σ_x at the cross section x=300 mm of model MO analyzed by using the corresponding inherent strains shown in Fig.4. The distributions of stress σ_x' and σ_x'' are calculated by using ε_x^* and ε_y^* separately. It should be noted that ε_x^* produces fundamental distribution of σ_x and ε_y^* influences the inclination of compressive part of σ_x near the ends.

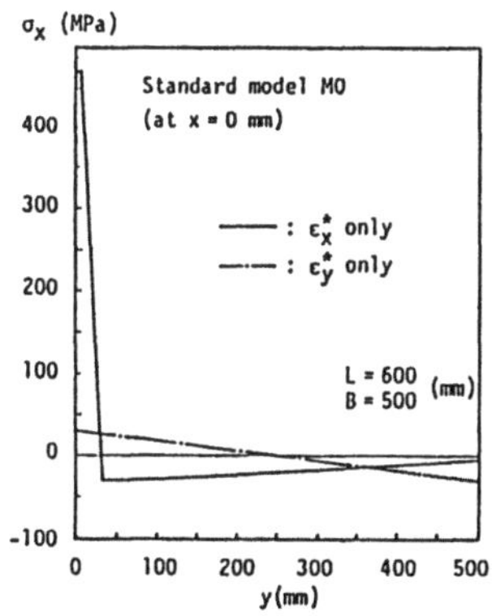

Fig.6 Stresses produced by ε_x^* and ε_y^*

In actual plate structures, such as ships, the length of plate is long enough in comparison with the width so that the important component of residual stress is σ_x caused by ε_x^* which distributes nearly in a trapezoidal form in transverse cross section(Fig.7). Such distribution was determined theoretically [11,12], assuming that heat input is provided to the weld line instantaneously so that the displacement along the cross section is uniform ($du_x/dy=0$) except near the free ends.

The width b of inherent strain zone and its magnitude $\hat{\varepsilon}_x^*$ in the mechanical heat-affected zone ($0\leq y \leq y_H$) are expressed as,

$$b = \xi b_0 \tag{4}$$

$$\hat{\varepsilon}_x^* = \zeta \sigma_{YW}/E \tag{5}$$

$$b_0 = 0.242 \frac{\alpha E Q}{c\rho h \sigma_{YB}} \tag{6}$$

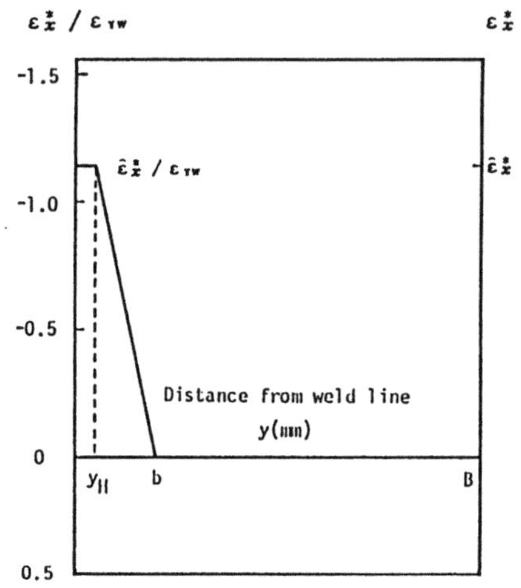

Fig.7 Idealized distribution of ε_x^*

where ξ is the normalized width of inherent strain zone with respect to b_0 which is the width for an infinitive width of a plate. ζ is the normalized magnitude of inherent strain. σ_{YB} and σ_{YW} are the yield stresses of the base plate and the weld metal, and E, α, c and ρ are Young's modulus, linear thermal expansion coefficient, specific heat and density respectively.
In the above expressions, ξ and ζ are given by complex equations. However, the following simple formulae were derived.

$$\left.\begin{aligned} \xi &\fallingdotseq 1 - \frac{0.27\alpha E T_{av}}{\sigma_{YB}} \\ \zeta &\fallingdotseq 1 + \frac{0.27\alpha E T_{av}}{\sigma_{YB}} \end{aligned}\right\} \tag{7}$$

where, $$T_{av} = \frac{Q}{2c\rho h B} \tag{8}$$

These can be applied to the yield stress σ_{YB} which varies from 230 MPa to 430 MPa.
In Fig.8, using the same material properties, the widths b and magnitudes $\hat{\varepsilon}_x^*$ of the inherent strains for different values of

T_{av} calculated by Eq.(7) are shown by lines and those obtained for a moving heat source by the thermal elasto-plastic analysis are plotted by solid circles. It is recognized that the two results have good agreement with each other. This indicates that the Eq.(7) are applicable not only to instantaneous heat source, but also to moving heat source, provided that the effective heat input is chosen to be the same in both cases.

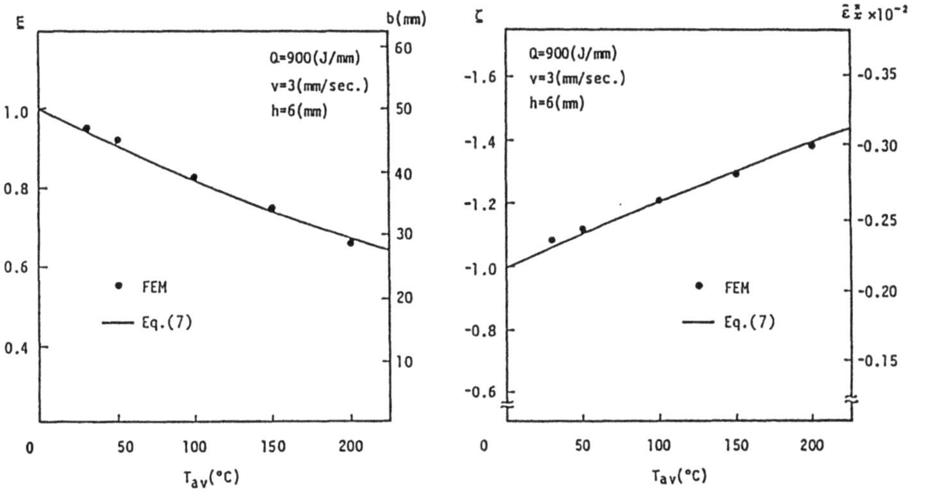

Fig.8 Width and magnitude of idealized distribution of ε_x^* by FEM and Eq.(7)

The estimated residual stress distributions in the middle cross-sections of six butt welded plates with different widths and lengths by the proposed method are shown in Fig.9 using several different marks. The residual stresses in the respective plates by thermal elasto-plastic analysis are also plotted in Fig.9 using lines. It is observed that the residual stresses in the butt welded plates with different aspect ratio L/B can be predicted accurately by the proposed method, provided that an inherent strain distribution in some butt welded plate had been known correctly from experiment or literature to calibrate the absolute values of b_0 and $\hat{\varepsilon}_x^*$ in the mechanical heat-affected zone.

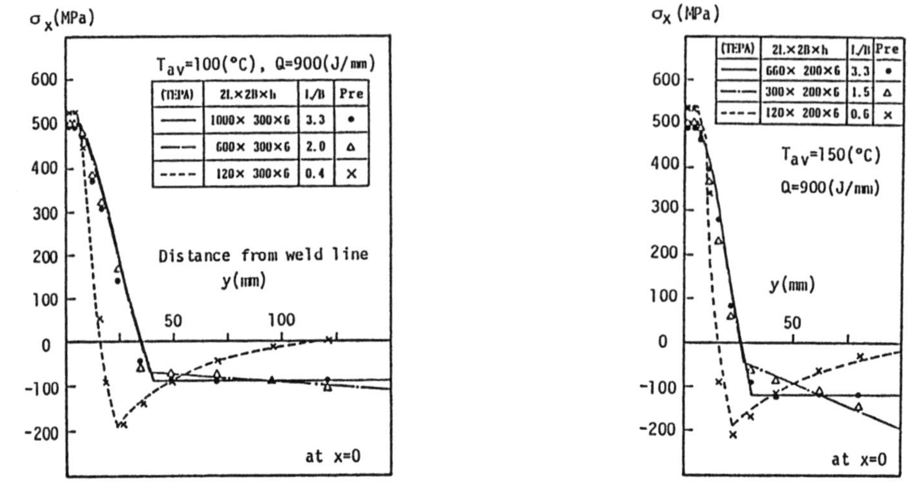

Fig.9 Welding residual stresses by FEM and the predicting method

New Measuring Method of Three-dimensional Residual Stresses[5-9]

In the new measuring methods, the effective inherent strains are estimated and residual stresses are calculated by imposing these effective inherent strains to the object in stress-free state. That is, residual stresses are measured using these inherent strains as parameters.

The object of measurement is uniformly distributed three-dimensional residual stresses produced in a long welded joint as shown in Fig.10(a). The source of residual stress (the inherent strain) is also uniform along the longitudinal axis except near the ends and symmetric with respect to the middle cross-section(x=0). Then, the whole effective three-dimensional inherent strain $\{\varepsilon^*\}$ is the function of the cross-sectional coordinates (x,y). Taking full advantage of this distribution, $\{\varepsilon^*\}$ can be separated into two groups: the longitudinal inherent strain $\{\varepsilon_l^*\}=\{\varepsilon_x^*\}$, and the cross-sectional one $\{\varepsilon_c^*\}=\{\varepsilon_y^*, \varepsilon_z^*, \gamma_{yz}^*\}$. This enables us to simplify the measurement of residual stresses.

In order to estimate these inherent strains separately, thin plates: one Specimen T and several Specimens L are cut out from the original Specimen R according to L_z method [6,9] (Fig.10). Specimen T is perpendicular and Specimens L is parallel to x-axis. For this process, the following assumptions are made.

(1) Cutting is accompanied by only elastic change of strain and does not produce any new plastic strain, that is, the inherent strain is not influenced by cutting.

(2) In a thinly sliced plate, inherent strain component being perpendicular to the section does not produce any stress. That is, in a thin plate, stress in the plane stress state are produced by only inherent strain component in that plane.

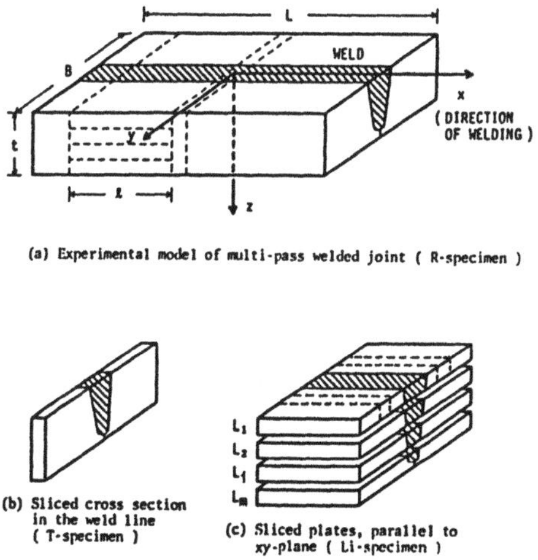

Fig.10 Experimental model and procedure of slicing Specimens T and L_i

Separation of Three-dimensional Inherent Strain Components

The inherent strain distributions in Specimens T and L are the same as that in Specimen R, since the cutting is made to satisfy assumption (1). ε_x being perpendicular to the cross section does not produce stress (assumption (2)) in Specimen T. Therefore, the source of residual stress remaining in Specimen T is the cross-sectional inherent strain $\{\varepsilon_c{}^*\}$.
Next, $\varepsilon_z{}^*$ being perpendicular to the section does not produce stress in Specimens L from assumption (2). $\varepsilon_y{}^*$ distributes

uniformly in the longitudinal direction of Specimens L. For this reason , stress by ε_y^* is released when Specimens L is cut out from Specimen R, and ε_y^* just expands or shrinks uniformly along the breadth. Accordingly, the source of the residual stress remaining in Specimens L is only ε_x^*. Three-dimensional inherent strain can thus be separated into the cross-sectional inherent strain $\{\varepsilon_c^*\}$ in Specimen T and the axial one $\{\varepsilon_l^*\}$ in Specimens L.
Then they can be estimated by the measurement of two-dimensional stresses remaining in Specimens T and L respectively. By this procedure, the measurement of three-dimensional residual stress is reduced into that of two-dimensional one.
Therefore, the three-dimensional residual stresses $\{\sigma\}$ produced in Specimen R can be obtained as the sum of the stress $\{\sigma^A\}$ produced by the cross-sectional inherent strain $\{\varepsilon_c^*\}$ and stress $\{\sigma^B\}$ by the longitudinal inherent strain $\{\varepsilon_l^*\}$.

$$\{\sigma\} = \{\sigma^A\} + \{\sigma^B\} \tag{9}$$

The actual measuring method and the procedure of three-dimensional residual stress components, $\{\sigma^A\}$ and $\{\sigma^B\}$, are described below.

Three-dimensional Residual Stress, $\{\sigma^A\}$, by Cross-sectional Inherent Strain, $\{\varepsilon_c^*\}$

As the cross-sectional inherent strain distribute uniformly along the longitudinal axis, the three-dimensional residual stress component $\{\sigma^A\}$ is in the plane strain state in specimen R. This is proven in [9]. The outline of the proof is as follows: The resultant stresses(normal force and moments) in plane strain perpendicular to the cross-section are expressed by a product of $\nu/(1-\nu^2)$ and those in plane stress, which must vanish since the stresses are self-equilibrating. In contrast with this, the stress remaining in the thinly sliced Specimen T cut out from Specimen R, $\{\sigma^{A0}\}$, is in the plane stress state, and can be directly observed by the stress relaxation method. Using the following equations, Eq.(3), directly observed $\{\sigma^{A0}\}$ can be converted into the stress $\{\sigma^A\}$ in the plane strain state. As a result, three-dimensional residual stress component $\{\sigma^A\}$ can be obtained without estimating inherent strains.

$$\sigma_y^A = \sigma_y^{A0} / (1-\nu^2)$$

$$\sigma_z^A = \sigma_z^{A0} / (1-\nu^2) \qquad (10)$$

$$\sigma_x^A = \nu\,(\sigma_y^{A0} + \sigma_z^{A0}) / (1-\nu^2)$$

$$\tau_{yz}^A = \tau_{yz}^{A0} / (1-\nu^2), \quad \tau_{zx}^A = \tau_{xy}^A = 0$$

where, ν : Poisson's ratio.

Three-dimensional Residual Stress, $\{\sigma^B\}$, by Longitudinal Inherent Strain, $\{\varepsilon_l{}^*\}$

For the measurement of $\{\sigma^B\}$, Specimens L are cut out from Specimen R as shown in Fig.10. Elastic strain remaining in each Specimens L can be observed when each Specimen L is sectioned into pieces in parallel to the longitudinal axis as depicted by the broken lines.
Using the relation between residual stress (strain) and inherent strain, the inherent strain distribution in each Specimen L can be determined. The entire distribution is composed of each ones. The residual stress $\{\sigma^B\}$ can be calculated by imposing the entire distribution of $\{\varepsilon^*\}$ on the stress-free specimen R.

Actual Measurements of Three-dimensional Residual Stress Distributions in Welded Joint

The material of Specimen R was mild steel and its initial residual stresses before welding were removed by stress relief annealing. Using submerged arc welding(current: 650A, voltage: 35V, welding velocity: 42cm/min.), Specimen R was made by multi-pass butt welding whose passes were accumulated from the bottom to the top of the specimen. The length, width and thickness of the specimen are L=200mm, B=200mm and t=50mm, respectively. One Specimen T and four Specimens L were cut out from Specimen R. The thickness of these sliced plates was 10mm. The length of each Specimen L, ℓ, was 70mm.
SR-4 gages with a gage length of 2mm were used. They were attached in pairs on both faces of the sliced plates which were Specimens T and L and the mean value of each pair of the observed strains was regarded the observed strain at the point.
According to the new measuring methods, three-dimensional residual stress $\{\sigma\}$ was obtained as the sum of residual stress

components $\{\sigma^A\}$ and $\{\sigma^B\}$ respectively. The results are shown in Figs.11 and 12 in which "○" and "●" indicate three-dimensional residual stresses directly observed by the check gages on the surface of Specimen R. The measured values and directly observed values are shown a good coincidence. Figure 12 represents the estimated welding residual stresses on the cross section of the joint.

Conclusion

The characteristics of inherent strain produced by welding was described. The new predicting method and measuring method of welding residual stress were introduced, in which, inherent strains are dealt as parameters. The new methods were developed without any approximation except the basic assumptions based on the theory of elasticity. Using these methods, the actual

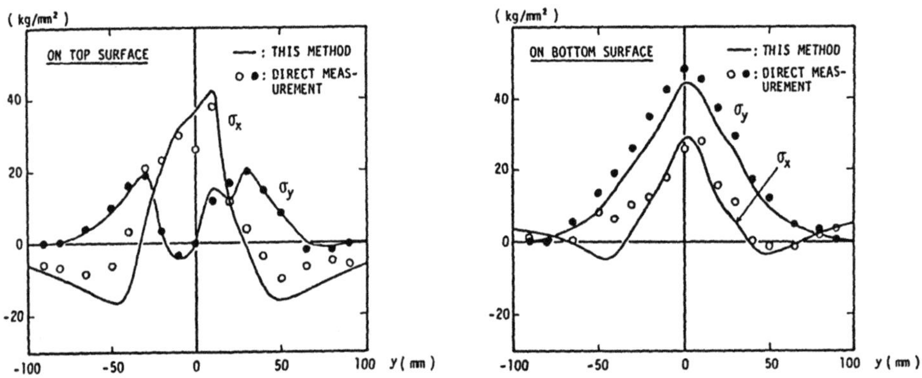

Fig.11 Welding residual stresses on the top and bottom surfaces in the middle of weld line

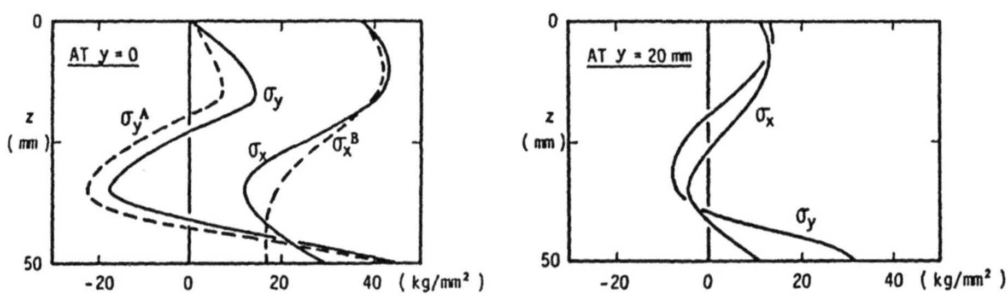

Fig.12 Estimated welding residual stresses on the cross section(x=0)

prediction and measurement of residual stresses produced in welded joints were performed and the practicality of these new methods and the validity of the theories were shown.

References

1. Ueda, Y.; Yamakawa, T.: Analysis of thermal elastic-plastic stress and strain during welding by finite element method, Trans. Japan Welding Soc. Vol. 2, No. 2 (1971), 90-100. Also IIW Doc. X-616-71 (1971).
2. Masubuchi, K.: Stress due to distributed incompatibility , J. Japan Soc. Naval Arch i.' /88 (1955) 189-200
3. Fujimoto, T.: A method for analysis of welding residual stresses and deformations based on the inherent strain, J. Japan Welding Soc., 39-4 (1970), 236-252 (in Japanese)
4. Satoh, K.; Terasaki, T.: Effect of welding conditions on residual stress distributions in welded structures materials, J. Japan Welding Soc., 45-2 (1976) 150-156 (in Japanese)
5. Ueda, Y.; Fukuda, K.; Nakacho K.; Endo, S.: A new measuring method of residual stresses with the aid of finite element method and reliability of estimated values, J. Soc. Naval Arch. of Japan, Vol. 138, (1975) 499-507 (in Japanese). Also Trans. JWRI(Welding Research Institute of Osaka Univ.) 4-2 (1975) 123-131.
6. Ueda, Y.; Fukuda, K.; Tanigawa, M.: New measuring method of 3-dimensional residual stresses based on theory of inherent strain, J. Soc. Naval Arch. of Japan, Vol.145, (1979) 203-211 (in Japanese). Also Trans. JWRI 8-2(1979) 249-256.
7. Ueda, Y.; Fukuda, K.; Fukuda, M.: A Measuring theory of three dimensional residual stresses in long welded joints, J. Japan Welding Soc., 49-12 (1980) 845-853 (in Japanese). Also 12-1 Trans. JWRI (1983) 113-122.
8. Ueda, Y.; Fukuda, K.: Simplified measuring method of three dimensional welding residual stresses -- proposal of nLy method and simple Ly method -- J. Japan Welding Soc., 52-2 (1983) 110-117 (in Japanese). Also Trans. JWRI 11-2 (1982) 95-104.
9. Ueda, Y.; Fukuda, K.: New measuring method of three-dimensional residual stresses in long welded joints using inherent strain as parameters-L_z method-, ASME Journal of Engineering Material and Technology, Vol.111, No.1, Jan. (1986) 328-334.
10. Ueda, Y.; Kim, Y.C.; Yuan, M.G.: A predicting method of welding residual stress using source of residual stress (Report 1), Trans. JWRI (Welding Research Institute of Osaka Univ.), Vol.18, No.1, (1989) 135-141.
11. Ueda, Y.; Yuan, M.G.: A predicting method of welding residual stress using source of residual stress (Report 2), Ibid., Vol. 18, No.1, (1989) 143-150.
12. Yuan, M.G.: A Predicting method of welding residual stresses in built-up plate structures using the characteristics of inherent strains, Dissertation for Dr. Engr., Osaka Univ., 1990.

Residual Stress Distributions After Welding as a Consequence of the Combined Effect of Physical, Metallurgical and Mechanical Sources

Th. Nitschke-Pagel, University Gh Kassel
H. Wohlfahrt, Technical University of Braunschweig, Germany

1. Introduction

As preceding papers /1–11/ pointed out various sources contribute to the residual stress state after welding and therefore extremely different residual stress patterns can arise as a consequence of welding. The aim of this paper is

- *to clarify the conditions for the predominance of one or the other of the residual stress generating processes,*
- *to show how these processes interact and thus*
- *to explain the influence of the type of the base material and the filler material and of welding parameters as heat input, welding velocity and cooling conditions on typical residual stress distributions after welding*

2. Different sources of residual stressses

2.1 Shrinkage stresses

The longitudinal shrinkage of highly heated areas is hindered by the colder and therefore less or not shrinking areas. If this hindered shrinkage of the highly heated areas would be the only stress generating process – or at least the dominating process – it would result in longitudinal tensile residual stresses in the weld, which would be compensated by compressive residual stresses in areas which are more remote from the weld seam. The magnitudes of the residual stresses due to the hindered shrinkage depend upon the degree of restraint. Only in connection with a strong restraint the tensile residual stresses can reach maximum magnitudes in the range of the yield strength.

At the end of the weld seam the longitudinal residual stresses must become zero and that is to say: the longitudinal residual stresses must show an inhomogeneous distribution along the weld seam. Therefore the transverse strains which are a consequence of the longitudinal residual stresses become also inhomogeneous along the weld seam indicating a hindered contraction in the transverse direction and resulting in transverse residual stresses of relatively small magnitudes – if shrinkage in the transverse direction is not hindered. The residual stress profiles as shown in Fig. 1 generated only by hindered shrinkage are typical for weldments in materials without phase transformations during welding and for soft soldered joints. In welding practice however, shrinkage in the transverse direction is hindered nearly always by external conditions – for instance during cooling of the last passes of each multilayer weldment. Then the tensile transverse residual stresses can reach the same magnitude as the longitudinal residual stresses. Maximum tensile residual stresses in the transverse direction are inevitably produced if broad weld gaps are clamped together by mechanical means. The mechanical stresses which

L. Karlsson, L.-E. Lindgren, M. Jonsson (Eds.)
Mechanical Effects of Welding
IUTAM Symposium Luleå/Sweden 1991

are necessary to bring together the abutting edges will be fixed within the structure after solidification, cooling down of the weld seam and removal of the constraint. In order to eliminate this additional stress source it is necessary to avoid the occurence of broad weld gaps by a good programme for the welding sequence combined with intermediate straightening if necessary.

2.2 Transformation stresses

The transformation of austenite, which occurs only in sufficiently high heated areas of welments or hard soldered joints, is connected with a volume expansion of the transforming areas, which is hindered in the longitudinal direction by the colder not transformating areas. The extent of the volume expansion depends upon whether the transformation results in a ferritic–pearlitic, bainitic or martensitic structure. If the phase transformation would be the only stress generating process – or at least the dominating one – the hindered dilatation would result in longitudinal compressive residual stresses in the transforming areas which must be compensated by tensile residual stresses in areas which are more remote from the transforming zone.

2.3 Model considerations on the interaction of shrinkage and transformation stresses

As reported in some foregoing papers (6–11) a realistic model on the formation of residual stresses as a consequence of welding has to take into account the interaction of shrinkage stresses and transformation stresses during cooling down of the weld seam. Whereas shrinkage occurs during the whole cooling process, the phase transformation occurs only in a temperature range. In order to understand the consequences of both stress generating processes model considerations are helpful, which simplify the complex real processes but which are able to highlight the most important influences on sign, magnitude and distribution of the resulting residual stresses. Such model considerations are presented in the following.

It depends upon the temperature–time curve during cooling and on the TTT–diagram of the steel whether austenite transforms to ferrite and pearlite in a relatively high and wide temperature range or the austenite transforms to bainite or martensite in a much lower and narrower temperature range. The stress–temperature diagram in Fig. 2 describes schematically the consequences of austenite transformations within different temperature ranges for the evolution of stresses in the weld seam during the cooling process and for the final residual stress state. The thick drawn lines of the diagram indicate the stresses which would originate in a simultaneously transforming zone as a function of the temperature. The thin straight lines illustrate schematically the yield strength of ferrite with pearlite and of bainite under tension and under compression as a function of the temperature. For the purpose of a clear representation of the main features the diagram is based on the following simplifying assumptions: a linear relationship between yield strength and temperature is supposed and Young's modulus E as well as the coefficient of thermal expansion α is taken to be independent of the temperature, resulting in linear relationships between stresses and temperature.

The diagram should be read from high to low temperatures. At high temperatures tensile stresses arise in the shrinking austenite which increase with decreasing temperature corresponding to the increasing yield strength of the austenite. The development of the tensile stresses is interrupted for instance in the temperature range of the transformation to ferrite with pearlite (example 1) or in the temperature range of a bainitic or a martensitic transformation (example 2 and 3). As a consequence of the hindered volume expansion connected with the phase transformation compressive stresses arise in the transforming zone. The compressive stresses are limited by the hot yield strength under compression of the resulting ferritic–pearlitic structure in example 1. In the case of a martensitic transformation (example 2 and 3) it is assumed that the resulting compressive stresses do not reach the relatively high hot yield strength of the bainite or the martensite.

After completion of the transformation the shrinkage of the transformed zones becomes again visible in the diagram as a decrease of the compressive stresses. During this period of the cooling process the slope of the stress-temperature line is proportional to E and α of the resulting structure and to the restraint of the welded joint. The stresses can change their sign again and even reach the hot yield strength under tension, if the transformation occurs at relatively high temperatures. Then tensile residual stresses of the magnitude of the yield strength should remain in the transformed zone after cooling down to room temperature – despite the occurence of the phase transformation. But, as can be seen clearly in the diagram, in the case of a transformation at relatively low temperatures, for instance a transformation to bainite, either low tensile residual stresses or even compressive residual stresses (ex ample 2 and 3) should arise in the transformed zone. The magnitude of the resulting tensile residual stresses is lower and the magnitude of the resulting compressive residual stresses is higher the lower the transformation temperature is.

According to this model-type consideration an influence of the phase transformation on magnitude or sign of the residual stresses can exist only if the transformation occurs in a sufficiently low temperature range. Then the temporary turnover of stresses into the compressive range, which is connected with the volume expansion, can remain effective and can be more effective the lower the temperature range of the transformation is.

The consequences of austenite transformations at higher or lower temperatures and of additional shrinkage for the entire distributions of transverse and longitudinal residual stresses after welding shall be described in principle in the following. For this purpose Fig. 3 shows a schematical diagram of the distributions of transverse stresses immediately after the phase transformation and also after another period of cooling and shrinkage under the assumption that shrinkage respectively expansion is hindered in the transverse direction. Fig 4 is the corresponding diagram of the longitudinal stresses.The diagrams represent the situation of a high heat input and consequently a relatively wide weld seam and heat affected zone. Then the cooling rate is relatively low and the transformation of austenite occurs in a relatively high temperature range. As illustrated by the dashed line 1 in Fig. 3 immediately after the austenite transformation transverse compressive stresses exist over a width which is distinctly broader than the transforming zone. After completion of the transformation shrinkage occurs mainly in a narrow zone on both sides of the weld centre line which shows still a high temperature. Thus the shrinkage process during further cooling results in a new stress peak which grows from the compressive stress level near the weld centre line. After an austenite transformation at a relatively high temperature shrinkage can be very effective and therefore the new stress peak can reach the tensile range (dash-dotted line 2) and finally a high tensile value close to the yield strength (fully drawn line 3). A lower heat input is connected with a narrower weld seam, a narrower transforming zone, a lower temperature of the austenite transformation and consequently with a sharper residual stress peak of a lower magnitude in the tensile range or even in the compressive range.

The dashed line 1 in Fig. 4 indicates that during the transformation of austenite compressive stresses are also produced in the longitudinal direction, but in contrast to the transverse stresses only within the transforming zone. In adjacent zones beyond the transforming zone tensile stresses arise due to equilibrium reasons. Shrinkage after the phase transformation is mainly concentrated on a zone close to the weld centre line and changes the stress distribution in the way that a tensile stress peak grows near the weld centre line and for equilibrium reasons compressive "peaks" arise on both sides of the tensile peak (dash-dotted line 2). In connection with a high heat input respectively a high temperature range of the austenite transformation a high peak of tensile residual stresses at the weld centre line has to be anticipated (fully drawn line 3). With a lower heat input and hence a lower temperature of the austenite transformation the magnitude of the tensile residual stress peak should also become lower and combined with a very small heat input compressive residual stresses have to be anticipated in the longitudinal direction in the narrow weld seam. In connection with any heat input the tensile stress maxima on both sides of the transforming area remain finally present as a typical consequence of the phase transformation.

The conclusion from the described model considerations is that the temperature range of the austenite transformation has the dominating influence on the formation of residual stresses due to welding and therefore that all welding and material parameters which have an influence on this temperature range are also influencing the residual stress distributions after welding. Consequently, on one side all parameters with an influence on the cooling rate respectively the cooling time between 800° C and 500° C, for instance heat input – and hence welding current, welding speed and preheating temperature – or weld seam geometry, are influencing the residual stress pattern. On the other side the chemical composition of the base and filler materials in connection with the type of the TTT–diagram and their yield strength and hot yield strength are determining parameters for the formation of welding residual stresses. Experimental results on the influence of all these parameters on sign, magnitude and distribution of residual stresses after welding shall be presented and discussed in the following sections.

3. Experimental procedure

The aim of the investigations which are described was to check systematically the influence of heat input variations on the residual stress state of steels with different TTT–diagrams and different yield strength values. Therefore widely varying heat input values – either realized by different welding currents and voltages or by different preheating temperatures – were used for pulsed arc TIG–welding or for producing a strip of molten metal (dummy seam) with the TIG–torch on one side of the 10 mm thick sheets. In the dummy seams transverse stresses can be appreciable as the depth of the weld pool is smaller than the sheet thickness and transverse shrinkage of the heated strip is externally hindered therefore. Furthermore dummy welds exclude the disturbing effect of mixing of filler and base materials. As a consequence dummy welds have the advantage to show the influence of the heat input on the residual stress state and the differences in the residual stress distributions of steels with different TTT–diagrams more clearly than real weldments. Therefore in the following residual stress distributions of dummy seams are mainly reported in order to show clearly the above discussed effects. Results of real weldments offer additional information how the residual stress distributions have to be modified if different filler materials – for instance filler materials with an extremely high yield strength in comparison with the base material – are used.

The exact cemical composition of the different structural steels St 52–3, StE 690 and StE 890 which have been used as base materials in the investigation is registered in /12/. Filler materials with the same chemical composition as well as with a different chemical composition have been chosen for the various weldments. The welding parameters of the pulsed arc TIG–weldments with different heat input values and further experimental details about the heat treatment of the sheets before welding, about the residual stress measurements by means of X–ray diffraction using Chromium K –radiation and about the measurements of the temperature–cooling time curves are also described in /12/.

4. Experimental results

The temperature–cooling time curves which have been measured for different welding heat input values H per unit length and for different preheating temperatures T_p are included in the TTT–diagram of the structural steel StE 690 in Fig. 5. As one can see from the temperature–cooling time curves with numbers 1 to 5 variations of the welding heat input per unit length, by means of current and voltage variation, and variations of the preheating temperatures have an influence on the cooling time with the same tendency: that is to say, the cooling curves for both kinds of parameter variation can be brought in one row. The temperature of the beginning of the austenite transformation increases from cooling curve 1 to cooling curve 5.

The experimental results in Fig. 6 and Fig. 7 illustrate the influence of a different heat input respectively of different transformation temperatures on the distributions of transverse and longitudinal residual stresses around dummy seams in sheets of the structural steel StE 690. These residual stress distributions verify the tendencies which are expressed in the schematical representations of Fig. 3 and Fig. 4. The residual stresses have been measured at the marked points on both sides of the weld centre line. In each case the stress distribution is completed as a dashed curve on one side under the assumption of symmetry to the weld centre line.

After welding with a small heat input the cooling rate is relatively high and hence the transformation temperature low. It can be concluded, for instance from hardness values, that in connection with a heat input per unit length of H = 2.88 kJ/cm 100 % of the austenite transforms into martensite, that is to say the transformation occurs completely below 460 °C. Consequently the transformation stresses are dominating and compressive residual stresses with a maximum magnitude of 120 N/mm^2 at the weld centre line can be found in the narrow transformed zone within the transverse residual stress field (fully drawn line in Fig. 6).If the heat input increases to 5.82 kJ/cm the austenite begins to transform at 510 °C and at 440° C – after a transformation of only 20 % of the structure into bainite – martensite begins to form (cooling curve 1 in Fig. 5). Under these welding and cooling conditions the transformed zone widens and shrinkage which occurs in the weld zone after the transformation of austenite gets a bit more influence on the residual stress state. As a consequence a new stress peak evolves near the weld centre line within the zone of compressive transverse residual stresses (dashed curve in Fig.6). This stress peak remains just below the tensile range and the measured distribution of the transverse residual stresses is in principal agreement with the residual stress distribution of curve 2 of the model consideration in Fig. 3. In connection with a heat input per unit length of 5.82 kJ/cm the distribution of the longitudinal residual stresses shows exclusively compressive stresses within the transformed zone without an additional stress peak (fully drawn curve in Fig. 7). However after welding with a heat input per unit length of 5.82 kJ/cm connected with preheating at 150° C (cooling curve 2 in Fig. 5) an additional stress peak with a relatively low magnitude in the tensile range could be observed at the weld centre line. This stress distribution (dashed curve in Fig. 7) is in full agreement with curve 2 of the model considerations in Fig. 4. The cooling curve 2 in Fig. 5 indicates that under this cooling condition the austenite begins to transform at 530° C and at 420° C 40 % of the austenite are transformed into bainite.

After welding with the heat input per unit length of 5.82 kJ/cm without preheating and with preheating relatively high tensile residual stresses have been observed in a larger distance on both sides of the weld centre line. These tensile stress maxima are a consequence of the volume expansion in the transforming areas in the longitudinal direction as described in the model considerations and illustrated schematically in Fig. 4.

In connection with the high heat input of 19.80 kJ/cm, corresponding to cooling curve 4 in Fig. 5, the transformation of austenite begins at 570° C and at 390° C 85 % of the structure is transformed into bainite. The dominating shrinkage process results then in a very high tensile peak of the residual stresses at the weld centre line – within the distribution of the transverse residual stresses (dash–dotted curve in Fig. 6) as well as of the longitudinal residual stresses (dash–dotted curve in Fig. 7). The magnitude of these stress peaks – 650 N/mm^2 respectively 600 N/mm^2 – reach nearly to the yield strength of the steel in the bainitic state. The remainders of the transformation induced transverse compressive stresses are found in a relatively big distance from the weld centre line (dash–dotted curve in Fig. 6) and the transformation induced longitudinal tensile stress maxima are clearly visible in a distance from the weld centre line which is only insignificantly larger (dash–dotted curve in Fig. 7). It seems plausible, that the magnitudes of the transformation induced tensile stress maxima increase with an increasing transforming volume, that is to say with an increasing heat input of the welding process. In the transverse direction the volume expansion during the transformation of austenite does not result in an elongation of adjacent, not transforming zones, but in a compression of these zones. Therefore no transformation induced tensile stress maxima can exist within the distributions of transverse residual stresses.

The following figures show examples for the distributions of transverse or longitudinal stresses in sheets of the steels St 52–3 and StE 890 with a dummy seam. In comparison with the equivalent residual stress distributions in sheets of the steel StE 690 one can see modifications which are typical for the altered chemical compositions respectively the altered TTT–diagrams or yield strength values.

Fig. 8 displays distributions of transverse residual stresses in sheets of the structural steel St 52–3 in which dummy seams have been produced with a heat input per unit length of 5.82 kJ/cm or 19.80 kJ/cm. According to the TTT–diagram of this steel and the cooling curve after welding with the lower heat input per unit length the transformation of austenite to bainite starts at a temperature of 440° C and the subsequent formation of martensite is finished at nearly 270° C. In connection with such low transformation temperatures the transformation stresses predominate even more than after welding of a dummy seam in a sheet of the steel StE 690 with the same heat input per unit length. Shrinkage after the austenite transformation can produce only a very weak peak within the zone with compressive residual stresses with a maximum magnitude of 250 N/mm^2 (fully drawn curve in Fig. 8). After welding with the high heat input of 19.80 kJ/cm 90 % of the austenite transform into bainite in the temperature range between 630° C and 480° C. Therefore shrinkage stresses become predominant and in a clear contrast to the residual stress distribution in connection with the lower heat input of 5.82 kJ/cm, but in principal agreement with the equivalent result for the steel StE 690, a relatively high tensile stress maximum appears at the weld centre line (dash–dotted curve in Fig. 8). The magnitude of the tensile stress maximum is equal to the yield strength of the steel St 52–3.

Fig. 9 exhibits distributions of longitudinal residual stresses in sheets of the steel StE 890 after welding of dummy seams with a heat input per unit length of 6.33 kJ/cm respectively 19.63 kJ/cm. Welding with the lower heat input produces transformation induced compressive residual stresses near the weld centre line in analogy to the result after welding of the steel StE 690 with a relatively small heat input. But the maximum magnitude of the compressive residual stresses is appreciably higher in sheets of the steel StE 890 than in the steel StE 690 (fully drawn curve in Fig. 9). In connection with the high heat input of 19.63 kJ/cm very high tensile stress maxima are induced in the not transforming zone as a consequence of the relatively large transforming volume and of the especially high yield strength of the structural steel StE 890. But the tensile stress peak which evolves as a consequence of post transformation shrinkage from the transformation induced compressive stress level reaches only a magnitude of 150 N/mm^2 at the weld centre line (dash–dotted curve in Fig. 9). This relatively low magnitude represents a strong contrast to the high magnitude of the corresponding tensile stress peak in the steel StE 690 (Fig. 7). The differences in the residual stress distributions after welding of the structural steels StE 690 and StE 890 can be explained with differences in the TTT–diagram. If equal temperature–cooling time curves are compared the transformation of austenite begins in the steel StE 890 always at a lower temperature than in the steel StE 690 and therefore the shrinkage stresses which develop after transformation of austenite cannot exert such a strong influence in the steel StE 890 as in the steel StE 690.

Fig. 10 is an extreme example for the modifications of features which have to be anticipated in the distributions of the longitudinal residual stresses if real weldments with a filler material are produced instead of dummy seams. Plates of the steel St 52–3 with a double–V edge preparation have been TIG–welded in 8 passes with a filler material of extreme strength (R_e = 830 N/mm^2), as used normally for structural steels with a correspondingly high strength. The length of the weld seam was 300 mm, the width of the plates after welding 200 mm and the plate thickness 20 mm. The heat input per unit length was 14.10 kJ/cm for the last pass. As one can see, welding with this heat input resulted in much higher logitudinal tensile residual stresses in the whole high strength weld metal zone than have been measured around a dummy weld seam in the same base material (heat input 19.80 kJ/cm). The magnitudes of the tensile residual stresses around the dummy seam are limited to a value of only 300 N/mm^2, which is somewhat below the yield strength of the steel St 52–3. This is a consequence of an especially broad irradiated area of the X– ray stress measurements, which had a width of 6 mm transverse to the dummy weld seam only in this special case instead of 1.5 mm or 2 mm in the normal experimental procedure /12/. The maximum magnitude of the tensile residual stresses in the real weldment with a high strength filler material reaches the extreme value of 750 N/mm^2. The transformation induced (com pare curve 3 in Fig. 4) side maxima of the tensile residual stresses beyond the heat affected zone are with a magnitude of nearly 400 N/mm^2 obviously limited by the yield strength of the base material (compare the maximum magnitude of transverse residual stresses at the centre line of a dummy weld seam in the steel St 52–3 in Fig. 8). The result that extremely high tensile residual stresses can develop in the weld metal zone of a weldment with a filler material of extraordinary high strength points at the fact, that the use of such filler materials can be critical – also for reasons of tensile residual stress magnitudes which can be dangerous. As usual the yield strength of the filler material should be adjusted to the yield strength of the base material.

To which extent the shrinkage of the highly heated areas can result in tensile residual stresses depends not only on the described interaction with the effect of the austenite transformation but also appreciably on the degree of restraint which is influenced directly by the welding sequence. Fig. 11 shows the distributions of transverse residual stresses in two hollow section joints welded with different welding sequences.

At first a hollow section joint as can be seen in Fig. 11 was welded with a correct welding sequence. After TIG–welding of the transverse V– type welds the four parts of the section were joined together by MAG–welding of the longitudinal seams and in that way the shrinkage of the transverse seam was not hindered significantly. Then tensile residual stresses with a low magnitude of about 300 N/mm^2 (das hed line in Fig. 11) have been found in the weld seam and in the heat affected zone. After welding with the wrong welding sequence – the transverse seams were TIG–welded at last so that their shrinkage was hindered very strongly – the transverse residual stresses (fully drawn curve in Fig. 11) around the transverse weld seams attain a maximum magnitude of about 700 N/mm^2 which corresponds exactly with the yield strength of the base material (high strength structural steel StE 690). This example evidently shows, that also in welded constructions it is possible to keep the tensile residual stresses on a low level if suitable welding parameters and a correct welding sequence are used. The application of wrong welding sequences produces high tensile residual stresses of the magnitude of the yield strength because of the strongly hindered shrinkage. Therefore the welding procedure has to be planned carefully if high tensile residual stresses must be avoided.

Another problem in conection with the generation of residual stresses in welded constructions which has to be considered is, that weldment distortions can appear after different steps of the production. Such distortions may appear, for instance, as a consequence of an inadequate heat input or because of an unfavourable welding sequence in connection with a low stiffness of the constructional unit. Therefore straightening of the different sections or the whole constructional unit by mechanical means or by an oxyacety-lene torch is necessary frequently to guarantee the dimensional stability of the unit. However it has to be considered that straightening by mechanical means can influence the residual stress system of a construction unit significantly in such a way, that for instance the mounting stresses, which must be applied to compensate inadmissible broad weld gaps are fixed in the construction so that these mounting stresses can be efficacious as high tensile residual stresses after completion.

Fig. 12 clearly shows the possible influence of a mounting process on the distribution of the residual stresses in a hollow section joint. The two parts of the construction have been joined with V– type wlds (weld seams (5) and (6)). At first the weld seam (5) was TIG–welded so that an inadmissible wide weld gap of about 25 mm appeared in the weld seam (6). After welding of the seam (5) the transverse residual stresses due to welding were measured by means of X–rays along the marked line in the weld seam (5). As can be seen in Fig. 13 the tensile transverse residual stresses in the cover pass of the weld seam (5) reach only a low magnitude of about 190 N/mm^2 (dashed curve in Fig. 13). With increasing distance from the weld centre line the tensile residual stresses decrease and change into compressive values. The compressive residual stresses in the base material are a consequence of the shot peening process which had been applied before welding to remove the rolling scale of the used sheets. Finally the weld gap of the seam (6) of about 25 mm was reduced to ca. 2 mm by mechanical means so that the seam (6) could be TIG– welded. After completion of the weld seam (6) the transverse residual stresses in the weld seam (5) have been measured again. Fig. 13 illustrates that the residual stress distribution in the weld seam (5) has completely changed after welding of the seam (6). The tensile residual stresses at the weld centre line reach about 680 N/mm^2 which is nearly the yield strength of the base material (StE 690) and also in a larger distance from the weld centre line (bet ween 2 and 20 mm) tensile residual stresses with magnitudes of 250 N/mm^2 to 550 N/mm^2 can be found. This extreme example makes evident, that in addition to favourable welding parameters, that is to say favourable coling conditions corresponding to the transformation behaviour of the used base material, and to a carefully designed welding sequence the avoidance of distortion during different welding steps is particulary important in view ofthe generated state of residual stresses in constructional units. Furthermore Fig. 12 shows, that high tensile residual stresses due to welding can be reduced efficiently by by stress relief annealing (dash–dotted curve in Fig. 12).

5. Summary

Model considerations on the interaction of shrinkage stresses and of stresses due to the transformation of austenite during cooling down and on the consequences of this interaction offer the possibility to explain a great variety of experimental results on residual stress distributions after welding. It turns out that the temperature range of the transformation of the austenite has a dominating strong influence on sign, magnitude and final distribution of the residual stresses after welding. Therefore all factors with an influence on this temperature range have also an influence on sign, magnitude and distribution of residual stresses due to welding. That is to say, these residual stresses depend on one side on welding parameters like heat input or geometry of the weld seam respectively geometry of the welded joint and on the other side on the type of the TTT–diagram of the filler material and hence on the chemical composition of these materials. The influence of the heat input per unit length is demonstrated clearly by experimental results : a relatively low heat input can produce compressive residual stresses in the weld seam, whereas a higher heat input results in tensile residual stresses in the weld seam. The magnitude of these tensile residual stresses in the weld seam can become as high as the yield strength of the filler material or the base material. Therefore, in order to avoid unnecessarily high tensile residual stresses, the yield strength of the filler material should not be unnecessarily high, but should be adjusted to the yield strength of the base material. On the other hand it could be shown that also in a material with an extremely high yield strength the magnitudes of the tensile residual stresses at the weld centre line remain quite moderate, if the TTT–diagram of the steel favours really low temperatures of the transformation of austenite. The experimental results indicate such a behaviour for the investigated structural steel StE 890. Finally attention must be paid to the fact that transformation induced longitudinal tensile stresses develop beyond the transforming zone and that the magnitude of these tensile residual stresses depends obviously less strong on the temperature of the austenite transformation and may reach the yield strength of the base material.

Furthermore the experimental results show evidently that the residual stress distributions which can be found in welded constructional units depend especially on the applied welding sequences. If wrong welding sequences are applied the restraint of the shrinkage of different parts of a construction can become maximum so that tensile residual stresses can arise in spite of low transformation temperatures due to the welding conditions. The avoidance of weldment distortions is as important as the observance of appropriate cooling conditions and a favourable welding sequence, as the mounting stresses which are necessary to compensate inadmissible wide weld gaps may be fixed in a constructional unit and extremely high tensile residual stresses can be created in that way which are fully independent on the cooling conditions.

References

/1/ H. Bühler, H.–A. Rappe, A. Rose : Archiv Eisenhüttenwesen 45 (1974), S.719–728

/2/ E. Macherauch, H. Wohlfahrt : Materialprüfung 19 (1977), S.272–280

/3/ H. Bühler, W. Jankowski : Archiv Eisenhüttenwesen 48 (1977), S.633–636

/4/ H. Christian, F.–X. Elfinger : Der Maschinenschaden 51 (1978), S.124–130

/5/ R. Prümmer, H. Wohlfahrt : Schweißeigenspannungen, in "Eigenspannungen und Lastspannungen", Hrsg. E. Macherauch u. V. Hauk, HTM–Beiheft 1982, S.73– 83

/6/ H. Wohlfahrt : Schweißeigenspannungen, Entstehung–Berechnung–Bewertung, in "Eigenspannungen", Hrsg. E. Macherauch u. V. Hauk, Deutsche Gesellschaft für Metallkunde e.V., Oberursel 1983, S.85–116

/7/ R. Poje : Schweißeigenspannungen in Elektronenstrahl– und Unterpulverschweißverbindungen von Stählen bei verschiedenen Blechdicken, Dissertation (1984) RWTH Aachen

/8/ H. Wohlfahrt : Härterei-Technische-Mitteilungen 41 (1986) Nr. 5, S. 248-257

/9/ J. Heeschen : Untersuchungen zum Dauerschwingverhalten von Schweißverbindungen aus hochfesten Baustählen unter besonderer Berücksichtigung des Eigenspannungszustandes und der Nahtgeometrie, Dissertation (1986) Gh Kassel

/10/ J. Heeschen, Th. Nitschke, H. Wohlfahrt : New Results on the Formation of Residual Stresses due to Phase Transformations in the Welded Structural Steels St 52-3 and StE 690, in "Residual Stresses in Science and Technology", Ed. by E. Macherauch and V. Hauk, DGM Informationsgesellschaft Oberursel, 1986, S. 1005-1013

/11/ J. Heeschen, Th. Nitschke, W.A. Theiner, H. Wohlfahrt : Schweißeigenspannungen - Grundlagen, Bedeutung und Auswirkung in geschweißten Bauwerken, DVS-Bericht 112, DVS-Verlag Düsseldorf 1988, S. 109-116

/12/ Th. Nitschke-Pagel; H. Wohlfahrt: The Generation of Residual Stresses due to Joining Processes. In "Residual Stresses, Measurement, Calculation, Evaluation", Ed. by V. Hauk, H. Hougardy and E. Macherauch, DGM Informationsgesellschaft Verlag Oberursel, 1991, S.121-133

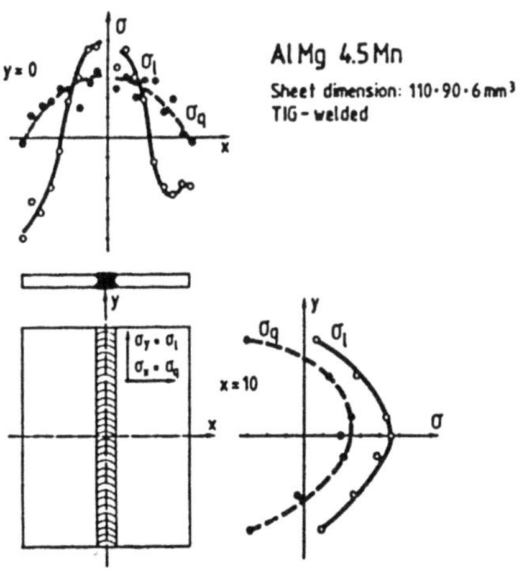

Fig. 1 : Distribution of longitudinal and transverse residual stresses along and perpendicular to the weld centre line of a two pass TIG-welded sheet of an Al Mg 4.5 Mn alloy

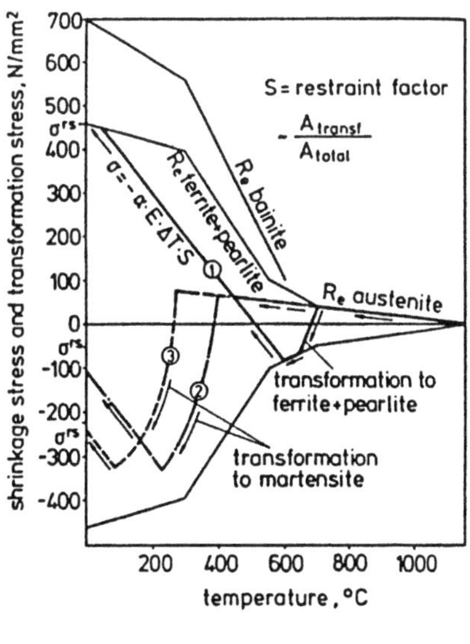

Fig. 2 : Schematical representation of the influence of the interaction of the shrinkage process and the transformation process on the resultung state of residual stresses

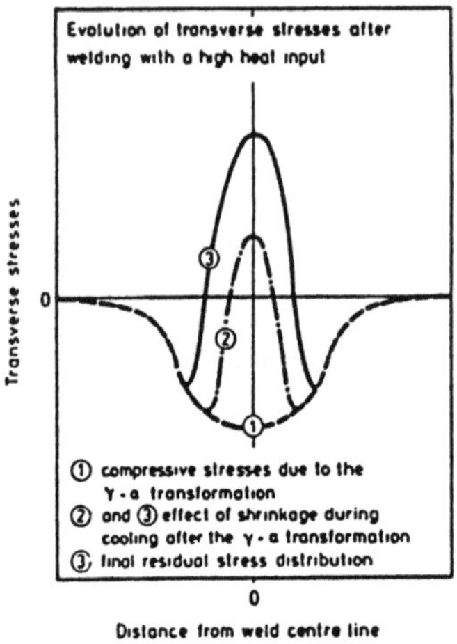

Fig. 3 : Schematical representation of the evolution of transverse stresses after welding with a high heat input

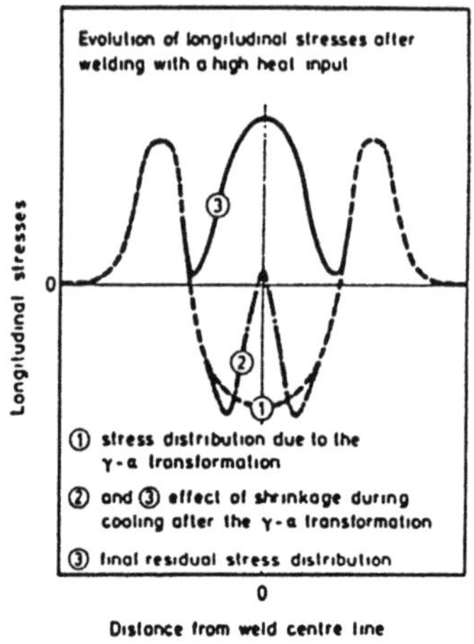

Fig. 4: Schematical representation of the evolution of longitudinal stresses after welding with a high heat input

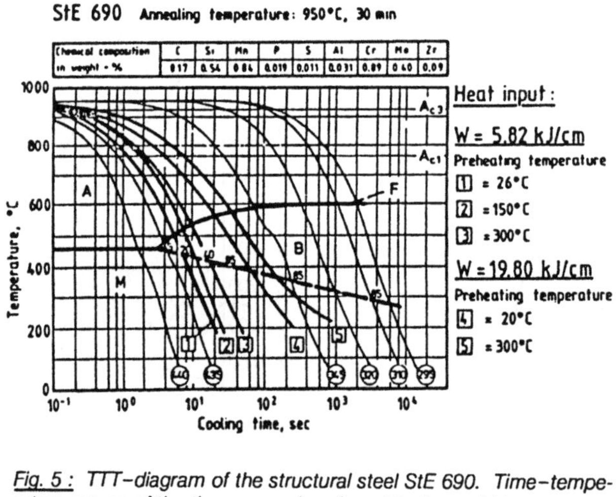

Fig. 5 : TTT–diagram of the structural steel StE 690. Time–temperature curves of the thermocouples closest to the weld toe

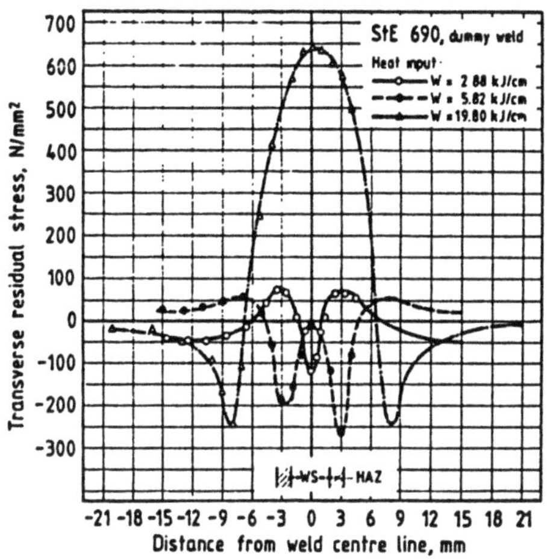

Fig. 6 : Distribution of transverse residual stresses at the surface of a dummy weld. Quenched and tempered steel StE 690 welded with different values of heat input

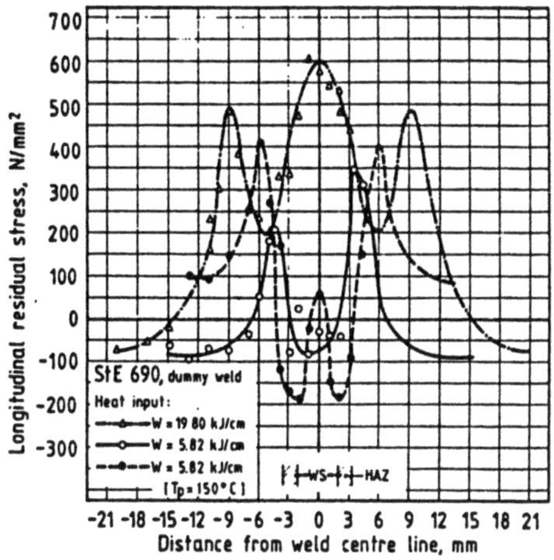

Fig. 7 : Distribution of longitudinal residual stresses at the surface of a dummy weld. Quenched and tempered steel StE 690 welded with different values of heat input

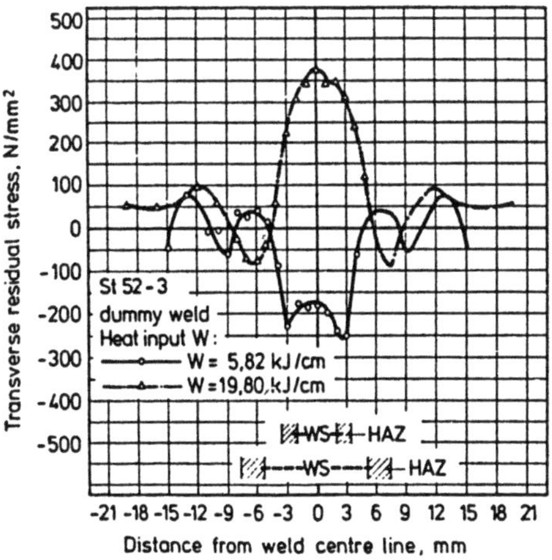

Fig. 8 : Distribution of transverse residual stresses at the surface of a dummy weld. Structural steel St 52–3 welded with different values of heat input

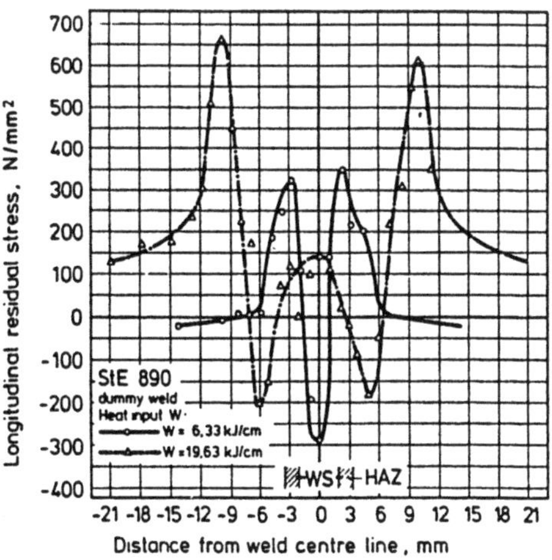

Fig. 9 : Distribution of longitudinal residual stresses at the surface of a dummy weld. Quenched and tempered steel StE 890 welded with different values of heat input

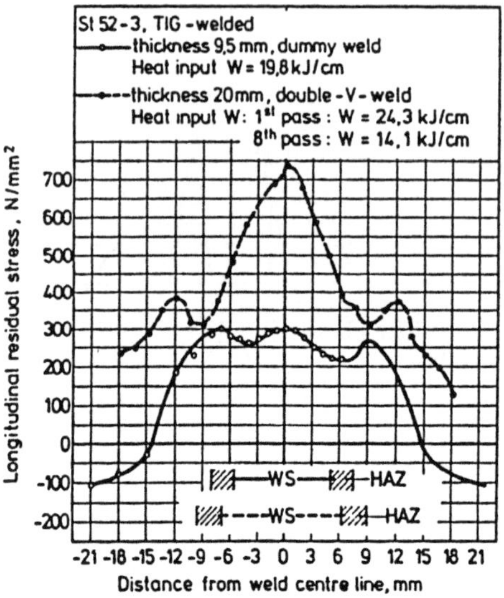

Fig. 10 : Distribution of longitudinal residual stresses at the surface of a dummy weld (heat input W=19.80 kJ/cm) and at the surface of a double-V-weld (8 passes, filler material : R_e = 830 N/mm²). Structural steel St 52–3

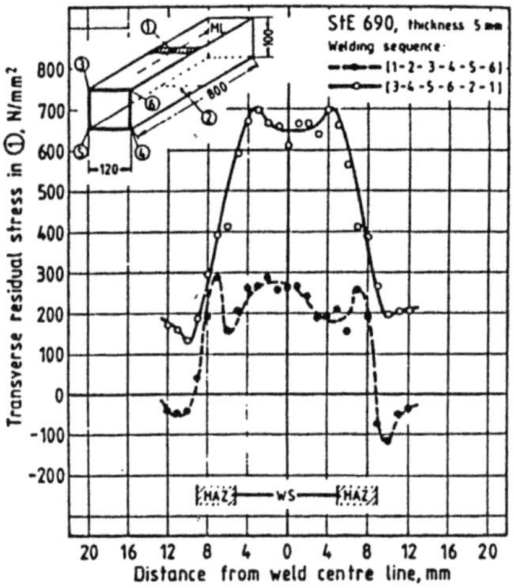

Fig. 11: Distribution of transverse residual stresses in TIG-welded transverse V-edges of hollow section joints after different welding sequences. Quenched and tempered high strength structural steel StE 690

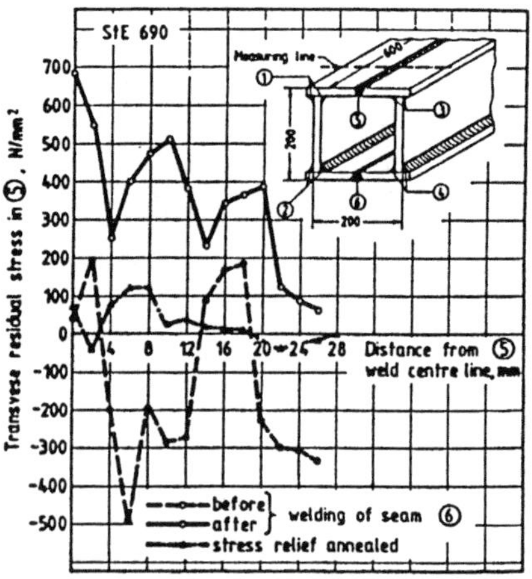

Fig. 12: Distribution of transverse residual stresses around a V–type weld (5) due to a mounting process in a hollow section joint before and after welding of the final weld seam (6). High strength structural steel StE 690

Measurement of Residual Stress Distributions Near the Toe of a Weld Between a Bracket and a Plate Using the Crack Compliance Method

W. Cheng and I. Finnie
Department of Mechanical Engineering
University of California
Berkeley, CA 94720 USA

Summary

The authors have developed a technique, referred to as the "crack-compliance method" to measure residual stresses using solutions based on fracture mechanics. In the present paper the method is extended to the measurement of residual stresses either at the toe of a weld joining a bracket to a plate or at the toe of a fillet weld which joins two plates. Results are presented for a specimen which simulates the attachment of a bracket to the wall of a nuclear reactor pressure vessel.

Analysis

The traditional technique for measuring residual stresses by layer removal is not easily applied to weldments because of the profile of the weld and the localized stress distribution [1]. Another technique which uses strains produced by a thin cut of progressively increasing depth has been used and validated for measurement of residual axial stresses in butt-welded cylinders, hoop stresses in quenched thick-walled cylinders [2] and bent beams [3].

At this time we consider a weld between a bracket and plate or a fillet weld as shown in Fig. 1. When a cut of depth a is made in a plate of thickness t, residual stresses are released, and the corresponding deformation can be estimated by applying the same stresses with opposite sign on the faces of the cut. If the width of the cut is small relative to the depth, the deformation due to the cut can be approximated by that due to a crack of the same depth. The general expression relating stress intensity factors to the normal strain ε_y on the surface at a distance s, measured from the cut in the y direction, has been given in [4], i.e.

$$\varepsilon_y(\hat{a},\hat{s}) = \int_0^a K_I \frac{\partial^2 K_I^f}{\partial F \partial \hat{s}} \frac{d\hat{a}}{E'} \tag{1}$$

in which $\hat{a} = a/t$, $\hat{s} = s/t$, $E' = E$ for plane stress and $E' = E/(1-\mu^2)$ for plane strain. The stress intensity factors K_I and K_I^f are those due to loading on the crack faces and due to a virtual force F at the location where the strain is to be measured. For a flat plate with an edge crack the stress intensity factor has been obtained for an arbitrary surface loading $\sigma(z)$ on the faces of the crack for all values of a/t [5], which, with some manipulation, may be expressed as

L. Karlsson, L.-E. Lindgren, M. Jonsson (Eds.)
Mechanical Effects of Welding
IUTAM Symposium Luleå/Sweden 1991

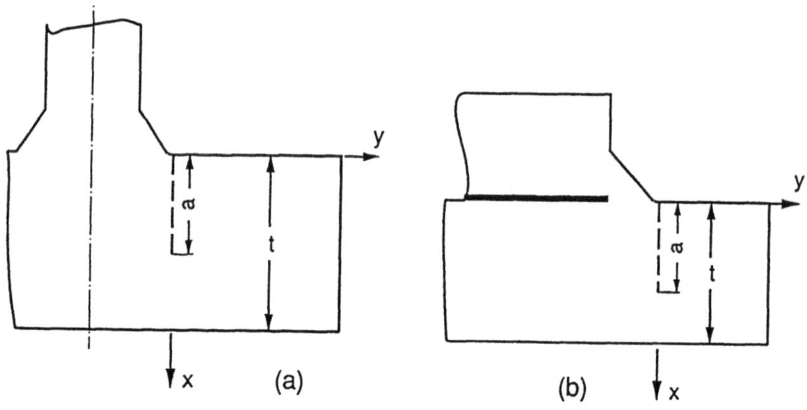

Fig. 1 (a) a weld between a bracket and a plate; (b) a fillet weld

$$K_I = \sqrt{\pi a}\,(\frac{2}{\pi})f_o(\hat{a}) \int_o^{\hat{a}} [(1-z)(2+G-3Gz)\cos^{-1}(\frac{Hz}{1-z})$$

$$+ \frac{H(1-Gz)}{\sqrt{1-(\frac{Hz}{1-z})^2}}]\sigma(z)dz = \sqrt{a}\, f_o(\hat{a}) \int_o^{\hat{a}} D(z,\hat{a})\sigma(z)dz \qquad (2)$$

where $z = x/t$ is measured from the mouth of the crack and

$$H = \frac{1-\hat{a}}{\hat{a}} \quad ; \quad G = \frac{3(1-7\hat{a})(1-\hat{a})^5}{28\hat{a}}$$

The stress intensity factor K_I^f can be obtained by substituting into Eq. (2) the stress field $\sigma^f(x)$ corresponding to the virtual force F [6].

For a plate with an attachment as shown in Fig. 1, the deformation of the plate due to making a cut is influenced by the compliance of the attachment. As a first step, we conceptually separate the attachment from the plate as shown in Fig. 2 and express the normal and shear stresses in terms of two polynomial series of order N

$$\sigma_y = \sum_{i=0}^{N} \sigma_i T_i(y) \quad ; \quad \tau_{xy} = \sum_{i=0}^{N} \tau_i T_i(y) \qquad (3)$$

in which σ_i and τ_i are coefficients to be determined for the ith order function $T_i(y)$. Now applying each term of the series as the loading condition to the plate and attachment, the displacement response at m locations along the boundary may be obtained by using the crack compliance function for the plate [6] and by a finite element computation for the attachment. Since the computation for the attachment is independent of the crack, it can be carried out readily on a personal computer and its results can be used for all crack sizes. Denoting the horizontal u and normal v displacements due to the loading given in Eq. (3) by a single and double prime for the plate and the attachment respectively, they may be expressed as

$$
\begin{aligned}
u'(y) &= -\sum_{i=0}^{N} [\sigma_i C^n_{iu'}(y) + \tau_i C^s_{iu'}(y)] \\
v'(y) &= -\sum_{i=0}^{N} [\sigma_i C^n_{iv'}(y) + \tau_i C^s_{iv'}(y)] \\
u''(y) &= \sum_{i=0}^{N} [\sigma_i C^n_{iu''}(y) + \tau_i C^s_{iu''}(y)] \\
v''(y) &= \sum_{i=0}^{N} [\sigma_i C^n_{iv''}(y) + \tau_i C^s_{iv''}(y)]
\end{aligned}
\tag{4}
$$

in which superscripts s and n are used to denote the compliances $C^s_{iu'}$, $C^s_{iv'}$, $C^n_{iu''}$ and $C^n_{iv''}$ for shear and normal stresses. From continuity the total displacements of the bracket and plate given in Eq. (4) must equal the displacements u^c and v^c due to a given surface traction on the crack faces for a flat plate, i.e.

$$
\begin{aligned}
u^c(y) &= -u'(y) + u''(y) = \sum_{i=0}^{N} [\sigma_i C^n_{iu}(y) + \tau_i C^s_{iu}(y)] \\
v^c(y) &= -v'(y) + v''(y) + \sum_{i=0}^{N} [\sigma_i C^n_{iv}(y) + \tau_i C^s_{iv}(y)]
\end{aligned}
\tag{5}
$$

in which

$$
C^n_{iu}(y) = C^n_{iu'}(y) + C^n_{iu''}(y) \quad , \quad C^s_{iu}(y) = C^s_{iu'}(y) + C^s_{iu''}(y)
$$

$$
C^n_{iv}(y) = C^n_{iv'}(y) + C^n_{iv''}(y) \quad , \quad C^s_{iv}(y) = C^s_{iv'}(y) + C^s_{iv''}(y)
$$

The unknown coefficients σ_i and τ_i can now be determined by imposing displacement continuity at m > N number of points along the boundary using a least squares fit which leads to

$$
\begin{aligned}
&\frac{\partial}{\partial \sigma_i} \sum_{j=1}^{m} \{[u_j^c - \sum_{k=0}^{N} (\sigma_k C^{nj}_{uk} + \tau_k C^{sj}_{uk})]^2 + [v_j^c - \sum_{k=0}^{N} (\sigma_k C^{nj}_{vk} + \tau_k C^{sj}_{vk})]^2\} = 0 \\
&\quad = \sum_{j=1}^{m} (u_j^c C^{nj}_{ui} + v_j^c C^{nj}_{vi}) - \sum_{j=1}^{m} \sum_{k=0}^{N} [(\sigma_k C^{nj}_{uk} + \tau_k C^{sj}_{uk}) C^{nj}_{ui} + (\sigma_k C^{nj}_{vk} + \tau_k C^{sj}_{vk}) C^{nj}_{vi}] \\
&\frac{\partial}{\partial \tau_i} \sum_{j=1}^{m} \{[u_j^c - \sum_{k=0}^{N} (\sigma_k C^{nj}_{uk} + \tau_k C^{sj}_{uk})]^2 + [v_j^c - \sum_{k=0}^{N} (\sigma_k C^{nj}_{vk} + \tau_k C^{sj}_{vk})]^2\} = 0 \\
&\quad = \sum_{j=1}^{m} (u_j^c C^{sj}_{ui} + v_j^c C^{sj}_{vi}) - \sum_{j=1}^{m} \sum_{k=0}^{N} [(\sigma_k C^{nj}_{uk} + \tau_k C^{sj}_{uk}) C^{sj}_{ui} + (\sigma_k C^{nj}_{vk} + \tau_k C^{sj}_{vk}) C^{sj}_{vi}]
\end{aligned}
\tag{6}
$$

for i = 0,...,N, in which the letter j is used to denote the location at y_j.

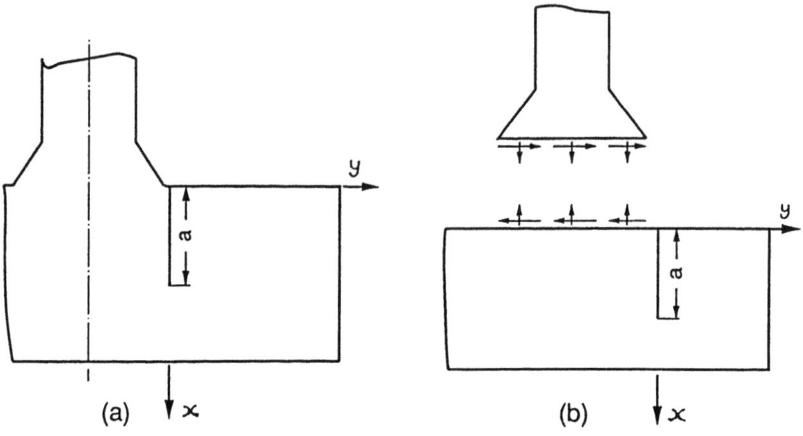

Fig. 2. Conceptual separation of the attachment from the plate

The additional deformation due to the attachment can now be obtained from the stress distributions $\sigma^a(x)$ and $\tau^a(x)$ on the plane of the crack corresponding to the boundary traction given in Eq. (3). It has been found that the normal strain at the location directly opposite the crack gives best overall strain response to release of normal stresses [6] and is not affected by any shear stresses on the plane of the crack [7]. Thus, from superposition, the stress intensity factor K_I for a plate with an attachment can be obtained by adding $\sigma^a(x)$ to the stress field $\sigma(z)$ in Eq. (2) from which the strain on the bottom surface can be found by using Eq. (1).

Estimation of Residual Stress Distributions From Measured Strains

To estimate a residual stress field through the thickness, we approximate it with an nth order polynomial series

$$\sigma(z) = \sum_{i=0}^{n} A_i P_i(z) \tag{7}$$

in which A_i is the amplitude factor for the ith order polynomial $P_i(z)$. For a body free of external loading Legendre polynomials are an appropriate choice since the force and moment equilibrium conditions can be always satisfied by setting A_0 and A_1 to zero.

Substituting Eq. (7) into Eq. (1) the strain at $\hat{s} = 0$ on the bottom surface may be written as

$$\varepsilon_y(\hat{a},0) = \sum_{i=0}^{n} A_i C_i(\hat{a})$$

$$= \sum_{i=0}^{n} A_i \int_0^{\hat{a}} \hat{a}[f_o(\hat{a})]^2 \{ \int_0^{\hat{a}} D(\hat{a},z)[P_i(z) + \sigma_i^a(z)]dz\}[\int_0^{\hat{a}} D(\hat{a},z)\sigma^f(z)dz] \frac{d\hat{a}}{E'} \qquad (8)$$

where $\sigma_i^a(z)$ is the stress field due to the presence of the attachment for the ith order function $P_i(z)$ and may be determined for a given crack size from Eq. (6).

Te determine the unknown amplitude factors A_i in Eq. (8), we obtain a number $m > n$ of strain readings by making progressively a cut of m increments. The method of least squares is then used to obtain A_i while minimizing the error over all data points. For an nth order approximation this leads to

$$\frac{\partial}{\partial A_i} \sum_{j=1}^{m} [\varepsilon(\hat{a}_j) - \sum_{k=0}^{n} A_k C_k(\hat{a}_j)]^2 \qquad i = 0,\ldots,n$$

$$= \sum_{j=1}^{m} \varepsilon(\hat{a}_j) C_i(a_j) - \sum_{j=1}^{m} \sum_{k=0}^{n} A_k C_k(\hat{a}_j) C_i(\hat{a}_j) = 0$$

Experimental Results

The crack compliance method was used to obtain the through-thickness residual stress distribution near the toe of a weld between a plate and a bracket with dimensions shown in Fig. 3. The bracket was made of Inconel. The plate which was made of A533-B low carbon steel clad with Inconel about 19 mm thick on the middle portion of the top surface where the bracket was located and with stainless steel about6 6.4 mm thick on the rest of the top surface.

For through-thickness measurement strain gages were attached at three locations at the bottom side of the plate directly opposite to the toe of the weld where a cut was to be made. Strain gages were also attached to the top surface near the weld for surface stress measurement. An electrical discharge machine was used to produce a cut of progressively increasing depth with a width of 0.33 mm. Noticeable strains were measured at the bottom face when the depth was only about 3 percent of the thickness. The residual normal stress distribution was estimated by 7th to 9th order Legendre polynomails with consistent agreement and the average value is shown in Fig. 4 as solid line. The estimated decrease of the residual stress near the surface was confirmed by the strains measured near the cut on the top surface. The residual stress distribution was also estimated by using the crack compliance functions for a flat plate and is shown in Fig. 4 as dashed line. It is seen that the effect of the bracket on the residual stress distribution is very significant for $x/t < 0.15$ while for $x/t > 0.4$ the effect becomes negligible.

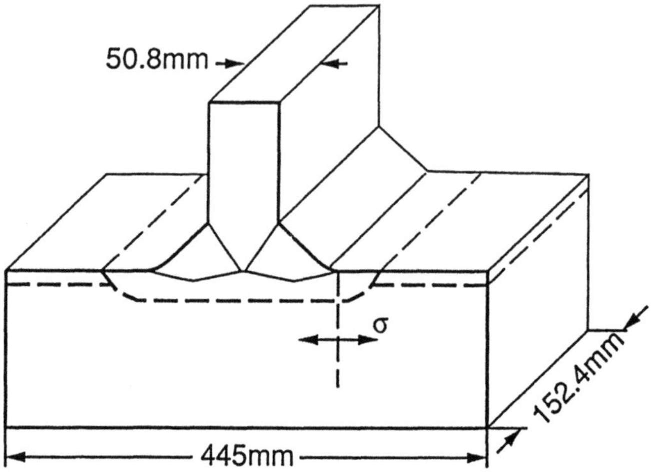

Fig. 3 Dimensions of the plate-bracket assembly

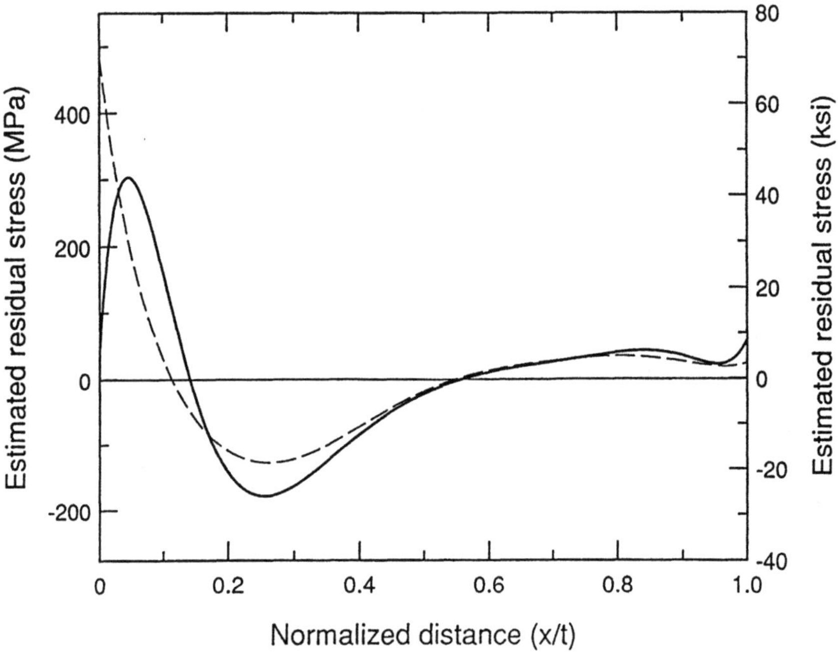

Fig. 4 Estimated residual stress distribution

Conclusions

The experimental procedure described appears well suited for through the thickness residual stress measurements on welded joints. We have applied the method, for example to study the effect of phase transformations on the residual stress distribution due to fillet welds. For surface stress measurement, the technique described in [4] has recently been improved to allow for the finite width of the cut [8]. Corrections may also be made to the surface stress solution to allow for attachment geometry.

Acknowledgment

This work was supported by the Electric Power Research Institute. We are grateful to the EPRI Technical Manager, Mr. J. Gilman for his assistance.

References

1. Cheng, W.; Finnie, I.: Examination of the Computational Model of the Layer Removal Method. Experimental Mechanics. 26 (1986) 150-153.
2. Cheng, W.; Finnie, I,: The Crack Compliance Method for Residual Stress Measurement. Welding in the World. 29 (1990) 103-110.
3. Cheng, W.; Prime, M., Finnie, I.: The Crack Compliance Method for Residual Stress Measurement. Presented at the 3rd International Conference on Residual Stresses, Tokushima, Japan, July 1991.
4. Cheng, W.; Finnie, I.; Vardar, O.: Measurement of Residual Stresses Near the Surface Using the Crack Compliance Method. ASME J. of Eng. Mat. and Tech. 113 (1991) 199-204.
5. Cheng, W.; Finnie, I.: K_I Solutions for an Edge-Cracked Strip. Eng. Fracture Mech. 31 (1988) 201-207.
6. Cheng, W.; Finnie, I.; Vardar, O.: Deformation of an Edge Cracked Strip Subjected to Normal Surface Traction on the Crack Faces. Submitted to Eng. Fracture Mechanics.
7. Cheng, W.; Finnie, I.: Deformation of an Edge-Cracked Strip Subjected to an Arbitrary Shear Surface Traction on Crack Faces. Submitted to Eng. Fracture Mech.
8. Cheng, W.; Finnie, I.: Surface Stress Measurement with Finite Width Slots (to be submitted to ASME).

The Neutron Strain Scanner: Measurements in Welds

P. J. WEBSTER

Department of Civil Engineering,
University of Salford,
Salford, M5 4WT, UK

Abstract

The neutron strain scanning technique is described and examples are given of non-destructive internal measurements made through a double-V weld in an aluminium alloy plate. The results reveal details of the residual macrostress distribution, microstrain and texture in individual weld beads and adjacent to the weld.

Introduction

Residual stresses arise in welds during fabrication as a result of differential cooling, phase changes and plastic deformation. They are one of the important factors contributing to the fatigue lives of welded components and a knowledge of their distribution and magnitudes is essential to any assessment of their integrity. Crystalline anisotropy and texture are also contributary factors to the properties of weldments.

Finite element modelling and other analytical methods are widely employed to calculate residual stresses in welds but it is essential that the theoretical calculations are validated by comparison with detailed, accurate and reliable experimental data. Traditional strain measuring techniques such as layering and hole drilling are usually, to a greater or lesser extent, destructive and the determination of stresses from the strain relaxation data is difficult for components with complex shapes. The X-ray diffraction technique is non-destructive but is effectively restricted, because of strong absorption, to near surface measurements which are often of limited value in welds due to irregularities in their surface finishes which are not typical of the interior.

Neutron strain scanning, using high resolution neutron diffraction, is a relatively new non-destructive technique that is being used to determine internal and through surface residual stresses in a wide range of crystalline materials and components [1-5]. It is in principle, subject to absorption and size constraints, possible accurately to determine the strain state at any point, in any direction, in a component and hence to derive the full stress tensor. This makes the technique uniquely suitable for the validation of theoretical calculations. Additionally information is provided relating to anisotropy and texture.

L. Karlsson, L.-E. Lindgren, M. Jonsson (Eds.)
Mechanical Effects of Welding
IUTAM Symposium Luleå/Sweden 1991

The neutron strain scanner

A beam of thermal neutrons incident upon a polycrystalline material is scattered, by the atoms on the atomic lattice, at specific "Bragg" angles to form a characteristic diffraction pattern. Measurements of the angles, intensities and profiles of the diffraction peaks provide information about the crystal lattice, its dimensions, site occupation and orientation.

The relationship between the diffraction angle θ, the interplanar spacing d and the neutron wavelength λ is given by the Bragg equation:

$$2d\sin\theta = \lambda \tag{1}$$

If a material is strained there are consequent changes, δd, in the dimensions of the lattice and, $\delta\theta$, in the angular positions of the Bragg peaks. Differentiation of the Bragg equation gives a relation for the lattice strain ε which may be expressed as:

$$\varepsilon = \delta d/d = -\,\delta\theta.\cot\theta \tag{2}$$

where the strain measured is that parallel to the scattering vector Q which bisects the angle between the incoming and outgoing neutron beams.

Strains are often classified as one of two types, microstrains or macrostrains. Microstrains vary over distances that are short relative to the microstructure and cannot be spatially resolved by most measuring techniques. They produce a continuous spread of lattice distortions in the range, $d \pm \Delta d$, and a corresponding broadening, $\pm\Delta\phi$, in the Bragg peaks, measurements of which are used to quantify the microstrain. Macrostrains vary throughout a component and may be spatially mapped by measuring the shift of the centre of the peak $\delta 2\theta$ at each location within the component. If there is texture and preferred orientation there will also be changes in the actual and relative intensities of the peaks.

In general, and almost certainly in welds, there will be microstrain, macrostrain and textural variation and all may be investigated using neutron strain scanners.

The principle of the neutron strain scanner is outlined in fig. 1. A small "gauge volume", on the axis of the instrument, is defined by neutron absorbing masks in the input and output beams and the component is located and moved relative to it using three automated orthogonal translators, two horizontal and one vertical, with rotation about the vertical axis. A suitable diffraction peak, preferably near 90°, is selected and measured as the component is scanned using a combination of translations and rotations. The peak profiles are recorded and their centre positions, widths, heights and intensities are determined by a computer fitting routine. Strains are calculated from the data and stresses are then derived from the strains using appropriate elastic moduli.

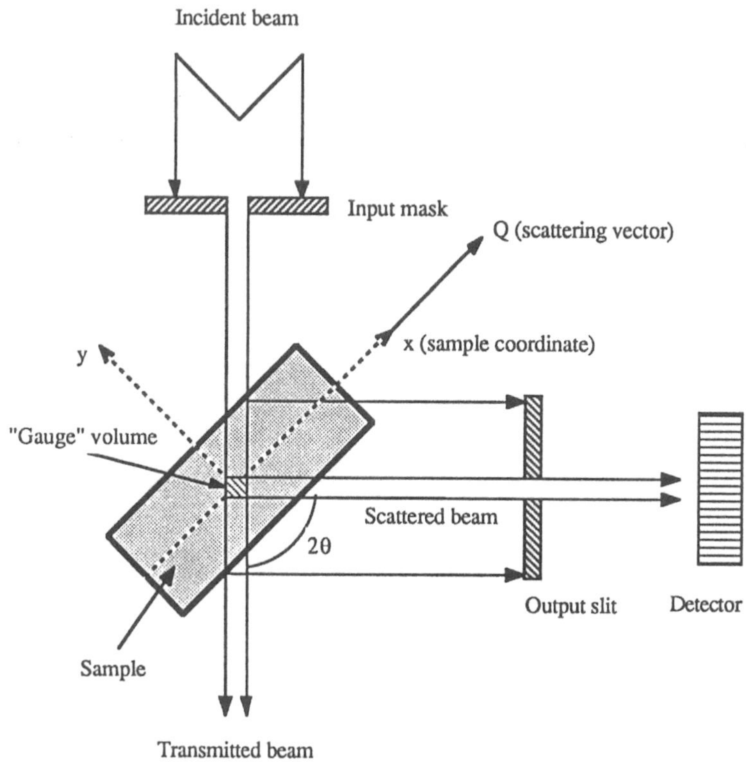

Fig. 1. Principles and outline of the neutron strain scanner.

As an intense neutron source is required scanners must be located at central neutron research facilities, with nuclear reactor or spallation neutron sources, such as at the Institut Laue Langevin (ILL) in Grenoble or ISIS at the Rutherford Appleton Laboratory in England. The size of component that can be surveyed is limited mostly by handling and absorption constraints. Absorption is material dependent and exponential but typically measurements may be made economically in most steels through a thickness of 30 mm, nimonic alloys 20 mm, titanium 50 mm and aluminium alloys 250 mm.

Measurements in welds

Measurements have been made in welds of several different types including conventional single and multipass welds and electron beam welds; in various engineering materials such as ferritic and austenitic steels, aluminium alloys, inconel and zirconium alloy and in a range of components with applications ranging from aerospace to civil engineering [6-11]. Results are described here of an investigation made on a double-V weld in a thick aluminium alloy plate.

The investigation was made on a 200 × 12 × 44 mm sample section, from the Welding Institute, cut transversely from a multi-pass double-V MIG weld across an aluminium alloy 5083 plate 44 mm thick by 900 mm square. Full details of the material and fabrication and other measurements are given in [6]. The approximate yield and tensile stresses for the parent material are 200 and 330 MPa respectively. The joint design, with the sequentially numbered bead pattern, and the co-ordinate system used are shown in fig 2.

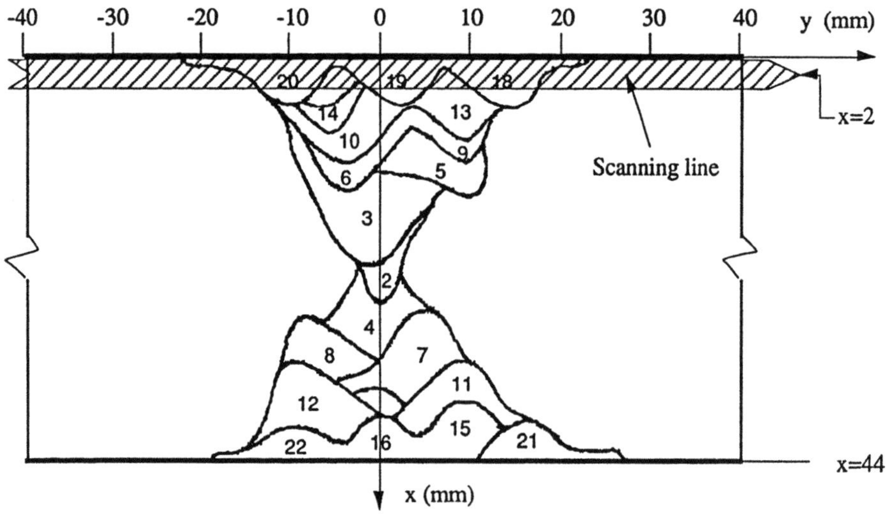

Fig. 2. Joint design and pass sequence for the aluminium alloy double-V MIG weld [6].

The strain scanner used was an adaptation of the neutron diffractometer D20 at the ILL with a neutron wavelength 0.14 nm. Data were collected for the (311) and (222) Bragg reflections at 51 points in linear transverse (y) scans from co-ordinates (2,-50) to (2,+50) centrally in the sample at a depth (x) 2 mm below the top surface, in steps of 2 mm for the X, Y and Z principal orientations. The sampling volume was 3 × 3 × 3 mm and the traverses passed through beads numbers 20, 19 and 18 and partially included beads 13 and 14.

The outlines of successive diffraction patterns, for the Y orientation traverse, are shown in fig. 3. The corresponding peak parameters for the (311) reflection are shown in figs. 4 and the stress pattern derived from the combined strain data in fig. 5.

Figs. 3 and 4 show how the peak profiles vary as the sample is scanned. In the host metal, from -50 to -24 mm and 26 to 50 mm the peak widths (microstrain) are constant at about 0.45° and the peak positions (macrostrain) change smoothly, but there is some variation in peak height and area, particularly on the positive side, due to grain size effects. In the weld the

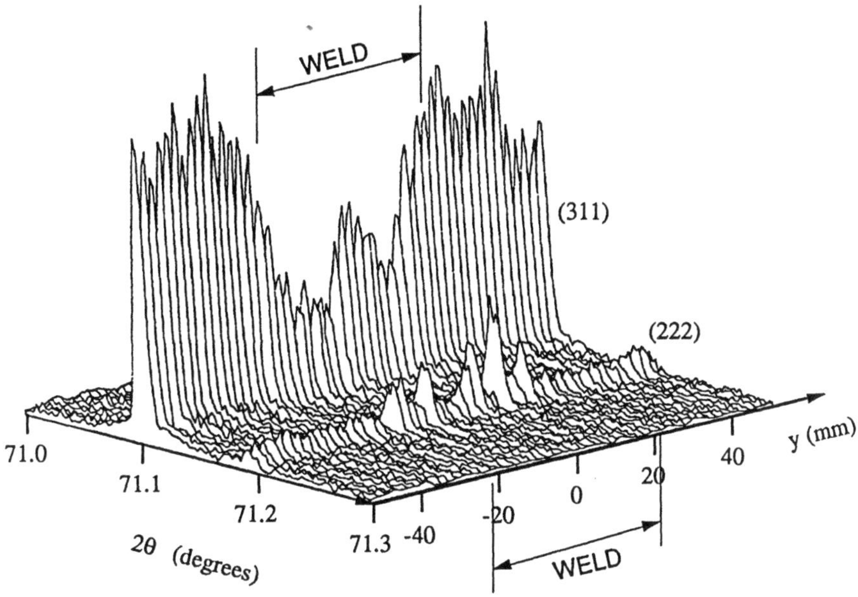

Fig. 3. Outlines of successive (311) and (222) diffraction profiles, for the Y orientation scan in 2 mm steps from y = -50 mm to +50 mm across the weld at a depth 2 mm.

average peak width has increased to about 0.6° which is an indication of the increase in microstrain in the plastically strained regions. The changes in peak position, peak height and area in the weld are related to the textural and macroscopic strain variations corresponding to the bead pattern. The sharp irregularities, and the changes in the relative intensities of the (311) and (222) reflections, are due to the particular orientations of a relatively small number of larger crystallites which have grown and reformed during the welding sequence.

If the few sharp perturbations due to the few large crystallites are disregarded the remaining peak positions are a direct measure of the macrostrain which rises to a maximum just outside the weld and varies from bead to bead inside.

Fig. 5 shows the transverse Y stress calculated from the strains assuming an average Young's modulus of 70 GPa.

As would be expected the maxima in tensile transverse stress are observed to be just outside the weld, at a distance of about 5 mm into the parent metal. The weld was made by depositing layers of weld beads alternately on opposite sides. The last beads deposited, 21 and 22, were laid in the other V. The stress variations shown correspond to the beads laid in the penultimate sequence 18, 19 and 20 on top of the beads 13 and 14.

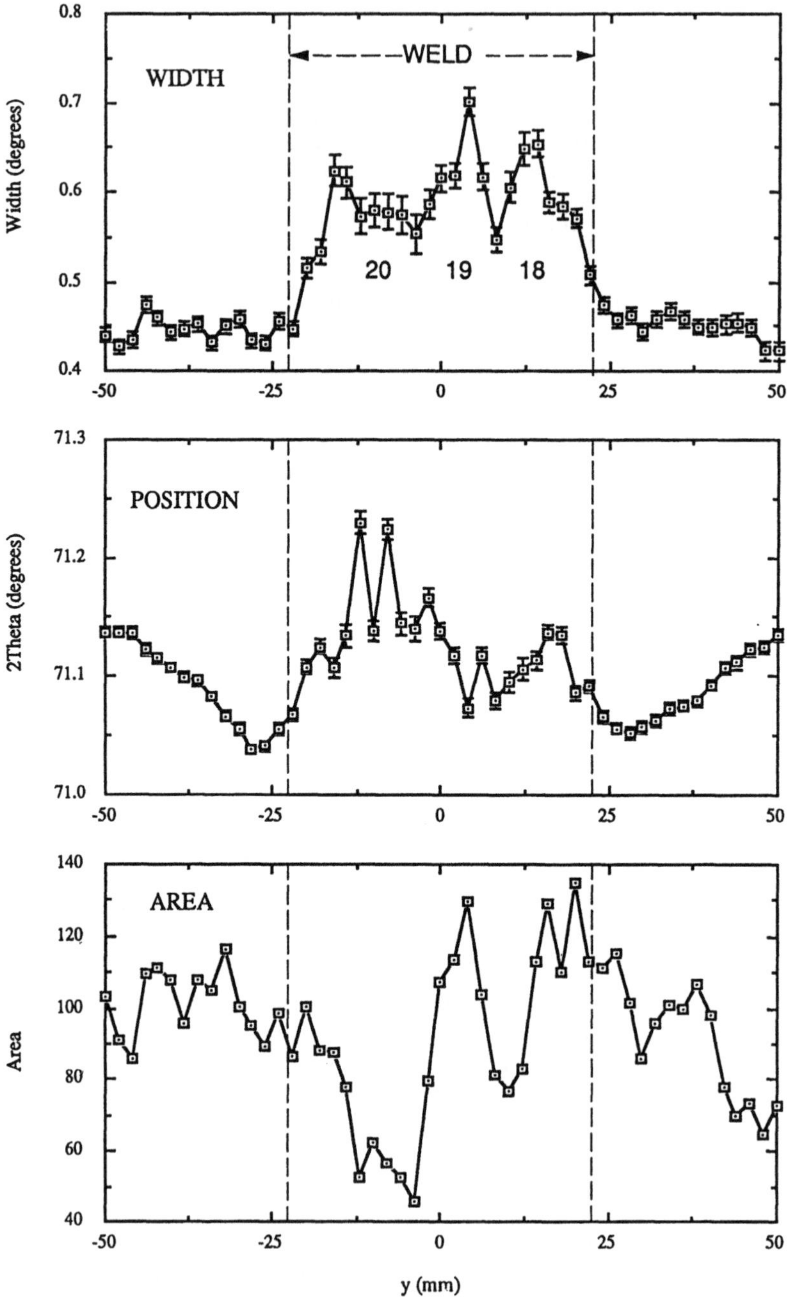

Fig. 4. Peak parameters for the (311) reflection, measured in the Y orientation, as a function of distance y across the weld, and position relative to the weld beads

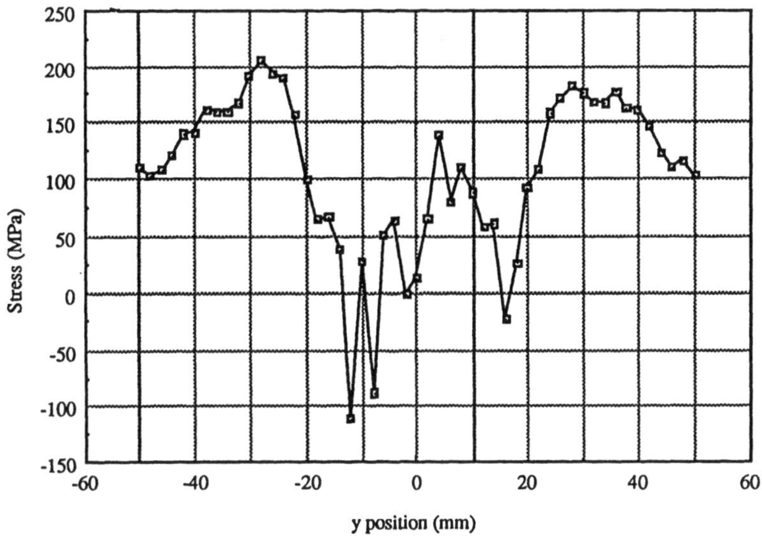

Fig. 5. Transverse residual stress distribution across the weld at a depth of 2 mm.

Summary

Neutron strain scanners are able non-destructively to measure internal and through surface residual strains in most polycrystalline materials. Their use is limited by absorption and portability constraints. Their ability to measure lattice strains accurately at many points and in chosen directions enables, in principle, the full stress tensor to be mapped throughout a component. The experimental results are particularly suitable for the validation of numerical calculations of residual stress and for the non-destructive determination of microstress, macrostress and texture in complex constructions. In welds they may be used to measure the stresses and texture within and between individual weld beads, in the heat affected zone and in the parent metal.

Acknowledgements

The author wishes to thank the Directors and staff of the Institut Laue Langevin for providing facilities and assistance for the development of strain scanners, and the UK Science and Engineering Research Council for financial support via grants GR/D63196 and GR/E55556 in co-operation with the Welding Institute who supplied the weld sample.

References

1. Allen, A.; Andreani, C.; Hutchings, M, T.; Windsor, C. G.: Measurement of internal stress within bulk materials using neutron diffraction. NDT International 14 (1981) 249-254.

2. Pintschovius, L.; Jung, V.; Macherauch, E. and Vohringer, O.: Residual stress measurements by means of neutron diffraction. Mater. Sci. Eng., 61 (1983) 43-50.

3. Stacey, A.; MacGillivray, H. J.; Webster, G. A.; Webster, P. J.; Ziebeck, K. R. A.: Measurement of residual stresses by neutron diffraction. J. Strain Analysis 20 (1985) 93-100.

4. Krawitz, A. D. and Holden, T. M.: The measurement of residual stresses using neutron diffraction. MRS Bull., XV (1990) 57-64.

5. Webster, P. J.: The neutron strain scanner: a new analytical tool for engineers. Steel Times 218 No. 6 (1990) 321- 323.

6. Smith, D. J.; Leggatt, R. H.; Webster, G. A.; MacGillivray, H. J.; Webster, P. J.; Mills, G.: Neutron diffraction measurements of residual stress and plastic deformation in an aluminium alloy weld. J. Strain Analysis 23 (1988) 201-211

7. Hosbons, R. R.; Ibrahim, E. F.; Holden, T. M.; Root, J. H.: The use of neutron diffraction to determine non-destructively the residual strain and texture in welds. "Recent trends in welding science and technology TWR'89", eds. S. A. David and J. M. Vitek; ASM International; Proceedings of the 2nd International Conference on Trends in Welding Research, Gatlinburg, May 1989, (1990) 103-106.

8. Webster, P. J.; Webster, G. A.; MacGillivray, H. J.; Smith D. J.; Ziebeck, K. R. A.: Comparison of residual stress distributions in thick section welds. ILL Annual Report No 5-25- 203,204 (1986) 119.

9. Webster, P. J.; Low, K. S.; Webster, G. A.; MacGillivray H. J.; Bourke, M. A. M.: Fatigue strength and residual stress in generic engineering components. ILL Annual Report No 5-23-311 (1987) 158.

10. Webster, P. J.; Webster, G. A.; Low, K. S.; Mills, G.; Bourke, M. A. M.; MacGillivray, H. J.: Residual stress in generic engineering components (b) An electron beam weld. ILL Annual Report No 5-23-325 (1988) 177.

11. Webster, P. J.; Webster, G. A.; Low, K. S.; Mills, G.; Bourke, M. A. M.; MacGillivray, H. J.: Residual stress in generic engineering components (ii) Residual stresses across an electron béam weld. ILL Annual Report No 5-23-349 (1989) 193.

Effect of Weld Metal Mis-match on Toughness Requirements: Some Simple Analytical Considerations Using the Engineering Treatment Model (ETM)

Condensed Version

Karl-Heinz Schwalbe
GKSS-Forschungszentrum Geesthacht GmbH
Max - Planck - Str.
2054 Geesthacht
GERMANY

Abstract

The Engineering Treatment Model (ETM) provides simple formulae for estimating the crack tip opening displacement (CTOD) as a function of applied load or applied strain for work-hardening materials.
Its application to a transverse welded wide plate shows that differences in the plastic properties of the weld metal and base material, respectively, significantly affect the CTOD. The expressions derived can serve for stating toughness requirements of the weld metal such that a weld metal crack would affect the toughness behaviour of the welded joint no more than a base material crack of the same size.

Scope of Investigation

The failure condition of a cracked component can be stated as

Driving Force = Toughness.

In the case of welds, countless investigations deal with the toughness side of the above shown equation, whereas very little has been done so far on the driving force side. The situation is not trivial because there are substantial gradients in the deformation properties across a weld which in turn affect the driving force at a crack. It is only recently that research recognizes lack of proper description and modelling of these inhomogeneities with respect to the fracture mechanics driving force. Some examples are given under Refs. [1-8] . In short, these experimental and finite element investigations demonstrate that overmatched weld metals exhibit a shielding effect due to their higher yield strength, i.e. they attract less strain than the base material, and that strain concentration occurs in undermatched weld metals, both with effects on the driving force acting on a weld metal crack accordingly.

L. Karlsson, L.-E. Lindgren, M. Jonsson (Eds.)
Mechanical Effects of Welding
IUTAM Symposium Luleå/Sweden 1991

In view of the capability of the Engineering Treatment Model (ETM) to estimate the CTOD in homogeneous materials quite reasonably [9-11] an attempt was undertaken to apply the ETM to a bi-material case such as a welded wide plate.

The following assumptions were made:

- The structure looked at was a transverse welded wide plate in tension, with a/W -> 0 .
- A through -crack was located in the weld metal.
- The wide plate was supposed to deform under prevailing plane stress.
- Residual stresses were not considered.
- The stress-strain curves were represented as power laws:

Base material

$$\frac{\sigma}{\sigma_{YB}} = \left[\frac{\varepsilon}{\varepsilon_{YB}}\right]^{n_B} \tag{1}$$

Weld metal

$$\frac{\sigma}{\sigma_{YW}} = \left[\frac{\varepsilon}{\varepsilon_{YW}}\right]^{n_W} \tag{2}$$

where the index Y refers to the yield point.

Furthermore, the mis-matching ratio

$$M = \frac{\sigma_{YW}}{\sigma_{YB}} \tag{3}$$

was defined.

Results

Due to the limited space the resulting expressions for the CTOD, δ, as a driving force parameter will be compiled without their derivation. A comprehensive study with a detailed derivation will be published in due course. The driving force in the weld metal, δ_W, will be presented in a normalized manner

$$\delta_R = \frac{\delta_W}{\delta_B} \tag{4}$$

i.e. it will be normalized by the driving force, δ_B, a crack of the same length in the base material would have. And also, δ_R will be expressed as a function of the applied global strain, ε, normalized by the base material yield strain, ε_{YB}.

Overmatching

1. Applied load, F, smaller than or equal to the base material yield load, F_{YB} (base material and weld metal elastic)

$$\delta_R = \frac{1}{M} \bullet \frac{1 + \frac{1}{2M^2}\left(\frac{\varepsilon}{\varepsilon_{YB}}\right)^2}{1 + 0.5\left(\frac{\varepsilon}{\varepsilon_{YB}}\right)^2} \tag{5}$$

2. Applied load between the base material yield load, F_{YB}, and the weld metal yield load, F_{YW} (base material plastic, weld metal elastic):

$$\delta_R = \frac{1}{1.5M}\left[\frac{\varepsilon}{\varepsilon_{YB}}\right]^{(2n_B-1)} \bullet \left[1 + \frac{1}{2M^2}\left(\frac{\varepsilon}{\varepsilon_{YB}}\right)^{2n_B}\right] \tag{6}$$

Since in the case of overmatching M>1, a shielding effect develops such that $\delta_R > 1$.To give an example: at the upper end of this loading range ($F = F_{YW}$, or $\varepsilon/\varepsilon_{YB} = M^{1/nB}$) δ_W is only one third of δ_B, i.e. $\delta_R = 1/3$, for $M = 1.2$ and $n_B = 0.15$.

3. Applied load greater than the weld metal yield load (both base material and weld metal plastic) :

$$\delta_R = \left[\frac{\varepsilon}{\varepsilon_{YB}}\right]^{\left(\frac{n_B}{n_W}-1\right)} \bullet M^{\left(1-\frac{1}{n_W}\right)} \tag{7}$$

Undermatching

1. Applied load smaller than or equal to the weld metal yield load (both base material and weld metal elastic) :

$$\delta_R = \frac{1}{M} \bullet \frac{1+\frac{1}{2M^2}\left(\frac{\varepsilon}{\varepsilon_{YB}}\right)^2}{1+0.5\left(\frac{\varepsilon}{\varepsilon_{YB}}\right)^2} \tag{8}$$

which is identical with Eq(5).

2. Applied load between weld metal yield load and base material yield load (weld metal plastic, base material elastic):

$$\delta_R = 1.5M^{\left(1-\frac{1}{n_W}\right)} \bullet \frac{\left(\frac{\varepsilon}{\varepsilon_{YB}}\right)^{\left(\frac{1}{n_W}-2\right)}}{1+0.5\left(\frac{\varepsilon}{\varepsilon_{YB}}\right)^2} \tag{9}$$

3. Applied load greater than the base material yield load (both weld metal and base material plastic):

$$\delta_R = \left[\frac{\varepsilon}{\varepsilon_{YB}}\right]^{\left(\frac{n_B}{n_W}-1\right)} \bullet M^{\left(1-\frac{1}{n_W}\right)} \tag{10}$$

which is the same as Eq(7).

Fig. 1 shows schematically how δ_R develops with $\varepsilon/\varepsilon_{YB}$. The diagram also shows the various loading ranges with the relevant equations. It can clearly be seen how the shielding effect of overmatching and the concentration effect of undermatching depend on the parameters M, n_W, and n_B/n_W.

δ_R greater than unity (concentration effect) means that the weld metal toughness must be higher than the base material toughness at least by the factor δ_R if the toughness performance of the welded joint is reqired to be no worse than that of the base material Similarly, the occurrence of shielding ($\delta_R > 1$) allows the weld metal to be less tough than the base material. Thus, a toughness requirement of the weld metal may be stated as

$$\frac{\delta_{cW}}{\delta_{cB}} \geq \frac{\delta_W}{\delta_B} = \delta_R \tag{11}$$

where δ_{cW} and δ_{cB} refer to the toughness values of the weld metal and base material, respectively.

Alternatively, the CTOD can be expressed in a Design Curve type way, i.e.

$$\delta_w^* = \frac{\delta_w E}{\pi a \sigma_{YW}} = \frac{\delta_w E}{\pi a M \sigma_{YB}} \tag{12}$$

This way of normalization results in the behaviour shown schematically in Fig. 2.

Conclusions

- The ETM provides analytical expressions for δ of a weld metal crack.
- Characterization of a weld metal crack by M is an oversimplification; work hardening plays an important role.
- The shielding effect of overmatching and the concentration effect of under matching can be quantified.

Acknowledgement

The financial support by the Deutsche Forschungsgemeinschaft is gratefully acknowledged.

References

[1] Lian, B., Denys, R., and van de Walle, R: An experimental assessment of the effect of weld metal yield strength overmatching in pipeline griths welds, 3rd Int. Conf. on Welding and Performance of Pipelines, London, 18-20 Nov. 1986.

[2] Koçak, M., Schwalbe, K.-H., Chen, L., and Gnirs, G.: Effects of notch position and weld metal matching U.K., 21-24 Nov. 1988 (GKSS 88/E/74).

[3] Koçak, M., Es-Souni, M., Chen, L., and Schwalbe, K.-H.: Microstructure and weld metal matching effects on HAZ toughness, 8th Int. Conf. on Offshore Mech. and Arctic Engineering 1989, Vol. III. The Hague, March 19-23.

[4] Koçak, M., Knaack, J., and Schwalbe, K.-H.: Fracture behaviour of undermatched welded joints, Proc. 9th Int. Conf. on Offshore Mechanics and Arctic Engineering, The American Society of Mechanical Engineers, 1990, pp. 453-459, (GKSS 90/E/23).

[5] Dong, P., and Gordon, J.R.: The effect of over and undermatching on fracture prediction models, Welding 90 (ed. M. Koçak), IITT-International, Gournay-sur-Marne, France (1990), pp. 363-370.

[6] Denys, R.M.: The future of wide plate testing in welding research, ibid., pp. 321-334.

[7] Lee, M.M.K., and Luxmoore, A.R.: The deformation and fracture characteristics of undermatched double-V butt welds in tension, ibid. pp. 371-380.

[8] Petrovski, B., and Sedmak, S.: Evaluation of crack driving force for HAZ of mismatched weldments using direct J-integral measurements in tensile panels, ibid., pp. 341-354.

[9] Schwalbe, K.-H., Cornec, A., and Heerens, J.: The Engineering Treatment Model (ETM) - a simple method for estimating the driving force under elastic-plastic and plane stress conditions, in: Defect Assessment of Components-Fundamental and Applications (ed. J.G. Blauel and K.-H. Schwalbe), ESIS/EGF Publications, London (1991), pp. 1111-1124.

[10] Schwalbe, K.-H., and Grüter, L.: Application of the Engineering Treatment Model (ETM) to the prediction of the behaviour of a circumferentially cracked pipe, ibid, pp. 1125-1134.

[11] Schwalbe, K.-H., and Cornec, A.: The Engineering Treatment Model (ETM) and its practical application, Fatigue and Fracture of Engineering Materials and Structures, Vol. 14(1991), pp. 405-412.

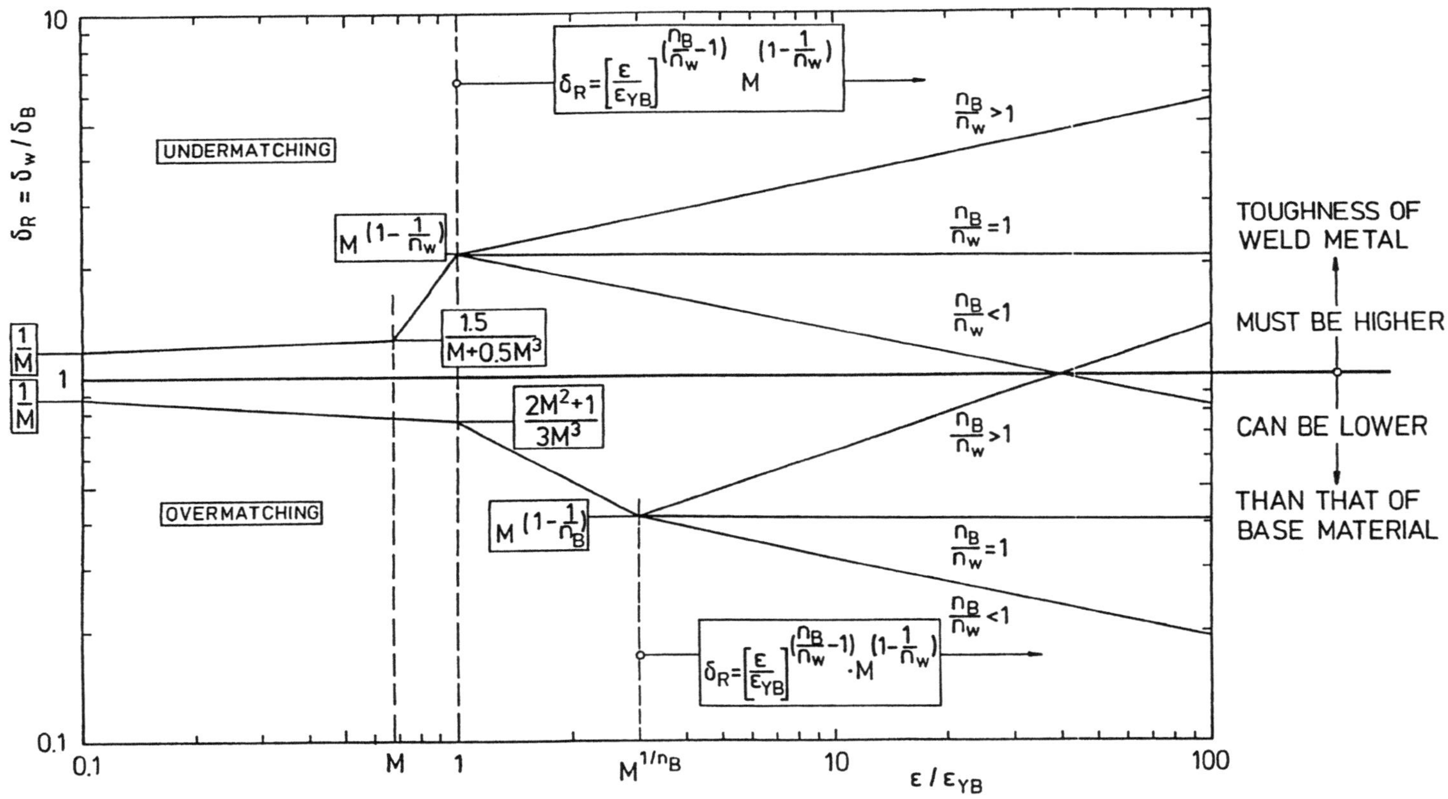

Fig. 1: Construction of δ_R as a function of $(\varepsilon/\varepsilon_{YB})$ in the various loading ranges.

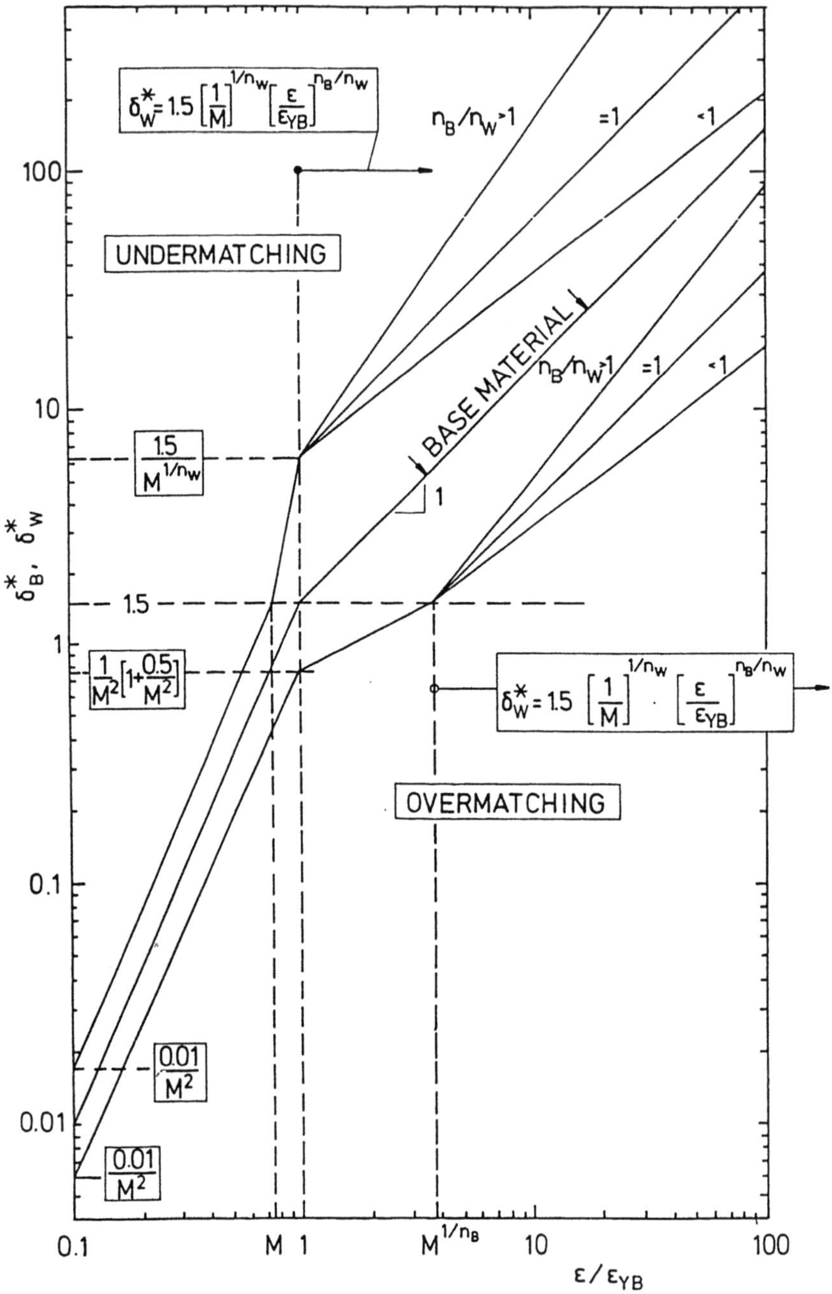

Fig. 2: Construction of δ_B^* and δ_W^* as a function of $(\varepsilon/\varepsilon_{YB})$ in the various loading ranges.

Compressive Strenght of Structural Members with Undermatching Weld Joints

Y. Ueda, H. Murakawa and H. Kimura

Welding Research Institute, Osaka University, Osaka

Summary

Undermatching weld joints [1] are often employed for high tensile strength steel to avoid crackings. The mechanical properties of weld joints, such as tensile strength [2-6] and fracture toughness [7] have been studied. However, its strength under compressive load [8] is not thoroughly understood yet because it involves buckling phenomena. To clarify the compressive strength of structural members with undermatching weld joints, the authors analyzed the behavior of columns, plates and pipes. F.E.M. and simple idealized models are employed and the effects of the joint location, slenderness ratio and strain hardening are clarified.

Strength of Undermatching Weld Joints

The buckling is usually observed in thin structures. The stress state in such structures can be treated as plane stress state. However, if soft weld joints exist, the elastic-plastic stress state near the joints becomes three dimensional stress and the yield strength and ultimate strength are increased due to so called plastic constraint. The plastic constraint can be separated into in-plane and

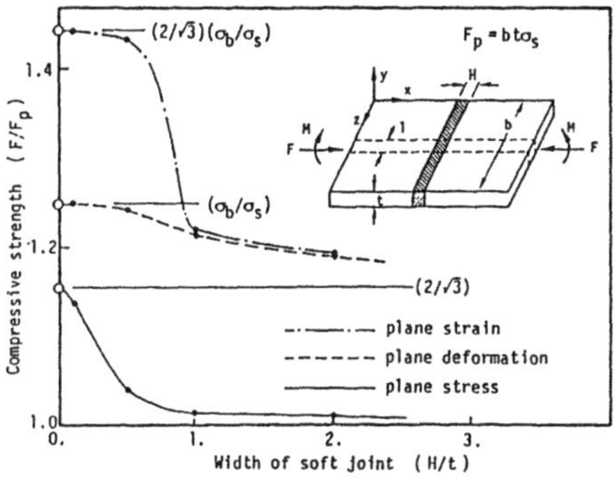

Fig.1 Effect of undermatching joint width on compressive strength.

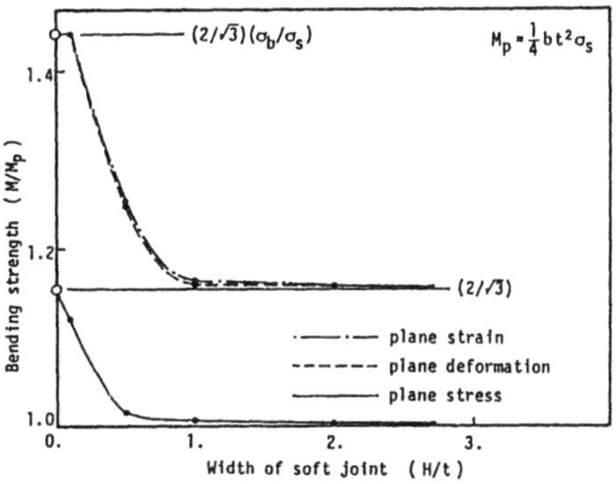

Fig.2 Effect of undermatching joint width on bending strength.

L. Karlsson, L.-E. Lindgren, M. Jonsson (Eds.)
Mechanical Effects of Welding
IUTAM Symposium Luleå/Sweden 1991

thickness directions. The illustration in Fig.1 shows a plate with a welding bead. The force is applied in the direction normal to the bead. The degree of the plastic constraint depends on H/t and H/b, which are relative width of the undermatching joint compared to the plate thickness t and width b. If the constraint in the width direction is considered, the deformed state becomes generalized plane strain in z direction when H/b is small. When the plate is fixed in the width direction, the deformation state becomes plane strain. On the contrary, if H/b is large, such as thin columns, it can be considered as plane stress state.

The effect of the plastic constraint in the thickness direction is examined for the forementioned three states by analyzing a strip with unit thickness as shown in Fig.1. In this analysis, two dimensional F.E.M. based on the plastic flow theory using Mises yield criterion is employed. The base metal and the weld metal are assumed to be perfect elastic-plastic materials and their Young's moduli and yield stresses are E=21000 kgf/mm², σ_b=80 kgf/mm² and $\sigma_s = 0.8\sigma_b =$ 64 kgf/mm². Though the welding residual stress exhibits significant effect on the behaviors of the joint or the member [9], it is not considered in this report. The computed strength of plate with undermatching joint is plotted against H/t in Figs.1 and 2. The strength is normalized by the full plastic compressive force or bending moment of the homogeneous plate with the yield stress which is same as that of the weld metal. The effect of the constraint in the width direction can be seen from the comparison among the three typical states. The plane strain state gives the highest strength and the plane stress state gives the lowest. On the other hand, the constraint in the thickness direction increases as H/t becomes smaller. However, there are upper and lower limits in the strength and they are shown in Figs.1 and 2. Theses limits can be derived from the relation between the stress state and the yield surface [10]. Since, the generalized plane strain state is the closest to the real state of the plate member in most structures, its detail is examined here. When H/t is large, the plastic constraint in the thickness direction is negligible and only constraint in the width direction is effective. Thus, the lower limits for both compressive and bending strength are $(2/\sqrt{3})$. However, the upper limits for the small value of H/t are different. The effect of the plastic constraint in the thickness direction may be taken into account by introducing a fictitious yield stress which is proportional to the strength ratio given by Figs.1 or 2. However, the width of the soft joint is assumed to be large enough and the plastic constraint in the thickness direction is ignored in the analyses discussed in the following chapters.

Buckling Strength of Columns

The column considered has a rectangular crosssection and its length is L as shown in Fig.3. It is assumed that the soft region, the width of which is H, is located ξ from the end of the column. Yield stresses of the base metal and the weld joint are assumed to be σ_b=80, σ_s=64 kgf/mm². Further, the sinusoidal form of initial deflection is assumed.

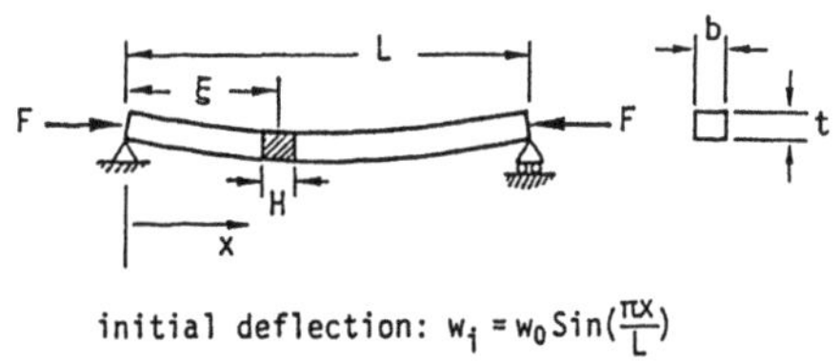

Fig.3 Column with undermatching joint.

Table 1 Dimensions of columns under compression.

	$\gamma=\lambda^2$	t (mm)	b (mm)	L (mm)
E.B.M.	2.0	20.0	20.0	415.6
T.B.M.	1.0	20.0	20.0	293.9
P.B.M.	0.5	20.0	20.0	207.8

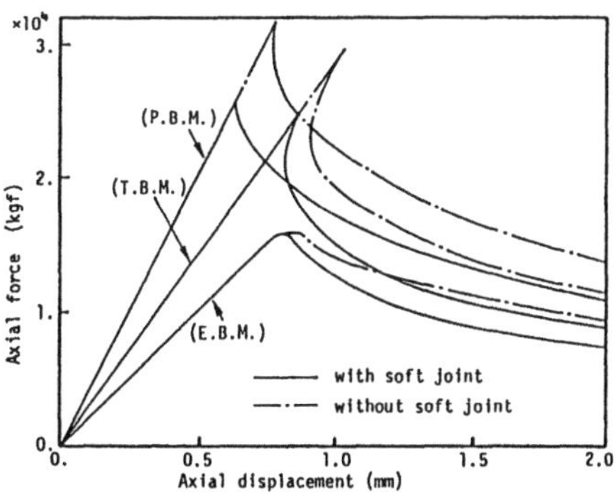

Fig.4 Load-displacement curves of columns with undermatching joint (H/t=1, w_0=t/1000).

(a) effect of slenderness ratio

The buckling phenomena can be categorized into elastic and plastic bucklings depending on the slenderness ratio λ, which is defined by the following equation,

$$\lambda^2 = \sigma_b/\sigma_E = (L/\pi)^2(A/I)(\sigma_b/E) \qquad (1)$$

where σ_b and σ_E are yield stress of the base metal and elastic buckling stress, and

$$A = tb,\ I = t^3b/12,\ \sigma_E = \pi^2 EI/L^2A$$

From the above definition, it can be seen that elastic

buckling occurs if $\lambda > 1$, and plastic buckling occurs if $\lambda < 1$. Three types of columns with $\gamma = \lambda^2 = 2.0, 1.0, 0.5$ are considered and these are referred to as elastic buckling model (E.B.M), transition B. M. (T.B.M.) and plastic B. M. (P.B.M.), respectively. Their dimensions are shown in Table 1.

Compressive load-displacement curves computed by F.E.M. using beam element considering elastic-plastic large deflection are shown in Fig.4. The solid and the chain lines represent columns with soft joint and homogeneous columns without it. The width of the joint is assumed to be equal to the thickness of the column. It can be seen that the compressive strengths are reduced roughly 20% due to the undermatching joint in the cases of transition (T.B.M.) and the plastic buckling models (P.B.M.). Whereas, small effect is observed for the elastic buckling model. Such differences in the effect of soft joint can be explained qualitatively by the following simple mechanical models.

The elastic and the plastic load-deflection curves are considered and the compressive strength is obtained as a intersection of these curves, as shown in Fig.5. The elastic load-deflection curve is given by the following equation which involves γ as a parameter.

$$w(x) = w_0 \sin(\pi x/L)/(1 - \gamma F/F_P) \qquad (2)$$

where

$$F_P = tb\,\sigma_Y$$

The broken lines in the figure show the load deflection curves of columns with initial deflection, $w_0 = t/100$, as an example. While, the plastic deformation after the plastic hinge is formed can be approximated by the rigid-plastic hinge model as shown in Fig.5. The load-deflection curve in this case is given by the following equation.

$$w(\xi) = (M_P/F)\{1 - (F/F_P)^2\} \qquad (3)$$

Two chain lines in Fig.5 are the plastic curves given by Eq.(3) assuming that the yield stresses are σ_b and σ_s, respectively. Thus, the intersecting points of the elastic and the plastic curves, shown by solid and open circles, represent the compressive strengths of column with soft joint and a uniform column. Then, the reduction of the strength due to the soft joint is given as the difference in heights of these two points. As seen from the figure, reduction of the strength is small for E.B.M. Whereas, it is large for T.B.M. and P.B.M. The reduction ratio is close to the ratio between yield stresses, i.e. $(\sigma_b - \sigma_s)/\sigma_b$. However, the above conclusions are valid only for columns for which one dimensional stress state is assumed. The behavior is different when the stress state becomes multi dimensional such as in plates or pipes.

(b) effect of the soft joint location

The effect of the joint location on the strength is examined using T.B.M. as an example. Three broken lines in Fig.6 show relations between the load and the deflection at the location of the weld joint for columns which have a joint at x=L/2, L/3 and L/8. These lines are plotted using Eq.(2). The chain lines show the load-deflection curves under plastic deformation for cases with plastic hinge formed at the base metal and the soft metal, respectively. The solid circles in the figure represent collapse points. The collapse modes are separated into two. Namely, collapse with a plastic hinge at the joint and that with a hinge at the center. If the joint locates near the center, the column collapse in the former mode and the reduction of the strength is larger as the joint approaches the center. On the other hand, if the joint locates near the end, it collapses in the latter mode and the undermatching joint shows no effect on the strength.

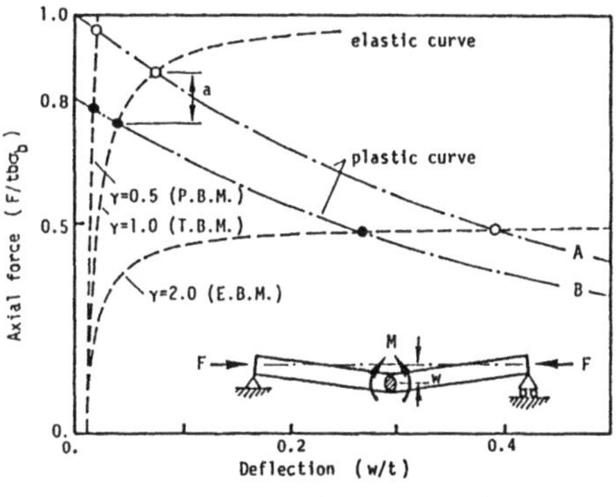

Fig.5 Prediction of compressive strength with undermatching joint at center (w_0=t/100).

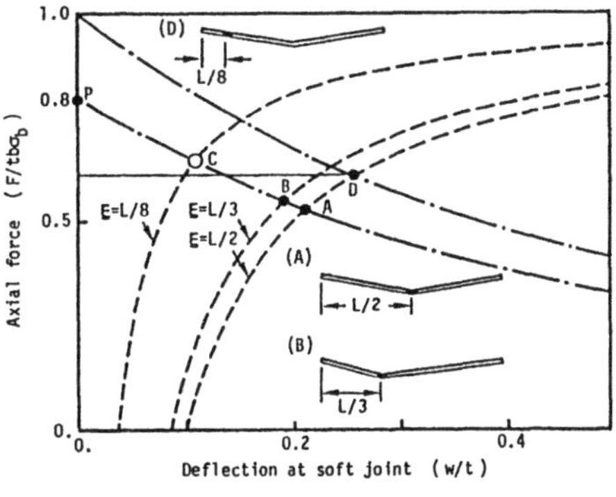

Fig.6 Effect of joint location on strength of column (T.B.M., w_0=t/10).

(c) effect of strain hardening

The plastic buckling model with a joint at its center is considered. Both the base and the soft metals are assumed to be linearly hardening materials and their hardening coefficient h is assumed to be h=E/10. The computed compressive strength of columns with various soft joint width are plotted in Fig.7. It is seen that the strength is increased by the strain hardening when the joint width is small.

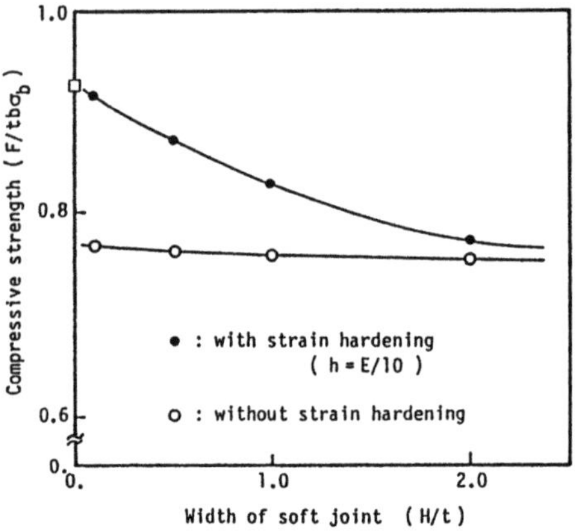

Fig.7 Effect of strain hardening on compressive strength (P.B.M., w_0=t/100).

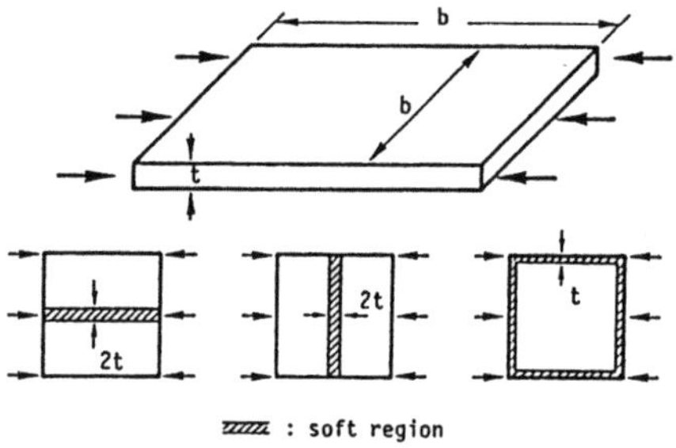

Fig.8 Fundamental patterns of weld joint location.

Compressive strength of square plates

The fundamental patterns of welding lines in plate members are shown in Fig.8. To clarify the effect of the slenderness ratio, elastic, transition and plastic buckling models are analyzed and their dimensions are shown in Table 2. The computed compressive strengths are normalized by that of homogeneous plates without soft region and summarized in Table 3. The figures in the parentheses are evaluated by a simple model which is presented in the reference [10].

Table 2 Dimensions of square plate under compression.

	Y	b (mm)	t (mm) simply s.	t (mm) clamped
E.B.M.	2.0	2000.	45.91	28.93
T.B.M.	1.0	2000.	64.92	40.92
P.B.M.	0.5	2000.	91.81	57.87

Table 3 Strength of plate with under-matching joint computed by FEM and idealized model (w_0=t/1000).

	s. s.	s. s.	s. s.	clamped	s. s.
E.B.M.	0.997 (1.000)	0.939 (0.949)	0.980 (0.941)	0.976	0.861
T.B.M.	0.973 (0.987)	0.862 (0.924)	0.913 (0.916)	0.915	0.815
P.B.M.	0.982 (0.982)	0.925 (0.924)	0.923 (0.912)	0.924	0.800

The reduction of the strength is relatively large when the joint is normal to the load. Especially in case of T.B.M., the strength is reduced by 14%. Among the three models, it is seen that the greatest reduction in strength occurs in T.B.M.

Compressive Strength of Pipes

Simply supported pipes with different radius to thickness ratios, D/t=25, 40, 100, are considered. Their dimensions are chosen so that they become transition buckling models when they are considered as simple columns. The dimensions are shown in Table 4. Concerning the joint location, a joint at the center normal to the load, which shows the greatest effect in the case of plates, is assumed. The base and the soft materials are assumed to be linearly hardening materials with same strain hardening coefficient h. The compressive strength is analyzed by F.E.M. using shell elements for two different strain hardening coefficients, h=E/100, h=E/10. The computed strengths are shown in Table 5. As seen from the table, the reduction of the strength due to the undermatching joint is small as compared to columns. There are two reasons. The first reason is the plastic constraint in the width or the circumferential direction. The second reason is the strain hardening. As shown for columns, the increase of strength depends on H/t. In general, H/D of pipes is much smaller than H/t of columns. Thus, the strength of a pipe is significantly increased by the strain hardening. Further, to compare the behavior after collapse, load-displacement curves

Table 4 Dimensions of pipes under compression.

D/t	25.0	40.0	100.0
D	1000.0	1000.0	1000.0
t	40.0	25.0	10.0
L	17291.	17552.	17817.

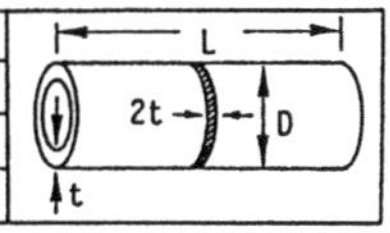

Table 5 Strength of pipes with and without undermatching joint.

	D/t=25	D/t=40	D/t=100
a) without soft joint h=E/100	0.874	0.884	0.842
b) with soft joint h=E/100	0.818	0.851	0.835
b) / a)	0.935	0.962	0.991
c) without soft joint h=E/10	0.875	0.885	0.844
d) with soft joint h=E/10	0.853	0.878	0.843
d) / c)	0.975	0.992	0.999

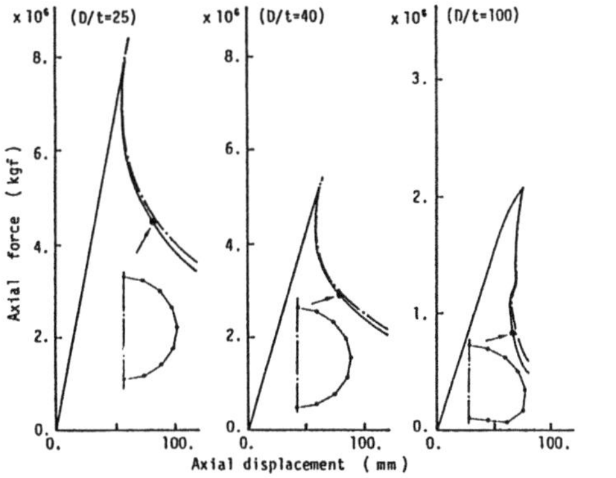

Fig.9 Load-displacement curves of pipes under compression (w_0=t/10).

are shown in Fig.9. It is seen that the effect of the soft joint on the post buckling behavior is not so significant.

Conclusion

The strength of the undermatching weld joint itself is analyzed and the relation between the strength and the plastic constraints in in-plane and thickness directions is clarified. Then, the compressive strength of columns, plates and pipes is studied. In case of plates and pipes, the plastic constraint in the width or the circumferential direction is effective to raise the strength significantly. In addition to this, if the strain hardening is taken into account, further increase of strength is expected.

References

1. Satoh, K. et al. : Welding Engineering, Rikogakusha 1979, (in Japanese).

2. Satoh, K. et al. : Size Effects on Static Tensile Properties of Welded Joints Including Soft Interlayer, J. Japan Welding Society, 37-11 (1968), 1214-1226 (in Japanese).

3. Satoh, K. et al : Effect of Mechanical Heterogeniety on the Static Tensile Strength of Welded Joints, J. Japan Welding Society, 40-9 (1971), 885-900 (in Japanese).

4. Satoh, K. et al. : Tensile Behaviors and Strength of Soft Welded Joints, J. Society of Naval Architects of Japan, 132 (1972), 381-393 (in Japanese).

5. Nitoh, H. et al. : Tensile Strength of Welded Joints for TMCP Type 50kgf/mm²-Class High Tensile Steel plates, J. Society of Naval Architects of Japan, 157 (1985), 304-311 (in Japanese).

6. Matsumoto, T. et al. : Investigations on Static Strength of Undermatching Welded Joint of HT60 Steel Plates, Quarterly J. Japan Welding Society, 4-2 (1986), 429-435 (in Japanese).

7. Satoh, K. et al. : Fracture Initiation Toughness of Materials with Locally Embrittled Region - with Reference to Electron Beam Welds of HT80 Steel, J. Society of Naval Architects of Japan, 153 (1983), 223-229 (in Japanese).

8. Kawano, H. et al. : Bending and Compressive Strength of TMCP Type HT50 Steel Welded Joints, Trans. of the West Japan Society of Naval Architects, 70 (1985), 187-198 (in Japanese).

9. Ueda, Y. et al. : Inelastic buckling of plates with residual stresses, IABSE, Zurich (1967).

10. Ueda, Y. et al. : Compressive Strength of Structural Members with Soft Weld Joints, Trans of JWRI (Welding Research Institute, Osaka University), 16-1 (1987), 177-187.

Effect of Mechanical Heterogeneity on the Absorbed Energy of Welded Joint

K.Seo[1], F.Nogata[1] and M.Kusaka[2]

[1] Department of Mechanical Engineering , Himeji Institute of Technology, 2167 Shosha, Himeji, Hyogo, Japan 671-22
[2] Steam Turbine Development Group Engineering Department, Mitubishi Heavy Industries, Ltd., 2-1-1 Shinhama Arai-cho, Takasago, Hyogo, Japan 676

Summary

The absorbed energy of the specimen with mechanical heterogeneity in Charpy test is studied. It is shown that the absorbed energy in Charpy impact test is affected not only by the fracture toughness of the material at notch tip but also by mechanical heterogeneity existing in the vicinity of notch tip. Furthermore, the strain energy release rate of the Charpy specimen with mechanical heterogeneity is analyzed under elastic-plastic condition. The analyzed result is shown to agree well with the proposed simple equation, which does not include the effect of mechanical heterogeneity.

And the fracture toughness of the materials at notch tip is estimated by using this analyzed strain energy release rate and the result of Charpy test. The estimated fracture toughness is not affected by mechanical heterogeneity.

1. Introduction

Charpy impact test has been routinely used in order to evaluate the notch toughness of the base metal, the weld metal or the heat-affected zone under various temperatures. In special, it is used for evaluating the ductile-to-brittle transition behavior of materials. This test determines the energy absorption in fracturing a specimen. The absorbed energy is considered to be a measure of fracture toughness of a material at notch tip[1]~[4]. Welded joint has more or less uneven hardness distribution, i.e. mechanical heterogeneity. This mechanical heterogeneity brings some difficulties in evaluating the fracture toughness.

L. Karlsson, L.-E. Lindgren, M. Jonsson (Eds.)
Mechanical Effects of Welding
IUTAM Symposium Luleå/Sweden 1991

In this study, it is shown that the absorbed energy does not evàluate a reliable fracture toughness of a material at notch tip if this test is applied to welded joint. Furthermore, the way of estimating fracture toughness, which is not effected by mechanical heterogeneity, is presented.

2. Experiment

2.1 Used material and preparation of specimen

600MPa class high tensile strength steel plate with 28mm thickness (MK484) was used in this experiment. Its chemical compositions and mechanical properties are shown in Table 1. The purpose of this study is to clarify the effect of mechanical heterogeneity on the absorbed energy in Charpy impact test. So three types of specimen shown in Fig.1 were prepared from this plate. One type was the standard V notch Charpy specimen, called ST Type. Another two types of specimen, W3 Type and W7 Type, had two beads of electron beam welding respectively. The width of bead was 3mm. The intervals between two beads in W3 Type and W7 Type specimens were 3mm and 7mm respectively. V notch of every specimen was located in the center of two beads.

Table 1. Chemical compositions and mechanical properties

C	Si	Mn	P	S	V	sol Al
0.13	0.44	1.46	0.012	0.009	0.049	0.026
Yield strength (MPa)		Ultimate strength (MPa)			Elongation (%)	
526		689			44	

Figure 2 shows the appearance of W3 Type and W7 Type specimens and the distributions of vickers hardness (500g) in specimens. Welded bead has high hardness in comparison with the base material. This is similar to the hardness distributions shown in the welded joint of common structural steels. Therefore, it can be considered that these specimens are models of welded joint with mechanical heterogeneity. According to the distribution of hardness and appearance of specimens shown in Fig.2, the materials at notch tip have no heat effect by electron beam welding at all and they still have the same mechanical properties as the base materials.

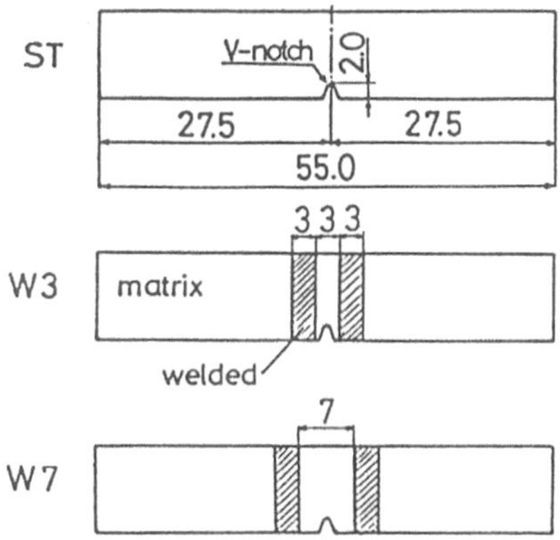

Fig.1. Charpy Specimen

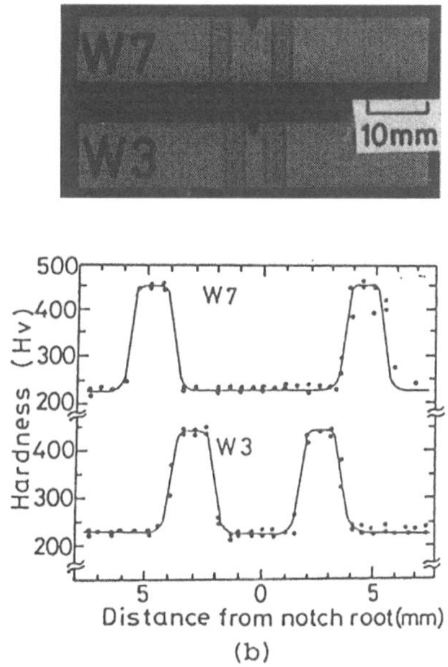

Fig.2. Appearance of Charpy specimen with weld beads, and distributions of hardness

2.2 Experimental Procedures

Instrumented Charpy impact testing machine, which had the maximum capacity 480J and was instrumented with strain gages to record the load-time profile during a test, was used. Impact tests were done on every type of specimens at temperatures ranging from -50℃ to 20 ℃. A dry-ice and alcohol bath controlled to ±1℃ was used for cooling specimens. The impact tests were completed within five seconds after withdrawing the specimen from bath. Initial impact velocity was 5.4m/s.

2.3 Experimental Results

A relationship between absorbed energy obtained from each specimen and test temperature is shown in Fig.3. In spite of the materials of notch tip being the same, transition temperature and the value of upper shelf energy differ with the types of specimens. Mechanical heterogeneity existing in the vicinity of notch tip decreases the upper shelf energy and raises the transition temperature. The results obtained in these tests are affected by mechanical heterogeneity existing in the vicinity of notch tip.

This shows that we must not simply consider that the transition temperature or the absorbed energy obtained by Charpy test should characterize the notch toughness of a materials at notch tip. To put it in other way, when we use a Charpy test to evaluate the notch toughness of weld metal or heat affected zone, we can not obtain their real notch toughness. This is because the plastic deformation of materials at notch tip is restrained by mechanical heterogeneity existing in the vicinity of notch tip.

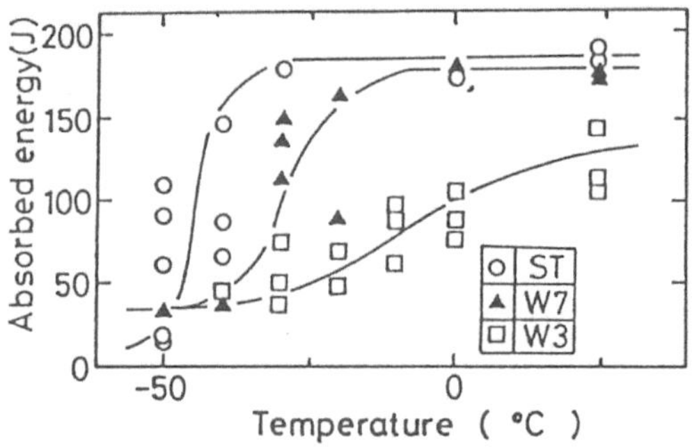

Fig.3. Temperature dependency of absorbed energy

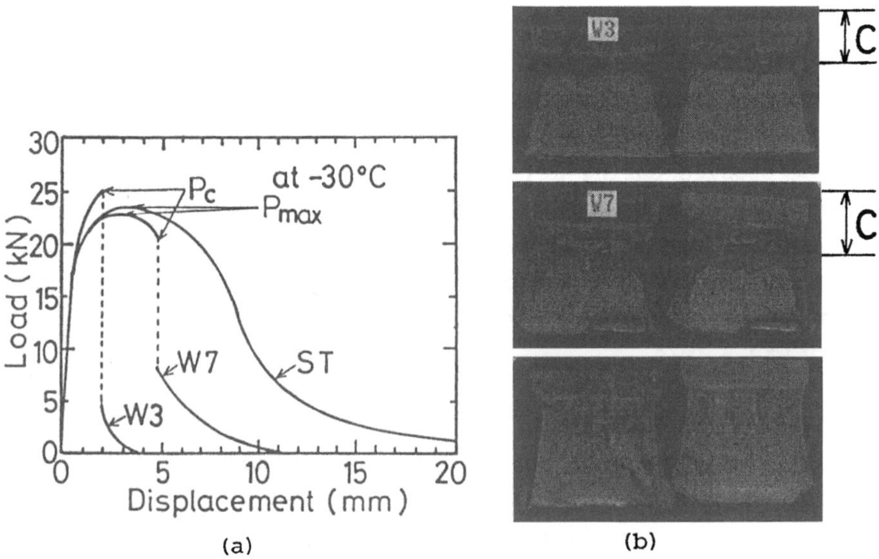

Fig.4. Example of load-displacement profiles and fracture surface

Figures 4(a)(b) show the examples of load-displacement curves and fracture surfaces of the specimen, which are obtained at test temperature -30℃. In ST Type specimen, after maximum load P_{max}

appears, load decreases gently accompanied with growing of ductile crack. The fracture surface of ST Type specimen shows 100% shear area. In W7 Type specimen, however, after showing maximum load, the load suddenly decreases largely at P_c. At this point, the ductile crack suddenly changes into a brittle crack. The fracture surface shows 65% shear area. In W3 Type specimens, the load P_c, at which brittle fracture occurs, appears before maximum load. Transition from a ductile crack to a brittle crack arises in earlier time than that of W7 Type specimen. W3 Type comes to behave more brittle and the specimen shows 29% shear area.

Although all of the three types of specimen have same material at notch tip, the fracture behavior largely differs with specimen types, that is, existence of mechanical heterogeneity.

3. Analysis of strain energy release rate in elastic-plastic fracture

For elastic-plastic behavior, COD and J-integral criterion have been used to evaluate the brittle fracture properties.

But in this study, these criterions can not be applied, because the plastic deformation is very large. So the strain energy release rate in elastic-plastic condition is analyzed.

3.1 Analytical method

In this study, numerical analysis of elastic-plastic stress under the plane stress condition was performed by FEM. Figure 5 shows an example of mesh division. Calculation was carried out for ST Type specimen and W3 Type specimen under static load. In calculation, mechanical properties of used material, which were obtained by impact tension test ($\dot{\varepsilon}$ =100/sec) [5], are as follows:

$\sigma = E\varepsilon$ (elastic)

$\sigma = B(\alpha + \varepsilon)$ (plastic)

$$\sigma_Y = 605\exp\left\{55.1\left(\frac{1}{T} - \frac{1}{T_0}\right)\right\}^{1/1-n}$$

$E = 200\text{MPa}$

$B = 1042\exp\left(\frac{55.1}{T}\right) \qquad n = 0.18$

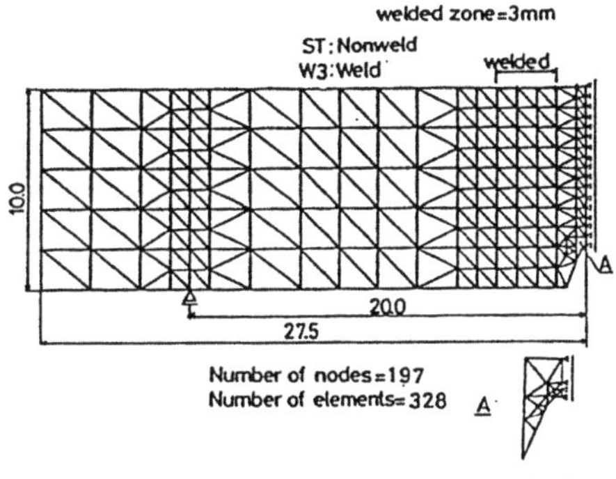

Fig.5. Element division

In analysis, as weld bead by electron beam has high yield strength, they are always considered to keep elastic condition. The calculation of strain energy release rate during a crack propagation was done by release the nodal force at crack tip, one after another[6].

3.2 Results of analysis

An example of load-displacement curve and strain energy variation under loading and crack propagating are shown in Fig.6(a)(b). Load increases in proportion to the displacement until P_Y, at which local yielding can occur in the element at crack tip. And after that, as displacement increases, load increases slowly A→ B accompaning with plastic deformation. When the displacement is fixed at point B and crack is propagated from the notch tip, the load decreases B→ C as crack length grows. During the crack propagation, elastic strain energy E_e decreases, but plastic strain energy E_p increases even if load decreases. So strain energy release rate during crack propagation in elastic-plastic condition can be calculated as follows:

$$G = \frac{\Delta E}{\Delta C} = \frac{E_{EX} - E_P - E_e}{\Delta C} \qquad (1)$$

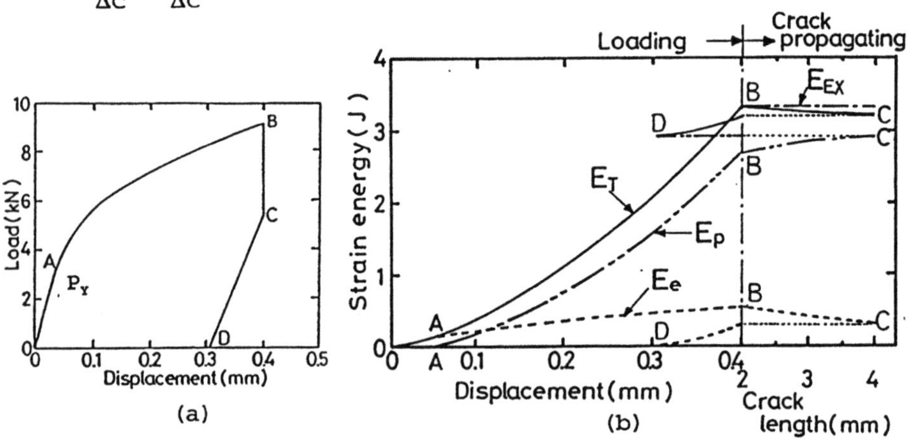

Fig.6. Load-displacement curve and strain energy variation

The strain energy release rate was calculated under various values of displacement δ. It is effected by crack length, type of specimen and displacement. According to linear elastic fracture mechanics, the strain energy release rate is given theoretically by Eq.(2) when load P is applied.

$$\frac{G}{G_Y} = \left(\frac{P}{P_Y}\right)^2 \qquad (P < P_Y) \qquad (2)$$

P_Y:the elastic limit load (local yielding can occure in the element at crack tip)

G_Y:strain energy release rate when load PY is applied

P:applied load

G:strain energy release rate when load P is applied

All calculated results are nondimensionalized by P/P_Y and G/G_Y as shown in Fig.7. The strain energy release rate in elastic-plastic condition is arranged approximately by the following equation (3), which does not have theoretical support.

$$\frac{G}{G_Y}=\left(\frac{P}{P_Y}\right)^m \tag{3}$$

In this equation, m differs with crack length C and can be equal to the value in Table 2.

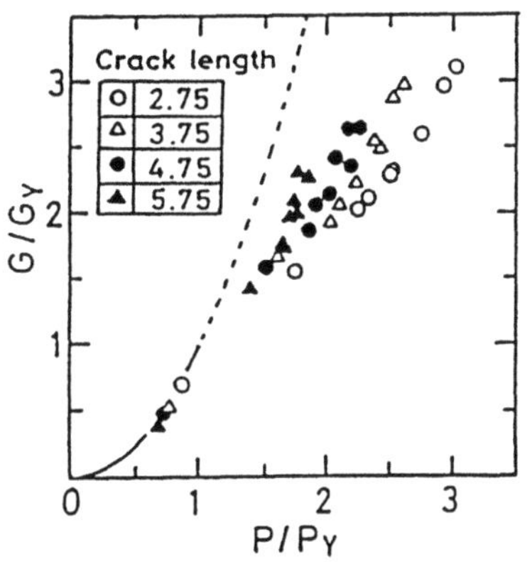

Fig.7. Relation between strain energy release rate and applied load

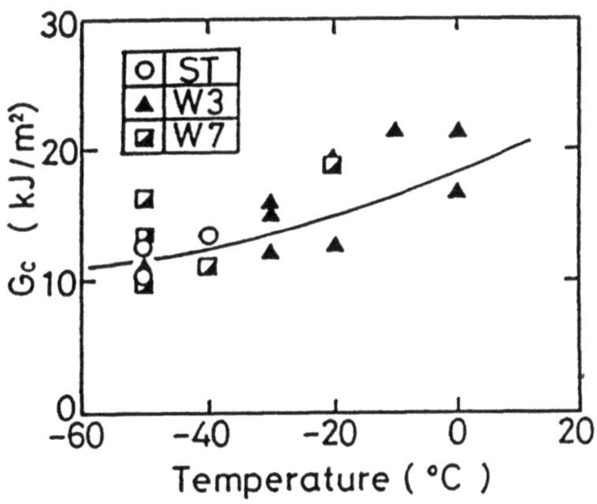

Fig.8. Estimated fracture toughness

Table 2. Values of m

Crack length	2.75	3.25	3.75	4.25	4.75	5.25	5.75
m value	1.00	1.15	1.16	1.25	1.30	1.30	1.36

4 Estimation of fracture toughness

We can get the load P_c, at which brittle fracture occurs, from the load-vs-displacement curve as shown in Fig.4(a), and ductile crack length C from the fracture surface as shown in Fig.4(b).

When applying P_c and C in the formula (3), fracture toughness G_c

(strain energy release rate at onset of brittle crack) can be estimated as follows:

$$G_C = G_Y \left(\frac{P_C}{P_Y}\right)^m$$

G_c values in each specimen were estimated and shown in Fig.8. Estimated fracture toughness is not affected by the mechanical heterogeneity. It can be decided only by temperature.

5.Conclusion

Obtained results are as follows:

1) The absorbed energy in Charpy impact test does not evaluate a reliable fracture toughness of a material at notch tip, if this test is applied to welded joint. It is affected not only by properties of material at crack tip but also mechanical heterogeneity existing near crack tip.

2) Strain energy release rate of the specimen with mechanical heterogeneity in the vicinity of notch tip is analyzed under elastic-plastic condition. It is shown that strain energy release rate can be obtained by the equation $G/G_Y=(P/P_Y)^m$.

3) The fracture toughness is estimated from the results of Charpy test and analyzed strain energy release rate. It is shown that the estimated value is not affected by mechanical heterogeneity.

Reference

1.Boyd, J.D.; Godden, M.J.: Evaluation of heat-affected zone toughness by the cross weld Charpy test. Symp. What Does Charpy Test Really Tell Us (1978) 74-89.
2.Canonico, D.A.; Stelzman, W.J.; Berggren, R.G; Nanstad, R.K.: Use of Instrumented Charpy Test to Determine Onset of Upper-Shelf Energy. Wele J.60(5) 85S-91S (1981).
3.Dalby, R.E.: Charpy V and COD correlation between test data for ferritic weld metals. Met. Constr. 13(1) 43-51 (1981).
4.Iyer, K.R.; Michot, R.B.: Instrumented Charpy Testing for Determination of the J-Integral. ASTM STP 563 146-164 (1974).
5.Seo, K.; Nogata, F.; Masaki, J.: The Absorbed Energy in Charpy Impact Test for Specimen with Mechanical Heterogeneity. J. of the Soc. of Materials Sience (in Japanese) Vol.35. No.399. 1405-1410 (1986).
6.Seo, K.; Masaki, J.; Nogata, F.: Brittle Fracture Initiation of Steel at the Temperature of the Ductile-to-Brittle Transition Region. Trans. Jpn. Soc. Mech. Eng. (in Japanese) Vol.58. No.489. 905-910 (1987).

On the Correlation Between Charpy Notch Toughness and Fracture Toughness of Base and Weld Metal

K. Eriksson

Department of Mechanical Engineering, Luleå University of Technology, S-951 87 Luleå, Sweden

Introduction

The Charpy test has been used since the beginning of the century (1905) for qualitative assessments of the brittle-to-ductile transition temperature (range) of structural steels.

Before the 2nd World War there was no rational basis for predicting structural behaviour. It was not known which criterion should be used, if any, for calibration against service performance.

By the early 1940's, following the Liberty ship disasters, it was recognized that the transition temperature of a cracked body was critical for the loss of strength [1].

Later investigations of the Liberty ship failures disclosed that fracture initiation, propagation and arrest plates featured on the whole different maximum values of Charpy energy at the service temperature [2]. The scatter is however large and data from individual plates overlapping.

In the 1950's it was found that the calibration against service performance is not universal but material dependent [1]. The Charpy energy for a given performance, e.g. arrest, is dependent upon the type of steel. Typically, quenched and tempered steels requires a much higher Charpy energy for arrest than ordinary carbon and carbon-manganese steels. For weld metal no systematic study of the relationship between notch toughness and defect tolerance as measured with fracture toughness has been carried through. It is more or less tacitly assumed that base metal requirements also apply to weld metal.

L. Karlsson, L.-E. Lindgren, M. Jonsson (Eds.)
Mechanical Effects of Welding
IUTAM Symposium Luleå/Sweden 1991

Fig. 1 features a further interesting property of an inhomogeneous steel: the transition temperature of the core is higher than that of the surface material. At -20°C the mean notch toughness of 15 specimens is 9J with very little scatter. This yields a lower bound of J equal to 11 kN/m, while in fact Jc = 120 kN/m was obtained experimentally at -30°C. This means that notch toughness grossly underestimates the effective toughness of a structural element. An assessment of the effective toughness based on notch toughness only would in this case not cause disastrous failure. The economic consequences of such an assessment might however be considerable. Suppose e.g. that a road or railway bridge, or even a suite of bridges were replaced just because standard notch toughness testing indicates a brittle material while in fact the effective toughness is quite satisfactory.

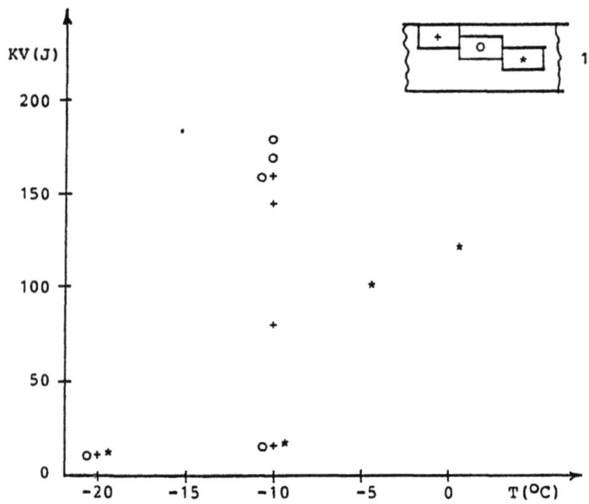

Fig. 1. Notch toughness of "surface" (+), "intermediate" (o) and "core" (*) material of an inhomogeneous structural steel.

The next example concerns the most common pressure vessel steel in Sweden, SS1330, in the form of a hot rolled and normalized fine grain 35 mm thick plate. Wåle and Bergh [9] have thoroughly studied the effect of prestrain and ageing upon various mechanical properties of this steel, notably the ductility.

Test pieces with the long axis parallel and perpendicular to, respectively, the rolling direction of the parent plate were prestrained to 5 and 10% and subsequently aged 30 min at 250°C.

Fracture toughness testing with compact specimens and notch toughness testing were performed. The test temperatures were 0 and 24°C. The data in Fig. 2 are taken from the investigation of Wåle and Bergh, where fracture toughness now is plotted against notch toughness. Also in this Figure is shown the lower limit estimation of fracture toughness based on notch toughness (the dotted line) obtained by using an expression from Sailors & Corten [8].

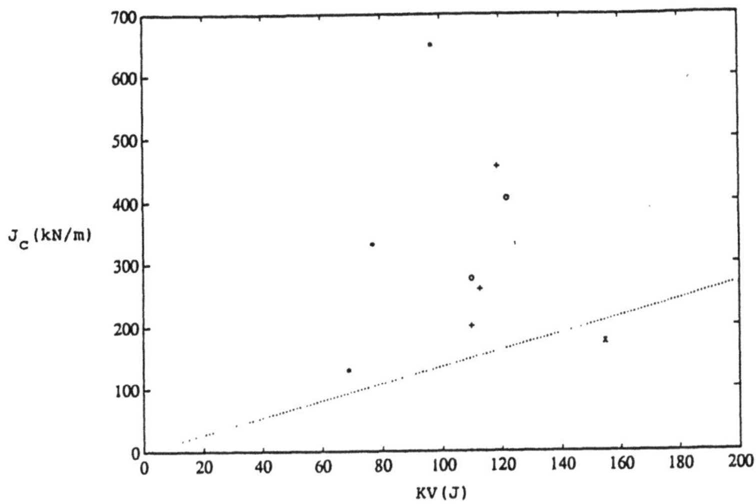

Fig. 2. Effect of prestrain and ageing upon notch toughness and fracture toughness [9]. Transversal 0°C (*), longitudinal 0°C (+) and transversal +24°C (o).

Both fracture toughness and notch toughness are reduced by the prestrain and ageing treatment. This means in practice that the steel is susceptible to embrittlement after cold forming and welding.

Notch toughness is however reduced very little in comparison to fracture toughness. In the longitudinal direction fracture toughness is reduced some 55% while notch toughness is reduced only 8%. The implication is that notch toughness is a very poor indicator of the reduced defect tolerance as measured with fracture toughness for this material.

The ideal behaviour is a proportional reduction of notch toughness and fracture toughness, that is, embrittlement should give data points on a straight line through the origin. Performing notch toughness testing only and using the lower limit estimation of fracture toughness gives very conservative estimates for unaffected and moderately embrittled base material.

The last example is a collection of weld metals. All welds are butt joints in 25 mm structural steel plate and produced with basic electrodes. Some further details are given in Table I.

Table I. Weld metal data

metal	Weld(ing) method	position	Edge prep	No of runs	Base mtrl	Note	Ref
B	GMA*	flat	X	22-28	E355	interpass	[10]
C	GMA*	-"-	X	-"-	-"-	temp 150°C	
D	MMA	vert up	V	7	SS2134		[11]
E	MMA	vert down	V	22	-"-		
J	GMA*/CO_2	flat	Y	7	-"-		

*Flux-cored wire

Heat input as recommended by the filler metal manufacturer.

Fracture toughness testing with Three-point-bend specimens and notch toughness testing were performed. Specimen orientation in a welded joint is shown in Fig. 3. The testing temperature varied between -40°C and +20°C.

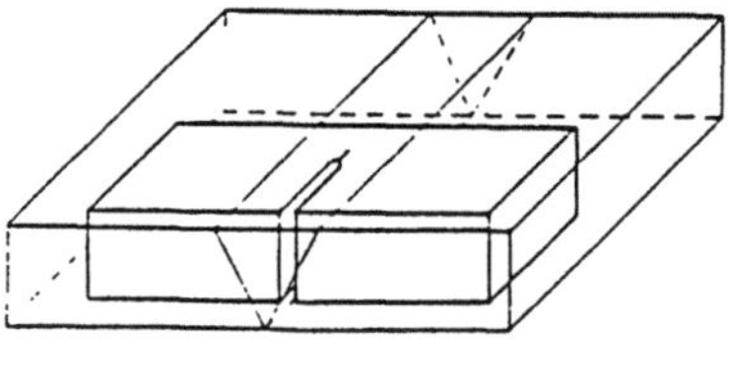

Fig. 3. Three-point-bend specimen and it's orientation in the welded joint.

The result of the testing, critical crack tip opening displacement δ versus notch toughness KV, is shown in Fig. 4. The dotted line is a lower bound estimate, $\delta = 67 \cdot 10^{-3}$ KV, with KV in Joule and δ in mm, according to [12].

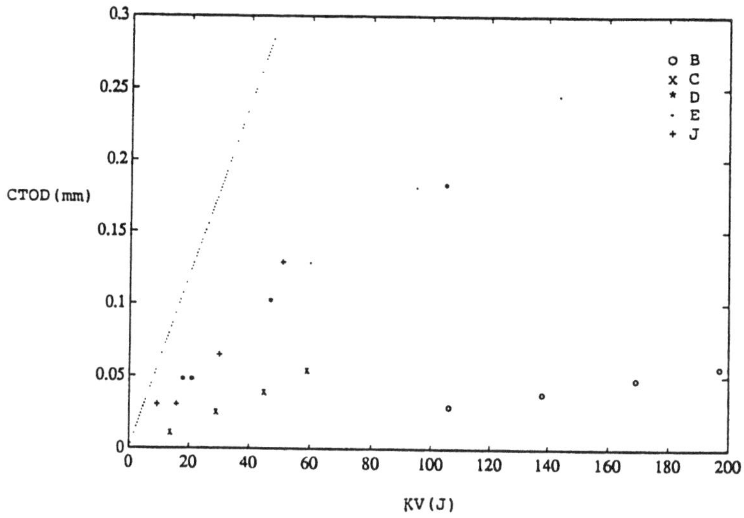

Fig. 4. Notch toughness and critical Crack Tip Opening Displacement for a set of manual metal arc and flux-cored wire weld metals.

In this case weld metals with acceptable and better notch toughness have very poor fracture toughness, far below the lower bound estimate. The defect tolerance of the weld metals is overestimated by the notch toughness to such an extent that the question is justified wether notch toughness is a measure at all of defect tolerance in this particular case.

Conclusion

This compilation of data from very different sources show that correlation between notch toughness and fracture toughness cannot be universal. It is shown that common empirical relationships grossly underestimates fracture toughness of inhomogeneous base material and worse, that toughness requirements on weld metal in terms of notch toughness do not necessarily guarantee the desired safety against fracture.

References

1. Pellini W.S. Evolution of Engineering Principles for Fracture-Safe Design of Steel Structures. Naval Research Laboratory, Report 6957, Washington D.C., 1969.

2. Williams M.L. & Ellinger G.A. Investigation of Structural Failures of Welded Ships. The Welding Journal 32(10), Research Suppl, 1953, pp 498-527.

3. Regulations for Steel Structures (BSK), National Board of Planning and Building, Stockholm, Sweden, 1987.

4. Recommendations for the choice and classification of Steels for use in Welded Structures. IIW Doc 22-59.

5. ASTM Standard E813-87, Standard Test for J_{Ic}, a Measure of Fracture Toughness, Annual Book of ASTM Standards, ASTM, Philadelphia, 1987.

6. PD 6493 Guidance on some Methods for the Derivation of Acceptable Levels for Defects in Fusion Welded Joints. British Standards Inst, London, 1980.

7. Eriksson K. Toughness Requirements for Older Structural Steels. Proceedings IABSE Workshop, Lausanne, Switzerland, 1990.

8. Sailors R.H. & Corten H.T. Relationship between Material Fracture Toughness using Fracture Mechanics and Transition Temperature Tests, ASTM STP 514, ASTM, Philadelphia, 1973, pp 164-191.

9. Wåle J. & Bergh S. Inverkan av kalldeformation på segheten hos stål. SA/FoU-Rapport 90/07. Statens Anläggningsprovning, Stockholm, 1990.

10. Norinder H. COD and Charpy-V Thoughness and Microstructure of Shelfshielded Wire Weld Metals. 4th Scand Symp on Mater Sci, NIT, Trondheim, Norway, 1986.

11. Eriksson K. J-integral Fracture Thoughness of Weld Metals. IIW Annual Assembly, Bratislava, Czechoslovakia, 1979.

12. de Meester B. Selection of Steels according to their Charpy-V Properties in Order to Avoid Brittle Fracture. IIW Doc 1020-88, Welding in The World, Vol 26, No 11/12, 1988, pp 308-324.

Three Dimensional Finite Element Modeling of Welded Structures

S. B. Brown and K. Christie

Department of Materials Science and Engineering
Room 8-106
Massachusetts Institute of Technology
Cambridge, Massachusetts 02139

H. Song

Anadrill Schlumberger
200 Macco Boulevard
Sugarland, Texas 77478

Introduction

Because of the geometric complexities of large structures, finite element welding simulation of such structures require commensurately large, three-dimensional finite element models, where the weld zone represents only a small portion of a structure. Examination of the local weld zone is not sufficient. Full three-dimensional models are required both since complicated structures normally involve elastic responses in all three dimensions and since the action of welding generates out-of-plane distortions and stresses.

These requirements of model size and dimension have hindered current finite element welding analyses of large structures. The primary constraint is that these models are too large and complex to be reduced to two dimensions. However, a full three-dimensional analysis increases complexity so much (due to the dramatic increase in total number of degrees of freedom associated with a model) that a simple three-dimensional model can overwhelm the computational capabilities of a mainframe which could handle a two-dimensional geometry with relative ease. Methods to simplify the complexity of finite element models are therefore necessary to permit reasonable analysis times.

The transient nature of welding can be exploited to reduce the size of finite element models that simulate the welding process. Welding is characterized by the motion of a relatively small, nonlinear zone traveling in a larger elastic body. The nonlinear zone represents the electrode-affected region where both nonlinear deformation behavior and large temperature gradients are concentrated. The rest of the structure away from the nonlinear zone remains linear and comparatively isothermal throughout the welding operation. Finite element models, therefore, can be simplified substantially by representing the local zone using a translating dense mesh and the overall structure using either a coarser mesh or substructures.

L. Karlsson, L.-E. Lindgren, M. Jonsson (Eds.)
Mechanical Effects of Welding
IUTAM Symposium Luleå/Sweden 1991

This article discusses the welding application of rezoning and dynamic substructuring, respectively. The combination of these two techniques dramatically reduces in computation time for simulation of welding.

Rezoning Applied to Welding

Welding simulation can exploit rezoning by modeling the local zone with a traveling dense mesh. The entire model is then remeshed at frequent intervals to track the motion of the local zone as the heat source progresses along its path. Only the local zone (not the entire weld path, as in earlier finite element analyses) needs to be modeled with a dense mesh. Rezoning also permits stepwise coupled thermal-mechanical analyses by updating the geometry and boundary conditions in the heat transfer analyses. Each rezoning step therefore is performed using the deformed mesh shape from the previous step.

We implemented rezoning in three, repeating steps. First, an incremental analysis produces a temperature field and deformed geometry. Second, a rezoning code produces a new mesh based on that deformed geometry with the local zone translated to capture electrode motion. Third, the field variables from the incremental analysis are mapped onto the new mesh. The next incremental analysis continues using initial conditions specified by these variable values which have been interpolated from the previous analysis.

Variables at each integration point on the new mesh must be obtained from corresponding data on the old mesh. First, the variables of the old mesh are extrapolated from the integration points to the nodes. Then the field variable values extrapolated to the node are averaged. The integration points of the new mesh are located in an element of the old mesh. Finally, the normalized coordinates of the integration point in that element are found and a variable value is interpolated in terms of its values at the old nodes.

During a typical welding simulation, such rezoning occurs many times. We use a shell program to accomplish this task automatically by exploiting the fact that most of these remesh procedures are only different from each other in the spatial position of the local zone and the transient distributions of variables such as temperature, displacement and stresses. This shell program will divide the simulation into a number of incremental analyses, generate new meshes between these increments according to the immediately preceding results, map the values of variables to the new mesh, and drive the finite element code to carry out the next incremental analysis.

Before any interpolation of a variable to a specific integration point in the new mesh, the exact location of this point in the old mesh needs to be determined. We use a variation of a strategy proposed by Murti and Valliappan [1] to determine if a point is inside or outside an element. The modified strategy works as follows:

For a two-dimensional, quadrilateral four-node element a point is inside an element if:

$$A_i = \frac{1}{2}\vec{a}_i \times \vec{b}_i \geq 0 \qquad (i = 1,2,3,4) \tag{1}$$

in which $\vec{b}_i$ is the ith side of the element. Its direction is always such that the element is on the left side of the vector. The vector $\vec{a}_i$ connects the point to the starting node of the ith side. If the point is not inside the element, there exists at least one negative product.

After the location of a new integration point within an old mesh element is determined, variable interpolation can be performed based on the normalized coordinates of that point inside the element. Different strategies can be used to obtain the normalized coordinates of a point inside an element [1,2]. The following procedure is straightforward and easy to implement. All variables can be interpolated through a set of interpolation functions based on the variable values at nodal points. The coordinates can also be transformed using the same rule for isoparametric elements:

$$\begin{aligned} x &= \sum_{i=1}^{n} N_i(\xi,\eta,\zeta)x_i \\ y &= \sum_{i=1}^{n} N_i(\xi,\eta,\zeta)y_i \\ z &= \sum_{i=1}^{n} N_i(\xi,\eta,\zeta)z_i \end{aligned} \tag{2}$$

If the nodal coordinates x_i, y_i, and z_i are known, these three equations uniquely determine a set of normalized coordinates ξ, η, and ζ associated with a spatial point with global coordinates x, y, and z.

Dynamic Substructuring

We use a dynamic substructuring procedure to reduce the finite element model size beyond what can be accomplished using rezoning. The linear portion of the model is reduced via static condensation to a set of equivalent boundary conditions on the nonlinear portion. An interval of analysis is performed, a set of displacements and stresses are obtained, and then the model is redefined in the distorted shape and rezoned. We then redefine the substructures to reduce the linear portion of the new mesh to equivalent boundary conditions on the new nonlinear, local zone. The analysis intervals repeat to track electrode motion as in the case of rezoning alone, but with dynamic substructuring the linear region of the mesh in each analysis interval is represented with substructures. Note that the term "dynamic" does not refer to a dynamic analysis associated with eigenmodes or inertial effects; it merely indicates that the substructuring procedure occurs repeatedly during the finite element analysis.

Substructuring assumes that temperature changes and gradients inside the substructure are small. We also assume that there are no other nonlinearities in the substructure such as contact points, inelastic deformation, and nonlinear boundary conditions. The dynamic remeshing is frequent enough so that both the displacement and strain increments during one incremental analysis are very small.

The dynamic substructuring technique involves the frequent redefinition of a substructure. At each redefinition, the accumulated stresses inside a substructure need special attention, because substructures can not accommodate initial stresses when they are redefined at the beginning of each incremental analysis. We use a special procedure to keep the substructure in an incremental analysis initially stress-free. In this procedure, we separate and store the accumulated stresses in a substructure before each incremental analysis. The stresses are updated after the incremental analysis by adding the stresses developed during the increment to the stored stresses. The boundary interaction forces between a local zone and its associated substructures can be determined from the traction forces derived from stress components along the boundary.

Rezoning and Dynamic Substructuring Applied to Plate Welding

We will now illustrate rezoning and dynamic substructuring techniques by showing how they may be applied to the simulation of plate welding. We also present a simulation of the plate-welding process using a full model without rezoning or substructuring to compare with the rezoning and dynamic substructuring results.

Figure 1 shows a two-dimensional finite element mesh of a plate. A heat source travels along the edge of the plate from one end of the denser mesh to the other. We do not include material deposition along the edge. The model then resembles edge heating rather than welding.

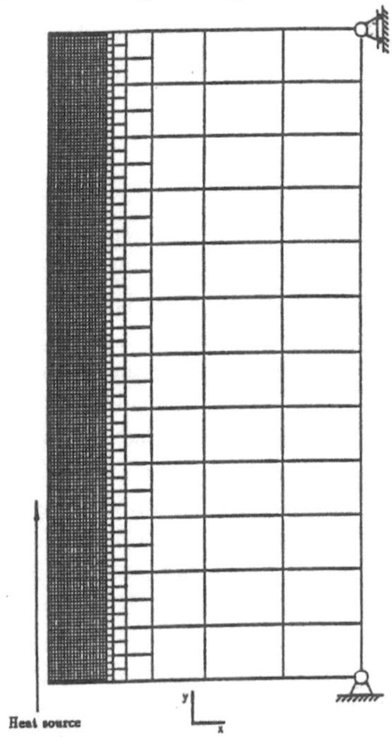

Fig.1. The full mesh of the plate heated on one edge

Heat is lost due to both convection and radiation primarily from the two surfaces of the plate. The two-dimensional model cannot directly simulate such losses through direct specification of boundary conditions. Instead, a volumetric heat sink is used over the plate to approximate convective and radiative heat losses from the upper and lower surfaces of the model.

Figure 2 includes the final distorted mesh predicted by a full model shown in Figure 1. The corresponding Mises stress distribution is illustrated in Figure 3. The stress distribution is very close to the reported stress distribution for a similar plate welding process modeled by Ueda *et al.* [3].

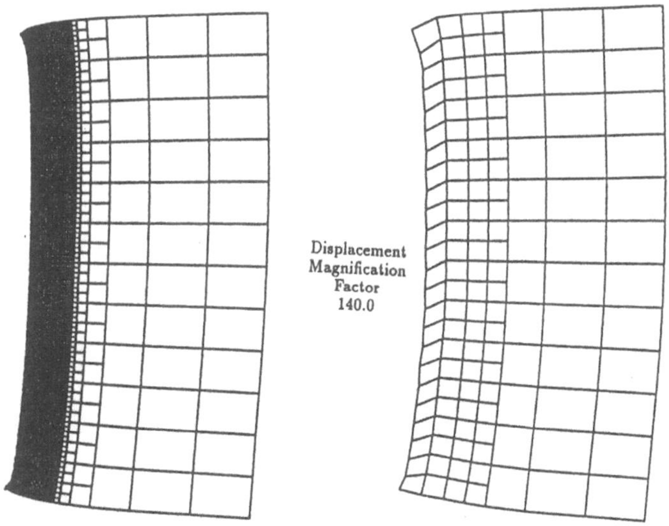

Fig.2. Final distortion of the plate, both full model and remeshed model.

Because the heat source must remain within the local zone, a smaller local zone implies a higher frequency of rezoning. To increase the efficiency of rezoning, the size of the local zone should be kept as small as possible so that the total number of degrees of freedom associated with the model is also small. However, higher frequency of rezoning requires more computational overhead since rezoning requires both computer time and storage space. The analyses described later show that the overhead cost of computer time is negligible compared to that of an incremental analysis in a typical simulation; thus a very small local zone can be selected.

The size of the local zone is also carefully chosen to include both high temperature gradients and thermal expansion. In the plate model, a temperature of 150 °C is used to delineate the boundary of the local zone.

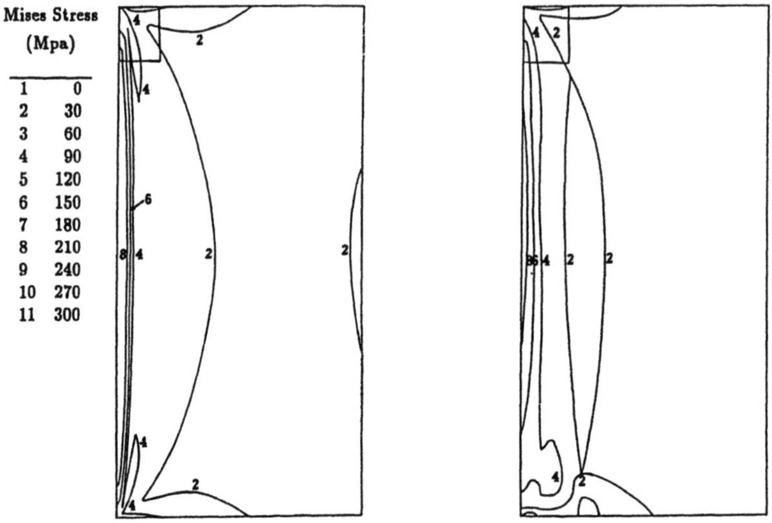

Fig.3. Final Mises stress distribution on the plate, both full model and remeshed model.

The rezoning and dynamic substructuring procedure is carried out by a shell program which drives a finite element code (ABAQUS) and our own subroutines. The finite element code performs the incremental analyses, and our own subroutines take care of the rezoning, variable mapping, and changes of boundary conditions. Specific details of the finite element analysis are described in another article[4]. Figure 4 shows the meshes used in nine incremental analyses for the same plate. The total mesh is redefined each time rezoning is performed.

The final distorted mesh is plotted in Figure 2. This final distortion pattern closely matches the one from the full model. The corresponding residual Mises stress contours are included also in Figure 3 which also closely resemble their counterparts resulting from the full model.

Comparison of the von Mises contours indicates that the magnitudes of the stresses are very close, and that the only major differences are the stress gradients along the weld wake. In the full model, a fine mesh is used all along the weld path; high gradients of both temperature and stress components are captured by the fine mesh density. The model with rezoning and dynamic substructuring, however, uses a much coarser mesh in the wake of the heat source.

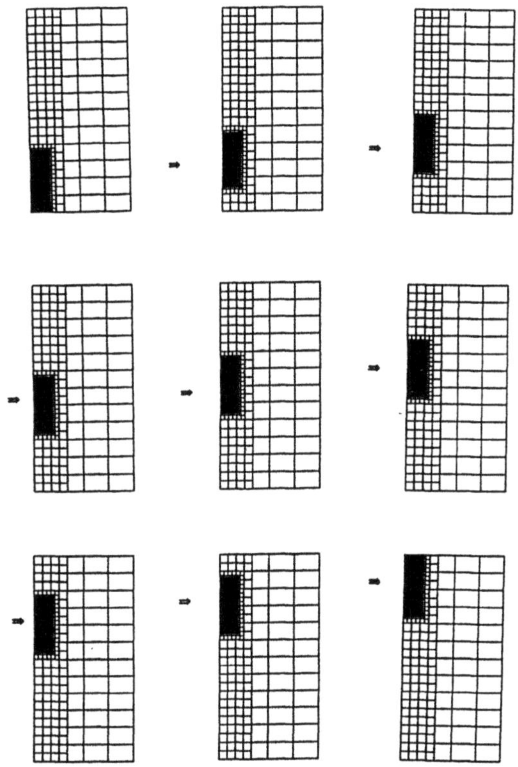

Fig.4. Different meshes for the incremental analyses

The total time needed for the simulation with both rezoning and dynamic substructuring represents a reduction of computer time by a factor of more than seven when compared with the full model simulation. It should also be noted that the plate model with a long slender weld path gives us a conservative assessment of the efficiency of rezoning. The computer time needed depends heavily on the bandwidth of the stiffness matrix, which in turn depends primarily on the number of nodes in the transverse direction. The bandwidth therefore does not change significantly when the model includes a shorter, denser mesh. Even higher efficiency of rezoning can be obtained with a high bandwidth structure. Much higher reductions are also expected when the structure is larger and more complex.

Conclusions

The transient character of welding, where a nonlinear, local zone translates through an otherwise elastic structure, suggests that model redefinition techniques can offer substantial reductions in analysis time. Since most of the thermal and mechanical gradients are concentrated within the local zone around the electrode, the rest of the structure can be modeled more coarsely than the local zone and then redefined as the local zone moves.

Implementation of model rezoning procedures on a simple welded plate model reduced the analysis time by over a factor of five. The addition of dynamically redefining the non-local portion of the structure with substructures provided an additional reduction in analysis time. The plate model presented here represents a lower bound on the reduction in analysis time, since more complicated two-dimensional structures should possess a larger stiffness matrix bandwidth. Three-dimensional weld simulations should demonstrate even larger reductions in time.

The model redefinition procedures presented here provide the additional advantage of permitting step-wise coupled thermo-mechanical analyses. Fully coupled analyses are still not readily implemented with current finite element codes. The incremental redefinition of the finite element model based on deformed states captures a major aspect of the thermal and mechanical coupling associated with welding without requiring a fully-coupled formulation.

Acknowledgements The work presented here was performed with funding from the Office of Naval Research, Grant number N00014-89-J-1187. The authors would also like to acknowledge the generous equipment and software donation by Apollo Computer Company and Hibbett, Karlsson, and Sorensen, Inc. Professor Tom Eagar provided substantive and useful advice. Sandra DeJong deserves particular thanks in providing editing assistance.

References

1 V. Murti and S. Valliappan. Numerical inverse isoparametric mapping in remeshing and nodal quantity contouring. *Computers and Structures*, 22(6), 1986.

2 K. Bathe. *Finite Element Procedures in Engineering Analysis.* Prentice-Hall, Inc., 1982.

3 Y. Ueda, Y. C. Kim, and M. G. Yuan. A predicting method of welding residual stress using source of residual stress (report 1). *Transactions of JWRI*, 18(1):135–141, 1989.

4 S.B. Brown and H. Song. Finite element simulation of welding of large structures. *submitted for publication*, 1991.

Influence of Weldmatching on Mechanical Behaviours

Huo Lixing, Zhang Yufeng, Wang Lijun

Welding Division, Department of Mechanical Engineering
Tianjin University. Tianjin, China

Summary

Different tests were carried out on the butt joints of several kinds of steels to study the effects of weldmatching on the fracture or strength, and fatigue behaviour of welded constructions. The effects of weldmatching on the strength of welded structures were investigated and appraised directly by experiments in accordance with the general yielding criterion based upon "Fitness for Purpose" philosophy. The test results show that overmatching of weld metal increases the possibility of having a general yielding and may help tolerate larger weld defects, and certain factors, such as strain ageing and heat treatment can change the degrees of weldmatching, thus affect a_{cr}. It is obvious from test results that great care must be taken, when the safety against fracture initiation of welded joints with different weldmatching is evaluated solely based on COD concept. Effect of weldmatching on the fatigue behaviour was appraised based upon four point bending test data of two kinds of welded joints. To analyse crack growth the fracture mechanics approach was used. Although the values of $\triangle K_{th}$ are equal in the two kinds of welded joints, the fatigue crack propagation rate da / dN for A 131 steel welded joint was faster than that for A537 welded joint.

Effects of Weldmatching on the Strength of Welded Structures

In order to study the effects of weldmatching on fracture behaviour the wide plate tests were carried out on the butt joints of A 131 and A 537 steels welded with the same kind of electrode LB52NS, on the butt joints of 16Mn steel welded with Chinese electrodes J427, J507, J606 respectively, on the butt joints of 14MnMoNbB steel welded with electrodes J707 and J107 respectively. The mechanical properties of these steels and weld metals are shown in Table 1.

1. Testing procedure

Specimen: For manufacturing specimens the plates corresponding the sizes of specimen (it is 100 × 500 × 20mm for A131, A537, 16Mn steels, 200 × 500 × 20mm for 14MnMoNbB steel) were cut into two half parts. After this, a × type 60 ° deg. edge preparation was made and welded together. After welding the weld metal rein-

L. Karlsson, L.-E. Lindgren, M. Jonsson (Eds.)
Mechanical Effects of Welding
IUTAM Symposium Luleå/Sweden 1991

forcement was removed and on the weld the notch of different length in each specimen was obtained by drilling a hole and sawing a slit (0.15mm) with a small hacksaw and orientated transversely to the direction of loading.

Table 1. Mechanical properties of testing steels and weld metals

Condition	Material	σ_sMPa	σ_bMPa	ψ%	$\delta_{5\%}$
As welded	A131	280	418	78.3	29
	A537	363	524	75	26
	Deposited metal of LB52NS	462	526	77.4	18.3
Strain aged	A131	412			
	A537	481			
	D.M. of LB52NS	460			
Strain ageing +heat treatment	A131	303			
	D.M. of LB52NS	360	479	79.8	18.3
Heat treatment	A131	270			
	A537	344			
	D.M. of LB52NS	360.6	479	79.8	18.3
As welded	16Mn	336	524		29
	D.M. of J427	412	490		28.5
	D.M. of J507	490	549		29.8
	D.M. of J606	591	659		25.5
As welded	14MnMoNbB	752	810	58.2	14.5
	D.M. of J707	588	756		>16
	D.M. of J107	>800	980		

Testing: The tests were carried out at room temperature in a 1000KN and 10000KN tension machine. During testing the COD was measured on a gauge length of 8mm bridged over the notch tips. Meantime, the curves of P against COD, COD against $\triangle L$ and P against $\triangle L$ were plotted directly on the X–Y recorder. In order to compare the test results with the yield spread on one side of specimens a grating of 10 lines / mm was fixed for Moire technique. From thses interference lines, the ligament or general yielding on the specimens was clearly shown.

Assessment criteria: In welded constructions there is a possibility to exist defects in the weld, that it is necessary to establish acceptable levels for these defects on the basis of Fitness for Purpose. It is evidently much more realistic to look for criteria which relate to practical behaviour and which are easy to use.

Loading a specimen with a central notch in tension, local yielding starts at the crack tips. Due to workhardening the load necessary to develop further yielding must increase and when the crack is short, the plastic deformation will be happened in the gross section under loading. Plastic deformation in the gross section, this condition is defined as general yield. But for long cracks and when workhardening is small, this plastic deformation will be confined to fans (with tops at the crack tips), fracture initiates before plasticity occurs in the gross section. This condition is defined as ligament yielding.

For structural steels the ligament yield case is not acceptable because of insufficient strength ($\sigma < \sigma_s$) even if ductility is sufficient. The general yield case is acceptable, since it guarantees a strength at least equal to or larger than the yield strength of the metal ($\sigma > \sigma_s$) and a good ductility.

As the shift between general yield and ligament yield will occur for a certain crack configuration, it is easy to find the critical crack dimensions by means of the wide plate test. Using the above defined concepts we may formulate an acceptability criterion for defects as follow: a defect (or a crack) will be acceptable if general yield occurs. For plates this requirement will be satisfied if the crack is subcritical. In this work for the sake of simplicity in the case of welded joints which contains a notch in the weld loaded transversely, the "general yield" concept was used when parent metal and weld metal deform plastic and the concept "ligament yield" was used when the weld metal alone behaves plastic.

2. Discussion and conclusion of the tests

General in the case of a transverse butt weld when the weld metal contained a plane sawcut notch directed parallel to the weld, the overmatching weld metal increases the possibility to meet a general yielding, in other words, it will be easier for the applied load on the notched specimen to reach the yield strength of the unnotched specimens. i.e plastic deformation will occur in the notch free parts of the specimen and a general yield will occur. Here it should be pointed out that:

For both A131 and A537 steels overmatching of weld metal increases the possibility of having a general yielding, however, from the test results, which are shown in Tab. 2, it can be seen that for A131 steel overmatching of weld metal with parent metal in a higher degree may help tolerate larger weld defects or critical crack size a_{cr}. In this case general yield spread will be facilitated, furthermore the danger of crack initiation at the tips of the notches is reduced.

The yield stress of the electrode J606 is much higher than the yield stress of the electrode J506 (see Table 1), so that in accordance with the general yielding criterion the critical crack size a_{cr} for the welded joints welded with J606 will be larger than that for the welded joints welded with J506 electrode. However, from testing results of 16Mn welded joints it is clear that the a_{cr} for the welded joints welded with J606 is as big as for the welded joints welded with J506 electrode. It seems to be that the high strength weld metal J606 is more sensitive to strain concentration than the low strength weld metal J506, this means that when choosing a weld metal it is important to know that with increasing strength of metal, the sensitivity of the metal to strain concentration is increased and the a_{cr} or fracture stress will be affected.

According to the general yielding criterion there are three cases in the fracture behaviours of undermatching welded joints. i.e, 1) Fracture with ligament yielding; 2) Fracture with ligament yielding, but fracture starts at the tips of the crack in the weld, however the stress field resulting from the thermal weld cycles strongly influences the crack path, the crack deflecting away from the weld and propagated into the parent metal. Because of the good arresting properties of the parent metal the crack will be arrested in it. After this, under loading the deformation will be taken place in the parent metal, until fracture happened; 3) Fracture with general yielding. From Fig 1 it can be seen that the curve of P (or σ) against $\triangle$L for undermatching weld in yield strength is different from the curve of P (or σ) against $\triangle$L for overmatching weld in yield strength. Looking at this diagram there are five parts instead of 3 parts. In the first zone the P value increases with $\triangle$L (the elastic deformation was taken place on the whole specimen), after this the plastic deformation is confined to the ligament of the notch in the weld.

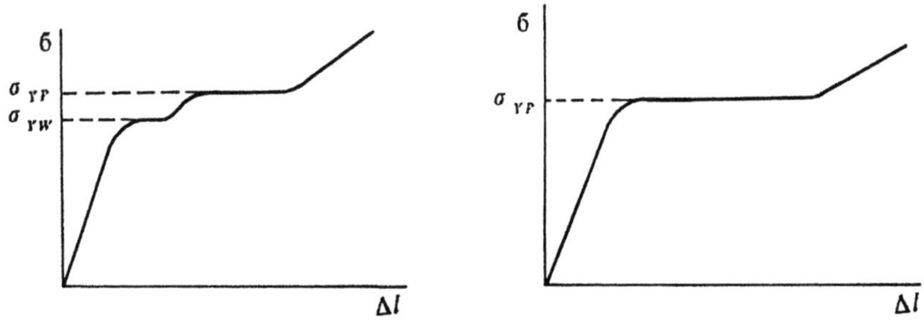

a. for undermatching weld b. for overmatching weld

Fig 1. P—$\triangle$L curve for fractive with general yielding

The P value remains constant until workhardening of the weld is obtained. i. e, the value of P increases again when the yield strength of weld is equal to the yield strength of the parent metal, resulting from workhardening of weld metal. This is the third part in the diagram. After this the plastic deformation spreads in the notch free parts of the plate. This will be happened in the fourth part in a form of horizontal line. This zone is characterized by the general yielding. The P increases again when the strain hardening in the parent metal starts.

Certain factors such as strain ageing and heat treatment, can change the degree of weld matching and thus according to the reason mentioned above can affect a_{cr} (see Table 2).

Table 2. Test results based on the general yielding criterion

Condition	Material		a_{cr}mm	a_{cr} / h(wide)
Parent metal	A 131		36~37	~0.33
	A 537		29~31	0.27
As welded	welded joint A	weld metal	49~50	0.45
		HAZ	49~50	0.45
	welded joint B	weld metal	40~41	0.37
		HAZ	39~40	0.36
Strain aged	W.M in welded joint A		19	0.17
	w.M in welded joint B		16~19	0.15~0.17
Strain ageing +heat treatment	W.M in welded joint A		31~39	0.28~0.35
	W.M welded joint B		/	/
Heat treatment	W.M in welded A		43~44	0.39
	W.M in welded joint B		39	0.35

joint A: A131 steel joint welded with electrode LB52NS
joint B: A537 steel joint welded with electrode LB52NS

There is an opinion that in the COD test, the crack initiation depends only upon the local property of the material surrounding the crack. Certainly, the crack tips position is very important in the COD test. But on the other hand, through the wide plate test (the specimen contains a transverse weld with a notch) we can find that COD depends not only on the local material surrounding the crack, but also depends on the whold condition of the specimen. In this case the COD parameter of the specimens depends both on the yield strength of the notch free parts and the notched part in the welded plate which contains a notch in the weld loaded transversely. That is to say that the P–COD pictures, the $\triangle$L—COD pictures and the values of the COD depend on the relations of strength between the notch free parts (parent metal) and the notched part (weld metal). The higher the yield strength of the notch free parts of the material, the larger the COD we have got before fracture. Or the higher the yield strength of the notched parts the smaller the COD we have got before fracture. In the COD design curve for material or constructions which possess a large critical value of COD the large crack size can be permitted, so that, great care must be taken when the safety against fracture initiation of welded joint with different weldmatching in yield strength is evaluated solely based upon COD concept.

Effect of Weldmatching on the Fatigue Behaviours

The second section describes the use of the four point bending test in estimating the fatigue behaviours of the two kinds of welded joints of A131 and A537 steels welded with LB52NS electrode, for in such case the effects of weldmatching considered are more similar to that in the wide plate test. In the present work to analyse crack growth the fracture mechanics approach was used. The shape, dimensions of the two kinds of specimens and loading system are shown in Fig 2. According to the research programme, first for the two kinds of welded joints the fatigue thresholds ΔK_{th} were measured and second, in the region where Paris law $da/dN = C(\Delta K)^m$ is obeyed, the relationships between da/dN, ΔK were treated with seven point incremental polynomial technique based on the a–N curves.

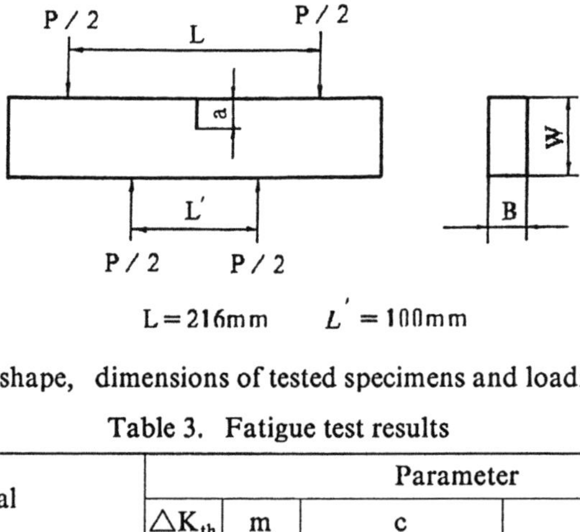

Fig 2. The shape, dimensions of tested specimens and loading system

Table 3. Fatigue test results

Material	Parameter			
	ΔK_{th}	m	c	da / dN
Welded joint of A131 steel welded with LB52NS	8.110	3.13	9.53×10^{-13}	$9.53 \times 10^{-13}(\Delta K)^{3.13}$
Welded joint of A537 steel welded with LB52NS	8.105	3.18	4.13×10^{-13}	$4.13 \times 10^{-13}(\Delta K)^{3.18}$

1. The determination of fatigue crack propagation threshold ΔK_{th}. The fatigue crack propagation threshold ΔK_{th} is a material property, although its measurement requires a tedious test. In our work based upon a continuous de-

crease of K the ΔK_{th} for the two kinds of welded joints were determined in accordance with Chinese Standard HB 5216–82. This method promises to reduce the time required to measure the threshold, while still maintaining reasonable accuracy. Studies were carried out on tension–tension testing at $R=1/3$ and at a frequency of 100Hz. The ΔK_{th} was obtained for average of each set of 2 specimens. It was recognized that the extrapolation of the log ΔK_aversus log $N_a$curve to a large N_a(10^6cycles) gave a good approximation of ΔK_{th} for the welded joints studied. This means that the threshold can be defined as the value of ΔK for which the fatigue crack propagation rate tend to be $da/dN<10^{-8}$mm / c. The test results are presented in Table 3.

From the test results it can be seen that the values of ΔK_{th} are equal each other in the two kinds of welded joints, in other words, there isn't any effect of the weldmatching on the propagation thresholds ΔK_{th}.

2. Effect on the fatigue crack propagation rate da / dN. In this case the tests were carried out in laboratory air at a frequency of 100Hz, and under R = 9800N / 49000N = 0.2 and R = 6537N / 32663N ≈ 0.2 conditions. i. e the relationships between da / dN and ΔK were obtained from 7 specimens of A131 welded joints and 8 specimens of A537 welded joints under R = 6537N / 32663N (ΔP = 26.13KN) and from each 4 specimens of A131 and A537 welded joints under R = 9800N / 49000N (ΔP = 39.20KN). Conditions.

For four point bending test the ΔK was calculated according to:

$$\Delta K=\frac{1}{4}\Delta P(L-L_1)\cdot\frac{F(a/w)}{BW^{3/2}} \tag{1}$$

where

$$\begin{aligned} F(a/W)&=11.94(a/w)^{0.5}-14.82(a/W)^{1.5}+77.82(a/W)^{2.5}\\ &\quad-139.02(a/W)^{3.5}+148.8(a/W)^{4.5},\ \textit{for}\ a/W\leqslant 0.5;\\ F(a/W)&=3.976/(1-a/W)^{1.5},\ \textit{for}\ a/W>0.5 \end{aligned} \tag{2}$$

The test results are presented in Table 3 and visualized in Fig 3. From testing results it is apparent that nonsignificant difference in the constant m between the two kinds of welded joints could be found, however there are some differences in constant C between them under the same loading conditions, i. e the value of C for A131 welded joint is two times more than that for A537 welded joint, there is therefore, some discrepancy between the static load and fatigue conditions, that the overmatching in high degree deteriorates the fatigue behaviour. The fatigue crack propagation rate da / dN for A131 welded joint was faster than that for A537 welded joint. To identify and explain this test results in our research programme the test on various welded joints with different rations of weldmatching will be completed in the near future.

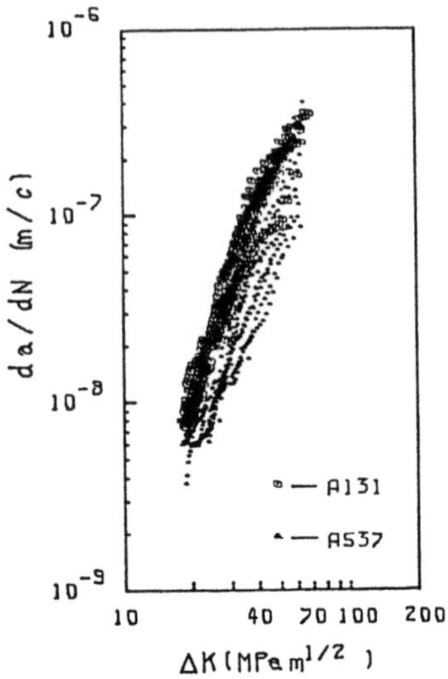

Fig 3. Fatigue crack growth curves for welded joints of A537 and A131 steel welded with electrode LB52NS tested in laboratory air, at $R=0.2$

Conclusions

Beneficial effect of overmatching of weld metal in yield strength on the strength of welded structures was found based upon General Yielding Criterion, however in the case studied by us overmatching in high degree deteriorates the fatigue behaviour of welded structures, it increases the fatigue crack propagation rate da / dN.

Acknowledgements

The authors wish to express their sincere thanks to Natural Science Foundation of China for their financial support, without which the present study could never have been accomplished. Sincere thanks are also expressed to IUTAM for supplying travel grant for attending this conference.

References

1. Soete W.,Huo L.X: The General Yielding Criterium. Revue De La Soudure 1981. No 3. 133–141

2. Huo L.X.,Zhang Y.F: The effect of some manufacturing factors on the strength of welded structures. Transactions of China Welding Institution Vol 9. 1988 No 1. 51–58

Combined Solid and Shell Element Modelling of Welding

M. Näsström L. Wikander L. Karlsson L. E. Lindgren
Department of Mechanical Engineering
Luleå University of Technology
S-951 87 Luleå, Sweden

J. Goldak
Department of Mechanical & Aeronautical Engineering
Carleton University
Ottawa, Ontario, Canada, K1S 5B6

Abstract

Finite element calculations of residual stress distribution in a welded component from a hollow square section inconel tube are presented in this paper, figure 1.
Shell element can be successfully used in finite element calculations of thin walled structures [1]. However, in the weld and the heat affected zone (HAZ) shell elements may not be sufficient, since the through thickness stress gradient is high in these regions. In the study presented here a combination of eight-nodes solid elements and four-nodes shell elements is used. The solid elements are used in and near the weld and shell elements are used elsewhere. This combination of solid elements and shell elements reduces the number of degrees of freedom in the problem in comparison with the use of solid elements only.

Introduction and statement of problem

In the design of welded structures it is of great importance to be able to calculate the residual stresses and the distortions due to welding. The residual stresses are needed to assess the strength of a welded structure. The distortions must often be known or avoided when welded components are mounted together to form a structure. One possible way to find the residual stresses and the distortions is to simulate the welding and the cooling to room temperature by use of the finite element method (FEM). Simulations have been performed of two-dimensional structures for nearly twenty years. Fully three-dimensional simulations of welding is very recent [2]. This is also the case for shell structures [1]. Fully three-dimensional simulations may be too cumbersome and simulations using shell elements may be too inaccurate in the weld and the heat affected zone.
The main purpose of this paper is to develope a method of simulation so that it is possible to combine the advantages of solid elements and shell elements in order to simulate realistic welding problems. Therefore a FE-code has been further developed for the combined solid and shell element modelling of welding. The problem used for the demonstration of the method is given in figure 1. This figure shows a part of a welded panel made of inconel 600. For demonstration purpose the part shown in figure 2 was chosen. This part was modelled using solid elements only, shell elements only and a combination of brick elements and shell elements, see figure 6. The simulations were divided into two parts, Thermal analysis and Mechanical analysis, see below.

L. Karlsson, L.-E. Lindgren, M. Jonsson (Eds.)
Mechanical Effects of Welding
IUTAM Symposium Luleå/Sweden 1991

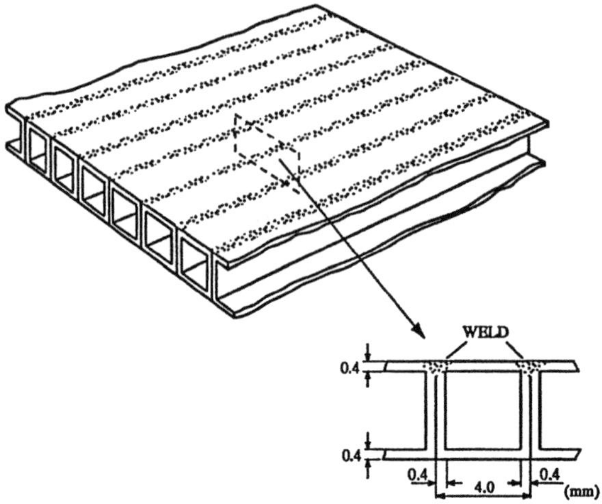

Figure 1. Inconel welded component.

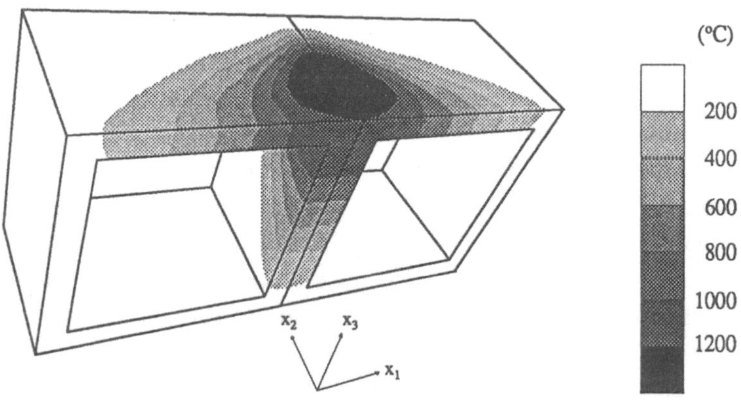

Figure 2. Analysed part of the structure of length (in x_2-direction) 2.5 mm and the temperature field at time 0.5 s.

Thermal analysis

Two hollow tubes with a square cross-section of 4x4 mm^2 and a wall thickness of 0.4 mm were joined side by side with a weld on the top surface along the x_2-direction, figure 2. The welding speed, v, was 4.25 mm/s, the gross heat input 39 kJ/m and the arc efficiency estimated to be 0.65. Temperature dependent material properties used in the thermal analysis are shown in figure 3. The peak in heat capacity, c, between 1354 °C and 1413 °C corresponds to the latent heat of 300 kJ/kg. The convective surface heat transfer coefficient was chosen to be 7.5 W/m^2°C and the radiative loss of energy corresponds to an emissivity of 0.65. The density is 8430 kg/m^3.

A cross section of the welded component was analysed. The three-dimensional time dependent temperature field was obtained from a two-dimensional analysis by the transformation below,the coordinate directions are seen in figure 2.

$$T(x_1, x_2, x_3, t) = \begin{cases} T_{2D}\left(x_1, x_3, t - \frac{x_2}{v}\right) & , x_2 - vt \leq 0 \\ T_{ambient} & , x_2 - vt > 0 \end{cases}$$

That is, the heat flow along the weld is neglected [3].

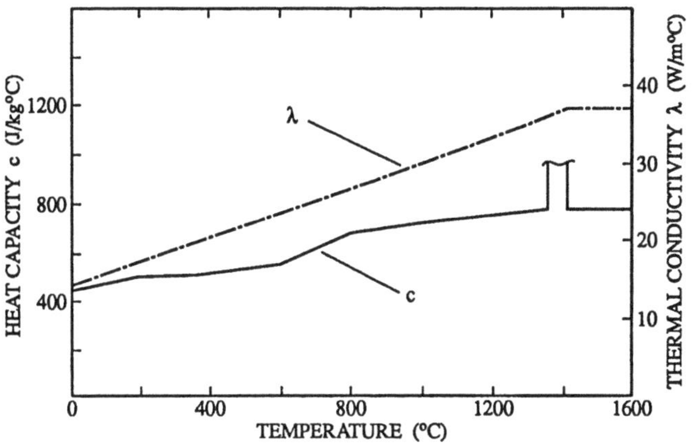

Figure 3. Thermal conductivity, λ, and heat capacity, c, as function of temperature.

An in house FE-code [4] was used in the two dimensional analysis. Due to symmetry it was sufficient to model only half the structure. It was divided into 695 four-node bilinear elements. The surfaces at x_3 = 0 mm, x_1 = 4 mm and x_3 = 4 mm were subjected to the above-mentioned radiative and convective boundary conditions, all other surfaces were considered adiabatic. The heat input was modelled by a line heat source with constant intensity on the top surface between x_1 = 0 mm and x_1 = 0.4 mm. The distribution of the heat input was elliptical in the x_1 - t plane with the duration 0.4 s and total width 0.8 mm. The simulation included 185 time/load steps, 80 for heating and the remaining for cooling. The temperature field was passed on to the mechanical analysis with the use of area coordinates.

Mechanical analyses

The shell element routines from [1] and the solid element routines from [2] were used in the mechanical analysis of the welded square pipe. They were combined in an in house code. Three different FEM calculations were performed. In the first only shell elements were used, in the second both brick and shell elements were used and in the third only brick elements were used. In all these cases the same temperature field and boundary conditions were used. Large displacements are accounted for in the the shell and solid element formulation, as well as large strains in the solid element. These features are not needed in the problem presented here. However these features are needed in the simulation of the welding of a large shell structure.
The nodes which were located on the x_2-x_3 plane in the weld were locked in the x_1-direction due to symmetry, the node located in x_1, x_2 and $x_3 = 0$ was locked in all directions and the node located at $x_1 = 0$, $x_2 = 2.5$ mm and $x_3 = 0$ was prevented from moving in the x_1 and x_3-directions. Finally the remaining two corner nodes on the x_1-x_2 plane were locked in the x_3-direction. The mechanical field is coupled to the temperature field only through the temperature dependent constitutive properties and the thermal strain. The temperatures were assumed constant in each element in order to obtain a compatible thermal strain field since linear elements are used [5] in the calculations The material is assumed to be thermo-elastic plastic with temperature dependent material properties. Inconel experiences no solid state phase transformations which is reflected in the ε^t curve in figure 4. The values of the thermal dilatation ε^t were obtained from the Swedish Institute of Metals Research. The following parameters were also needed in the mechanical analysis; elastic modulus E, Poisson's ratio ν and yield stress σ_y. The temperature dependence of these parameters shown in figures 4 and 5 were taken from [7]. The hardening modulus used is shown in figure 5. In the calculations the thermal strain increment was set to zero at temperatures above 900 C°. Results from another calculation where the cut of temperature was set to 1200 C° showed a maximum difference in maximum residual stress of at most five percent. The mechanical analysis was performed for times t = 0 to t = 10000 using 210 time/load increments for the welding process and 55 time/load increments for the cooling process.

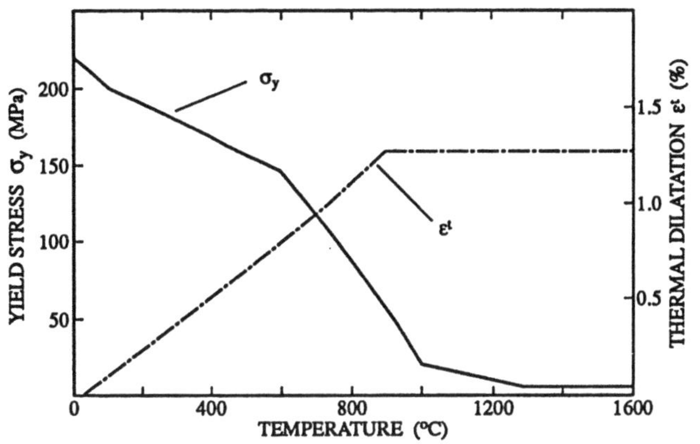

Figure 4. Yield stress σ_y, thermal dilatation ε^t, as functions of temperature.

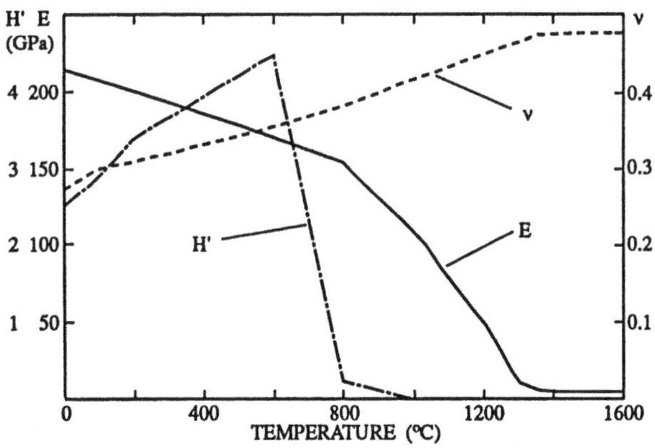

Figure 5. Hardening modulus H', Young's modulus E, Poisson's ratio ν, as functions of temperature.

The finite element model used in the combined analysis (case 2) is shown in figure 6a. This model has 5350 degrees of freedom. In this model the parent solid elements are integrated in the usual manner for this type of element. The only changes that are necessary are that the shell assumption of no strain energy normal to the shell element must be evoked [6].

The solid model, shown in figure 6b, used in the solid analysis (case 3) had 6846 degrees of freedom.
The shell element model shown in figure 6c was used in case 1. This problem had 3425 degrees of freedom.

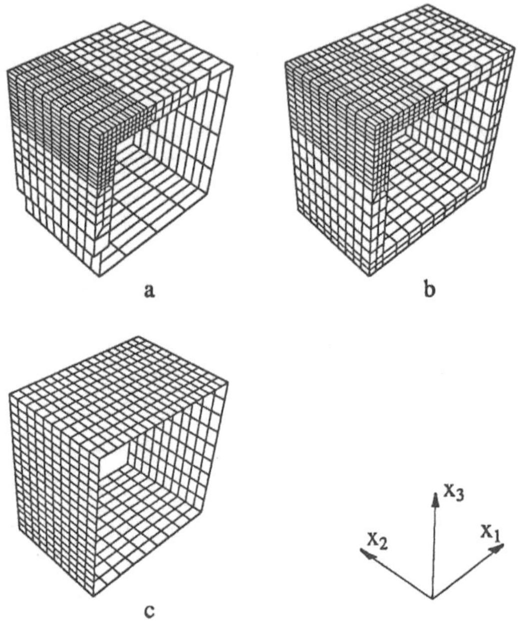

Figure 6. Finite element meshes used a) combined solid and shell element model b) solid element model c) shell element model.

<u>Results and evaluation of models</u>

The results in figures 7 and 8 show that the combined model can predict the residual stress state if compared to the solid model, while the shell model fails, especially close to the weld. This follows partly from the fact that the modelled structure can not really be considered as a shell structure, particularly using a fine mesh. It should be noted that there are no contours on the part containing shell elements in figures 7a and 8a.
Surprisingly, the residual x_2-stresses on the top surface close to the weld are mainly compressive, both in the combined and the solid analyses, see figure 7 a,b. This effect is mainly due to bending.
The residual x_2-stress outside the heat affected zone are similar in all three models see figure 7 a-c. The mesh used in the shell analysis, see figure 6c, was too coarse in the heat

effected zone to model the mechanical behaviour as the temperature gradients were large in this region. The shell model was also unable to capture the behaviour due to through thickness temperature variation since all gauss points in each element were exposed to the same temperature. This may explain the great difference between the shell model and the models with solid element.

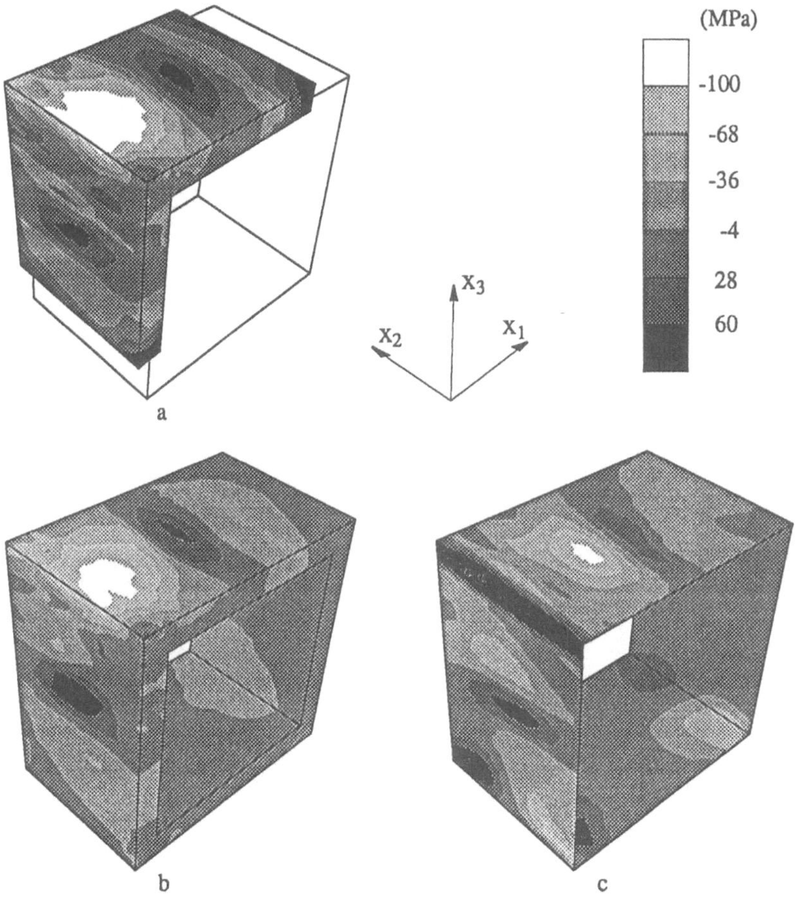

Figure 7. Contours of residual x_2-stress a) combined solid and shell element model b) solid element model c) shell element model.

In the x_1-direction, see figure 8 a-c, there are somewhat larger differences between the stresses in the combined and the solid model. This difference can be derived from the solid to shell transformation which does not allow relative movements in the shell normal direction of the connecting solid and shell master nodes. For the shell model the residual

x_1-stress deviates more than the x_2-stress. Note that the displayed x_1-stress in figure 8c should be regarded as a circumferential stress, i.e. it is only the top surface x_1-stress which can be compared with the other models.

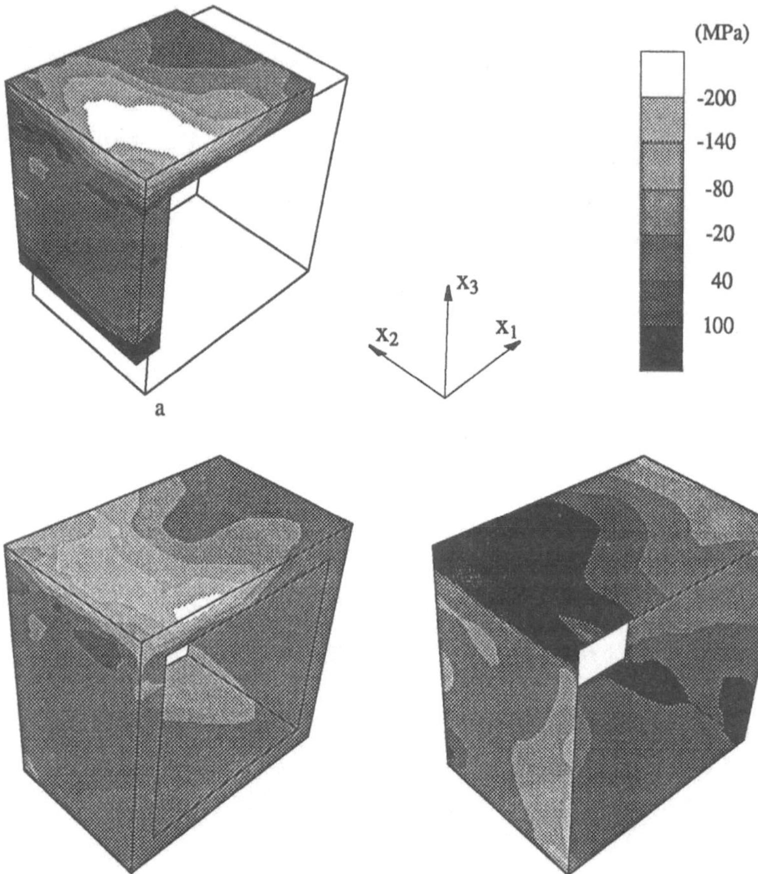

Figure 8. Contours of residual x_1-stress a) combined solid and shell element model b) solid element model c) shell element model.

Discussion and conclusion

The residual stress results show that the combined model can be successfully used in structures where high stress and temperature gradients are localized to a narrow region i.e. a fine solid mesh in the heat effected zone and a coarse mesh of shell elements elsewhere. In the investigation presented here no mesh refinement was made in the weld direction, which if used would reduce the number of degrees of freedom further. In order to avoid stress concentration in the shell normal direction in the solid to shell connection, relative movements of the connecting solid nodes should be allowed. Such a modification would permit the use of shell elements closer to the weld and reduce the number of degrees of freedom. It would also be of interest to investigate an extruded model in the x_2-direction to reduce the influence of end effects.

References

1. L. E. Lindgren, and L. Karlsson "Deformations and stresses in welding of shell structures" International Journal for Numerical Methods in Engineering, Vol. 25, 635-655 (1988)

2. M. J. McDill, J. A. Goldak, A. Oddy, M. J. Bibby "Isoparametric Quadrilaturals and Hexahedrons for mesh grading elements" Communications in Applied Numerical Methods, Vol 3, pp 155-163, (1987)

3. M. Jonsson, L. Karlsson and L. E. Lindgren "Deformations and Stresses in Butt-Welding of Large Plates" in R. W. Levis (editor): Numerical Methods in Heat Transfer, Volume III, pp 35-58,Wiley,London,1985.

4. L. Karlsson and L. E. Lindgren "Combined Heat and Stress-Strain Calculations" in M. Rappaz, M. R. Ozgu and K. W. Mahin (editors): Modelling of Casting, Welding and Advanced Solidification Processes V pp 187-202 (1991), (Proc. of the Int. Conf.on Modeling of Casting, Welding and Advanced Solidification Processes, Davos,Switzerland, September 16-21, 1990).

5. J. Goldak "Modeling Thermal Stresses and Distortions in Welds", in S.A. David and J. M. Vitek (editors): Recent Trends in Welding Science and Technology, ASM INTERNATIONAL, (Proc. of the 2nd Int. Conf. on Trends in Welding Research, Gatlinburg, Tennessee, USA, 14-18 May, 1989.) Materials Park Ohio, 44073 USA 1990, pp. 71-82.

6. J. D. Chieslar and A. Ghali "Solid to Shell element Geometric Transformation" Computers & Structures, Vol 25. No. 3, pp. 451-455, (1987).

7. Volvo Flygmotor AB, 461 81 Trollhättan, Sweden (Mr Ronny Jonsson).

Mathematical Modelling of Molten Pool in Arc Welding

Takayoshi OHJI, Akira OHKUBO and Kimiyuki NISHIGUCHI

Department of Welding and Production, Osaka University
2-1 Yamada-oka, Suita, Osaka, 565 JAPAN

A mathematical model of arc weld pool has been discussed in this paper. In the model analysis, the heat flow in welding is numerically analyzed to estimate the molten pool size, and a surface tensional balance equation for the molten pool is solved numerically to obtain a weld geometry.
The model of finite difference, developed in the present work, make it possible to simulate some typical processes such as a thin plate arc welding and a fillet welding of thick plate.

I INTRODUCTION

One of the important problem to be solved in the welding engineering is to construct a mathematical model for the computer simulation of welding process.
As the knowledge of the mechanism of heat and mass transfer in a molten pool under welding is limited, the following two fundamental assumptions have been introduced in this work,
(1) Heat flow in weld pool is assumed to be conductive, i.e., the influence of the metal flow[1-3] in molten pool on the heat flow is neglected.
(2) Molten pool is assumed to be in a static equilibrium[4,5] under gravity, arc pressure and surface tension, and it is considered to be in a quasi-stationary state.
The purpose of this work is to construct a mathematical model of arc weld pool under these assumptions, and discuss the possibility of numerical simulation of arc welding process by the model.

II MODEL OF MOLTEN POOL

As shown in Figs.1 and 2, two types of model, Model I and II, have been developed to predict the weld geometry in arc weld-

L. Karlsson, L.-E. Lindgren, M. Jonsson (Eds.)
Mechanical Effects of Welding
IUTAM Symposium Luleå/Sweden 1991

ing. Figure 1, typical of Model I , shows the model for a thin plate GTA welding, and Fig.2, typical of Model II , is for a fillet welding. The heat source in these models is assumed to be uniformly distributed in a circular region, as shown in these figures.

On the model analysis, the heat flow equation[6] in quasi-stationary state and the surface tensional balance equation of molten pool[5,7] are numerically solved.

As an example, the model[7] for thin plate GTA welding of full penetration are constructed from following equations,

$$\frac{\partial}{\partial x}\left[K(T)\frac{\partial T}{\partial x}\right] + \frac{\partial}{\partial y}\left[K(T)\frac{\partial T}{\partial y}\right] + \frac{\partial}{\partial z}\left[K(T)\frac{\partial T}{\partial z}\right] = -c(T)\,\rho(T)\,v\,\frac{\partial T}{\partial x} \tag{1}$$

$$\sigma\left[\frac{(1+(\Phi_Y)^2)\Phi_{XX} - 2\Phi_X\Phi_Y\Phi_{XY} + (1+(\Phi_X)^2)\Phi_{YY}}{(1+(\Phi_X)^2+(\Phi_Y)^2)^{3/2}}\right] = \rho g\Phi - Pa - \lambda \tag{2}$$

$$-\sigma\left[\frac{(1+(\Psi_Y)^2)\Psi_{XX} - 2\Psi_X\Psi_Y\Psi_{XY} + (1+(\Psi_X)^2)\Psi_{YY}}{(1+(\Psi_X)^2+(\Psi_Y)^2)^{3/2}}\right] = \rho g\Psi - \lambda \tag{3}$$

$$\Phi_{XX} = \frac{\partial^2\Phi}{\partial x^2}, \quad \Phi_X = \frac{\partial\Phi}{\partial x}, \quad \cdots\cdots$$

where, T:temperature, K(T):thermal conductivity, ρ (T):mass density, c(T):specific heat, v: welding speed, Φ (x,y):top surface of molten pool, Ψ (x,y):bottom surface of molten pool, σ :surface tension, Pa:arc pressure, λ :const.

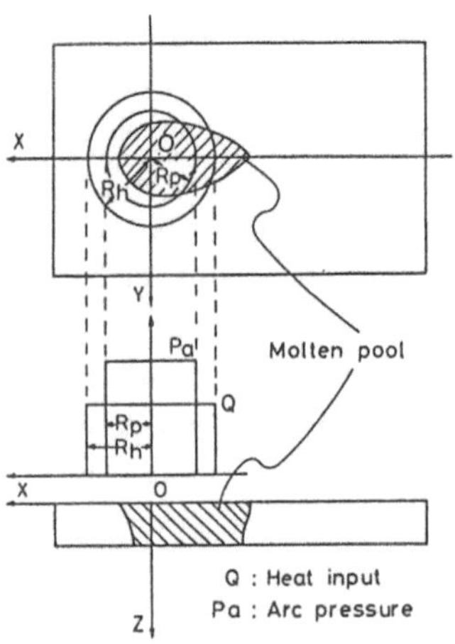

Fig.1 Model of molten pool for thin plate welding (Model I)

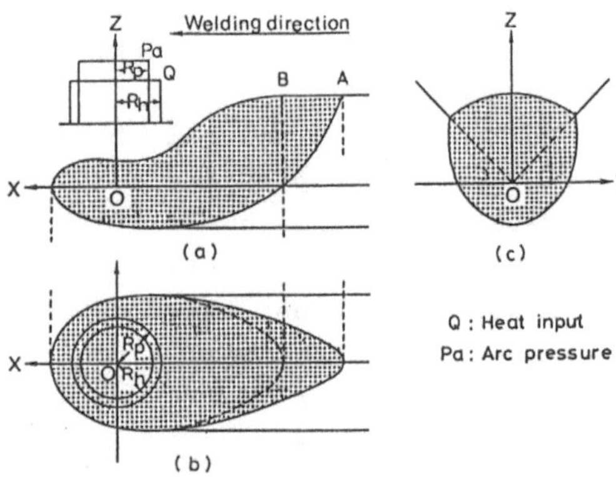

Fig.2 Model of molten pool for fillet welding (Model II)

Equations (2) and (3), the surface tensional balance equation, are concerned with the profiles of top and bottom surfaces of molten pool in quasi-stationary state. These non linear partial differential equations have been numerically solved by an iterative method[5,7] of finite difference.

The heat source in Model II is situated on the calculated weld pool surface. Accordingly, Model II make it possible to discuss the influence of weld pool configuration on the heat flow in arc welding. Another type of iterative method[8] has been applied to solve the free boundary problem, because the heat flow in the model depends on the molten pool configuration and the pool configuration depends on the heat flow.

III RESULTS AND DISCUSSION

① Thin Plate GTA Welding (Model I)[7]

Figure 3 is an example of calculated molten pool in thin plate welding of full penetration. (a) shows the longitudinal cross section of weld pool. (b) is the top view of molten pool, and (c) is the weld cross section after solidification. Though the calculation has been carried out under the condition of Pa=0, the molten pool is slightly distorted downwards by gravity. Figure 4 is another example of calculation in flat position, which shows the relation between weld cross sectional area and weld surface profile. As shown in the figure, the weld shape becomes concave with decrease of weld cross sectional area.

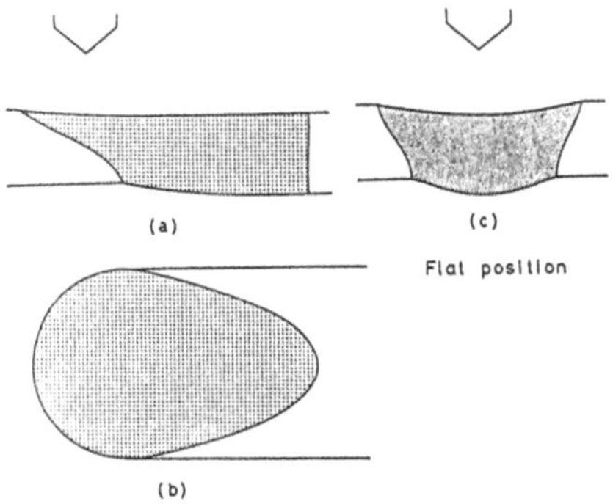

Fig.3 An example of computed molten pool in flat position. (Model I ,q=1500J/s,v=20cm/min H=3mm,Pa=0,Rh=4mm)

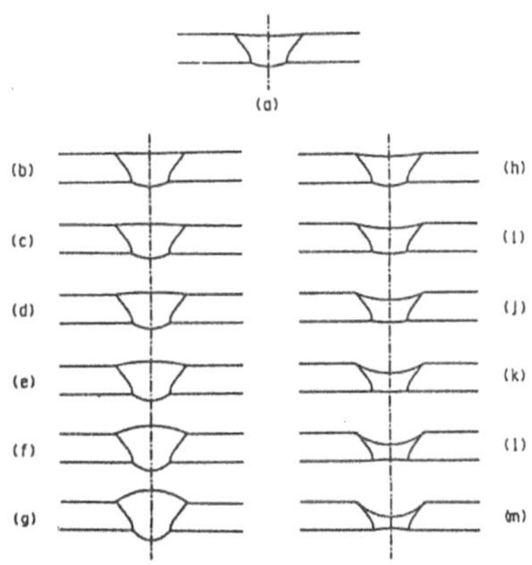

Fig.4 Relation between bead surface profile and weld cross section in flat position (Model I ,H=3mm)

The calculated molten pool in vertical position are shown in Fig.5, where the effect of gravity is clear. In the vertical up position (b), a "center-rich" type of weld is obtained as a result of hanging down of molten pool. The dotted zone in the figure (b) means the area where the solid bottom may appear because the calculated displacement of pool surface is over the bottom of molten pool, estimated by the heat conduction analysis. In this report, this area is called "gouging

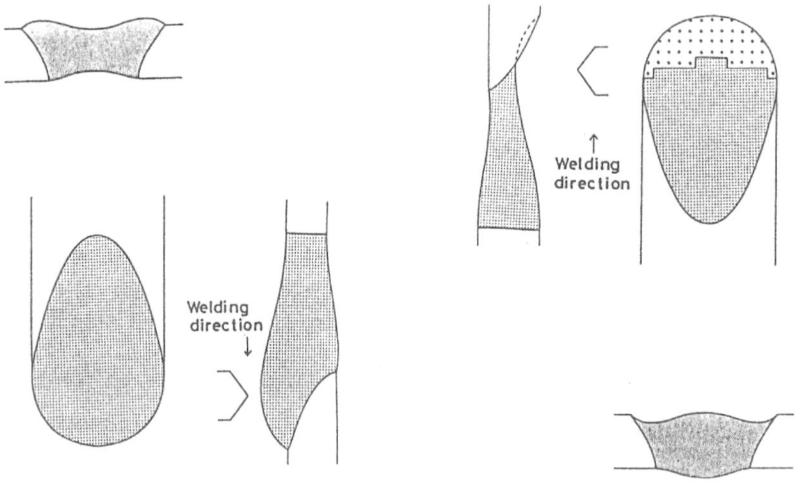

(a) Vertical down position (b) Vertical up position
Fig.5 Molten pool in vertical position (Model I , H=3mm, q=1500J/s, v=20cm/min, Pa=0, Rh=4mm)

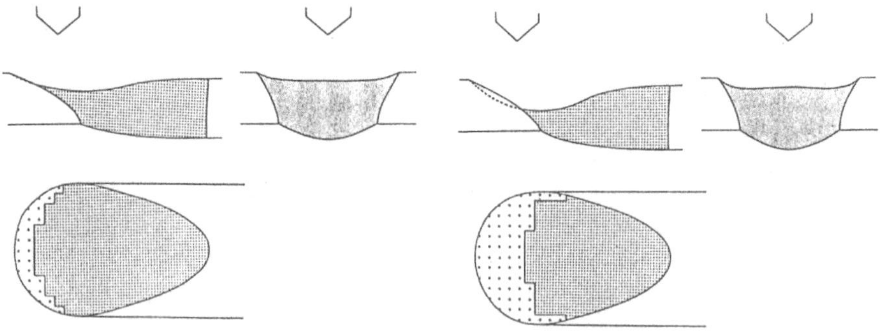

(a) Pa = 245 [Pa] (b) Pa = 490 [Pa]
Fig.6 Computed gouging zone in flat positional molten pool (Model I , H=3mm, q=1500J/s, v=20cm/min, Rp=4mm, Rh=4mm)

zone". The calculation suggests that the development of gouging zone by gravity causes undercutting or other defects in vertical up position.

Figure 6 shows the effect of arc pressure on the molten pool in flat position. It is clear that the gouging zone is due to the arc pressure, and the zone is broadened with increase of arc pressure.

② Fillet Welding

Figure 7, a typical example of the calculated weld pool for a fillet welding by Model II , shows the effect of arc pressure on the weld penetration. As is shown by the solid curve in Fig.7, a deep penetration is obtained as a result of the digging action of arc pressure when the arc pressure is high.

Figure 8, another example of calculated weld pool in fillet welding, shows the effect of wire melting rate on the weld

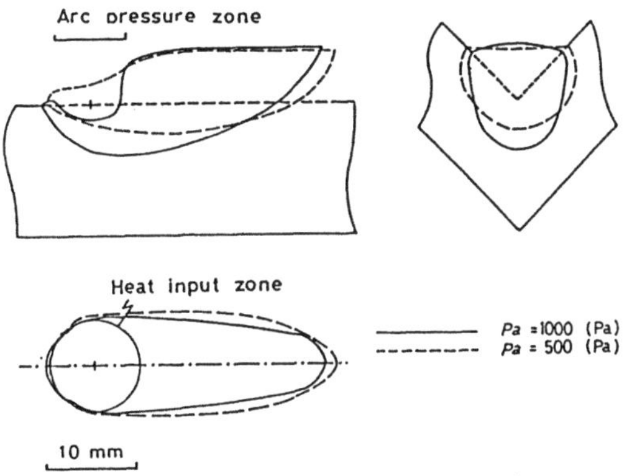

Fig.7
Effect of arc pressure on weld penetration
(Model II , H=1cm, q=7.5kJ/s, v=20cm/min, Rp=4mm, Rh=5mm, Mr=2mg/A·s)

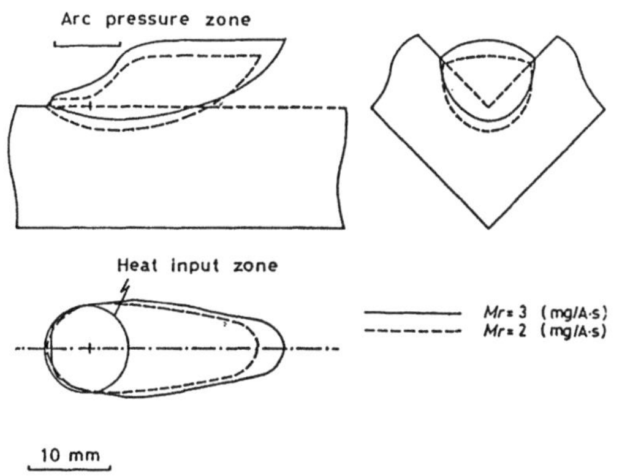

Fig.8
Effect of wire melting rate on weld penetration
(Model II , H=1cm, q=6kJ/s, v=20cm/min, Rp=4mm, Rh=5mm, Pa=560Pa)

penetration. The penetration is shallow when the deposition rate is high, as is shown by the solid curve in Fig.8.
The effect of arc pressure on bead surface profile of horizontal fillet welding is shown in Fig.9. Each of figures (a), (b), (c) and (d) is the cross section at each point of the weld pool. The broken curves in these figures are the surface profiles for non-arc pressure, and the solid curves show the effect of arc pressure. As can be seen in the figure (d), the center-rich type of surface profile in fillet welding is due to arc pressure.

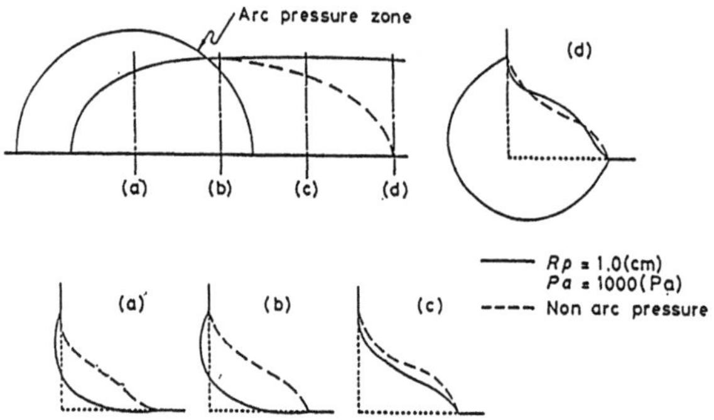

Fig.9 Effect of arc pressure on weld geometry in horizontal fillet welding (Model I, H=1cm, q=7.5kJ/s, v=40 cm/min, Rp=5mm, Rh=4mm, Mr=2mg/As)

③ Prediction of Process Parameter

Figure 10 shows the computed tolerance zone for thin plate GTA welding of mild steel. The solid curve for each plate thickness is the calculated critical condition of burn-through[7]. The broken curve is the condition of lack of fusion at the bottom side of the plate, i.e., the bottom bead is not formed at the lower side of the broken curve.
In Fig.11, the computed result for mild steel of 3mm thickness has been compared with the experiment by GTA. The solid curves in Fig.11 (a) show the calculated tolerance zone. In the figure (a), ◎ and ○ mean the normal bead, and × is for burn-through and ▲ is for lack of fusion bead in experiment. In Fig.11 (b), the experimental weld geometry, indicated by ◎ in the figure (a), is directly compared with the computed weld geometry. The correspondence between the experiment and the calculation is quite satisfactory.

In the case of copper, a high heat input and low velocity condition is required to get a full penetration weld. A computed tolerance zone for copper plate of 3mm thickness has been compared with the experiment, as shown in Fig.12. A good correlation between the model analysis and the experiment has been obtained in this case too.

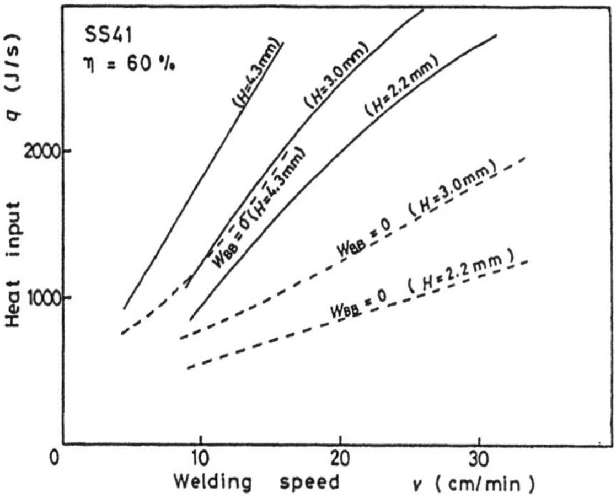

Fig.10 Computed tolerance zone for thin plate GTA welding (Model I ,Mild steel, Rh=4mm, Pa=0)

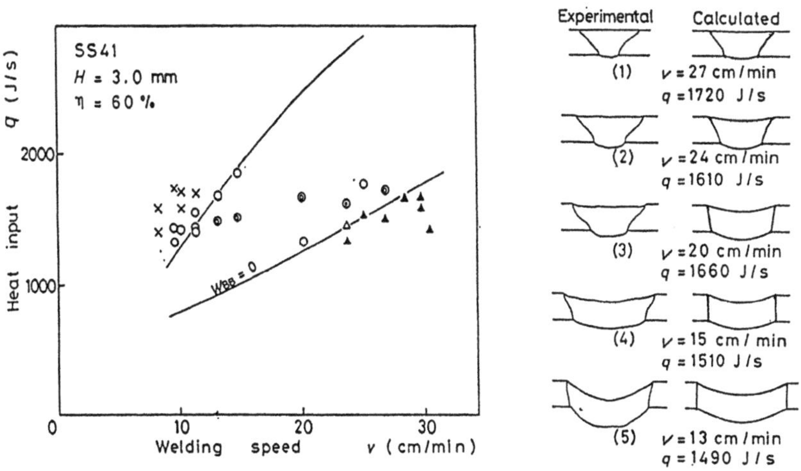

(a) (b)

Fig.11 Comparison between computed and experimental tolerance zones for GTA welding (Mild steel, H=3mm)

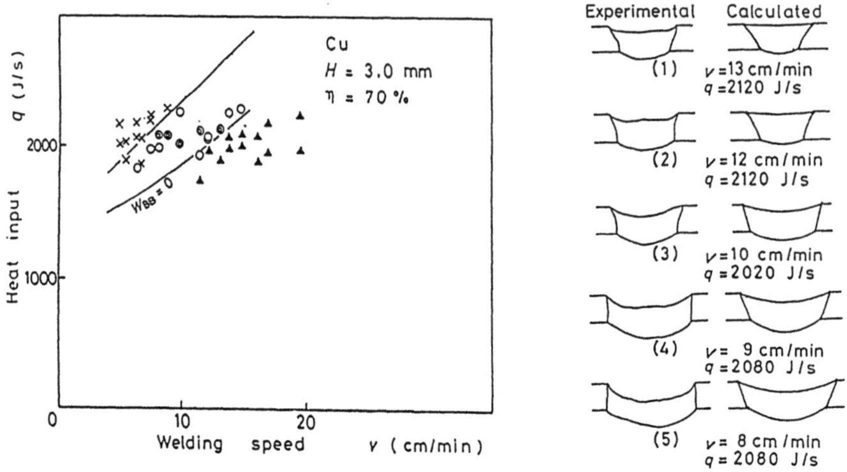

(a) (b)

Fig.12 Comparison between computed and experimental tolerance zones for GTA welding (Copper, H=3mm)

IV CONCLUDING REMARKS

In the present work, a mathematical model for arc welding has been developed. The result simulated by the model is quite satisfactory. Accordingly, it is concluded that the model is a useful tool for the simulation of arc welding process.

REFERENCES

(1) Lancaster,J.F.,Physics of Welding, Chap.4, Pergamon, 1986
(2) Heiple,C.R. and Roper,J.R., Welding J., Vol.61(1982), 97s
(3) Ohji,T.and Miyake,A. et al.,Quart. J. Japan Weld. Soc., Vol.8(1990), 54 (in Japanese)
(4) Nishiguchi,K.,Ohji,T.and Matsui,H.,Int.Inst.of Welding, IIW Doc.212-240-72 (1972)
(5) Ohji,T.,Ph.D. Dissertation, Osaka Univ.,(1978) (in Japanese)
(6) Rosenthal,D., Welding J., Vol.20(1941), 220s
(7) Ohji,T.and Nishiguchi,K., Technol. Repts.of Osaka Univ., Vol.33(1983), 35
(8) Nishiguchi,K.,Ohji,T.,Yoshida,Y.and Shirakashi,M., Quart. J. Japan Weld. Soc., Vol.4(1986), 673 (in Japanese)

Computation of Temperature Distribution due to Welding in Piping Systems

T. T. Öberg

Metals Laboratory, Technical Research Centre of Finland,
SF-02150 Espoo, Finland

Summary

This paper deals with the analytical and experimental analyses of the welding induced thermal fields in the girth and the overlay welds in piping systems. The knowledge of the temperature distribution forms the basis of further studies of the thermal stress fields, which in turn are used in crack growth analyses. The temperature analysis was done by using analytical methods based on known solutions for a moving heat source in a solid. For numerical analysis the mathematical formulae were implemented in a computer program TEDIW. The computer program was tested using two case studies from the literature. In these examples, also the measured temperature distribution was available. Both examples showed good agreement with predicted and measured values. In order to illustrate the capabilities of the program, an overlay welding case was also analysed.

Introduction

The prevention of intergranular stress corrosion cracking (IGSCC) near the heat affected zone of girth welds in piping systems of nuclear reactors is a task of considerable interest. The factors influencing the onset of IGSCC are welding-induced sensitization of the material, a corrosive environment and tensile residual stresses due to the girth welding of a pipe. One of the most effective approaches against IGSCC is to try to change the tensile residual stresses to compression at the inner surface of the pipe. This can be done by using different stress improvement methods such as induction heating, heat sink welding, overlay welding or mechanical shrinking of the pipe.

In this paper the thermal effects of the welding are investigated. For the analysis, a computer program, which is based on the analytical solution of moving heat source originally given by Rosenthal [3], was developed and tested. The program is similar to that of Solomon and Levy [5]. However, in overlay welding along the circumference of the pipe, the influence of the weld continuation is also taken into account for both transverse and circumferential direction of welding.

L. Karlsson, L.-E. Lindgren, M. Jonsson (Eds.)
Mechanical Effects of Welding
IUTAM Symposium Luleå/Sweden 1991

Moving heat source in an infinite solid

The use of Rosenthal's [3] model in the simulation of welding is based on the assumption that the heat conductivity λ is not a function of temperature. In that case the fundamental equation of heat conduction is given as [2]

$$\rho c \frac{\partial \theta}{\partial t} = \lambda\left(\frac{\partial^2 \theta}{\partial^2 x} + \frac{\partial^2 \theta}{\partial^2 y} + \frac{\partial^2 \theta}{\partial^2 z}\right) \quad , \qquad (1)$$

where θ is the temperature, ρc is the specific heat, t is the time and x,y and z are the cartesian coordinates.

The movement of the welding arc is taken into account by using a coordinate system (w,y,z), which moves at the same speed v as the welding arc. This condition is illustrated in Fig. 1. The temperature distribution is assumed to be quasi-stationary meaning that the temperature derivative with respect to time in equation (1) can be neglected. This leads to the temperature solution

$$\theta - \theta_0 = \frac{Q}{4\pi\lambda R} EXP(-vw/2\kappa - vR/2\kappa) \quad , \qquad (2)$$

where $\kappa = \lambda/\rho c$ is the thermal diffusivity and $R=(w^2 + y^2 + z^2)^{1/2}$ is the distance from the heat source to the point where θ is calculated and Q is the heat input from the welding arc.

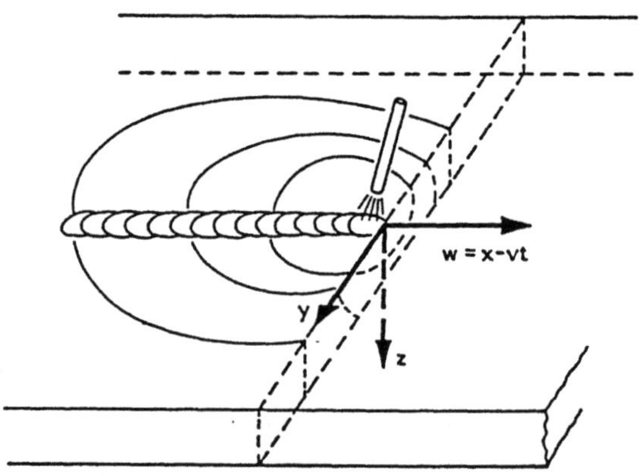

Fig.1. Temperature isotherms in a coordinate system, which moves with a welding arc.

Moving heat source on a plate of finite thickness

The solution (2) can be extended also to cover the temperature distribution in a plate of finite thickness H by adding a series of heat sources to equation (2). This method is described mathematically as [2]

$$\theta-\theta_0 = \frac{Q}{4\pi\lambda} EXP(-vw/2\kappa)\left[\frac{EXP(-vR/2\kappa)}{R} + \sum_{n=1}^{\infty}\left(\frac{EXP(-vR1/2\kappa)}{R1}+\frac{EXP(-vR2/2\kappa)}{R2}\right)\right] , \tag{3}$$

where $R1=(w^2+y^2+(2nH-H_0)^2)^{1/2}$ and $R2=(w^2+y^2+(2nH+H_0)^2)^{1/2}$ are the distances from the fictitious heat sources.

The solution of the moving heat source in a plate of finite thickness may also be treated as a two dimensional problem [1]. Leaving out the initial temperature, one obtains

$$\theta(w,R) = \frac{Q/H}{2\pi\lambda} EXP(-vw/2\kappa)\, K_0(vR/2\kappa) , \tag{4}$$

where K_0 denotes the modified Bessel function of the second kind and zeroth order and $R=(w^2+y^2)^{1/2}$ is the radial distance from the heat source.

Multipass welding along the circumference of a pipe

In multipass welding the temperature distribution during the second and subsequent passes is influenced by the welding of the previous pass. In order to account for this phenomenom, two methods have been suggested:

The method proposed by Rybicki et al. [4] uses superposition of heat sources similar to equation (3) to compute the required time-temperature curves for a welding pass. These curves are calculated by treating the pipe as a plate. After that, the curves are numerically integrated to obtain an average of the temperature field. For the subsequent pass the temperatures are also computed by using equation (3), but, in this case, the total temperature will be obtained by adding the average temperature of the previous pass and the experimentally determined ambient temperature to the temperature field.

The second method for the simulation of multipass welding was proposed by Vaidyanathan et al. [6]. In this approach, the welding passes are considered to be moving heat sources, which are separated in space by a distance equal to the circumference C (= π x diameter) of the pipe. The previous heat source (i) should vanish at the moment the subsequent source (i+1) appears. However, it is expected that the error is negligible if one regards the prior sources as continuous in time. Using equation (4), one can then write the temperature distribution as

$$\theta(w,R) = \sum_{i=1}^{n} \frac{Q/H}{2\pi\lambda} EXP(-vw_i/2\kappa)\, K_0(vR_i/2\kappa) \quad , \tag{5}$$

where n is the number of passes and w_2-w_1= w_3-w_2= ...w_n-w_{n-1} = C are the distances of separation in circumferential direction.

Overlay welding along the pipe circumference

The mathematical model developed here for the overlay welding deviates from the solution of equation (5) such that after each pass, in addition to the x-coordinate, also the y-coordinate will be separated in space by a distance which, in this case, equals to the distance B between each pass. Futher, in this work, equation (3) instead of (4) was used in the calculation of temperature distribution within a single pass. This was done because the distribution in z-direction may also be important, particularly with thick plates. The flow diagram of computer program TEDIW developed is shown in Fig. 3. The notations used in Fig. 3 refer to Fig. 2.

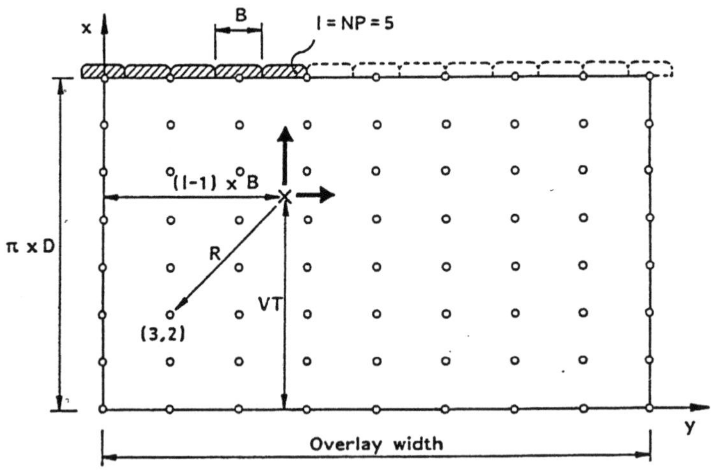

Fig.2. Geometric model of the overlay welding.

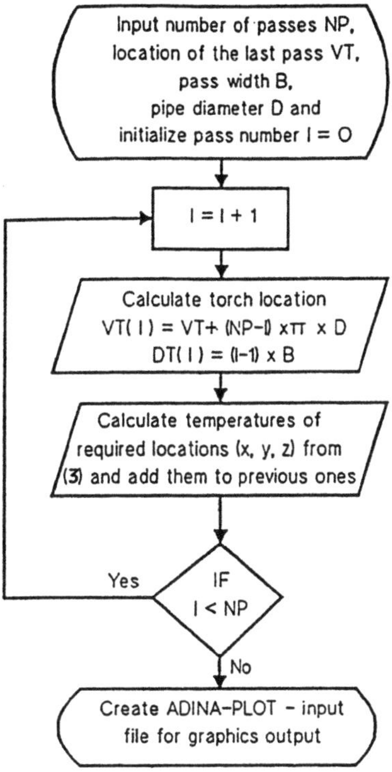

Fig.3. Flow diagram of the computer program TEDIW.

Program verification

In order to verify the basic equation (3), a relatively simple welding condition, reported by Solomon and Levy [5], was analyzed. The case concerns the analysis of SMA welding along the surface of a stainless steel plate. The results are shown in Fig. 4 where the horizontal axis indicates the x -direction distance from the welding arc. Since the basis of this analysis is the same than that of Solomon and Levy's [5], the results are identical to their solution. The measured points in Fig. 4 are results from the experimental work of Nippes et al., as quoted by Solomon and Levy [5]. Good agreement with those results is obtained.

The second verification example for the TEDIW program deals with girth welding of a type 304 stainless steel pipe reported by Rybicki et al. [4]. The girth weld consisted of 6 passes shown in Fig. 5. The figure also shows the positions of three thermocouples employed for experimental temperature measurements during the welding process.

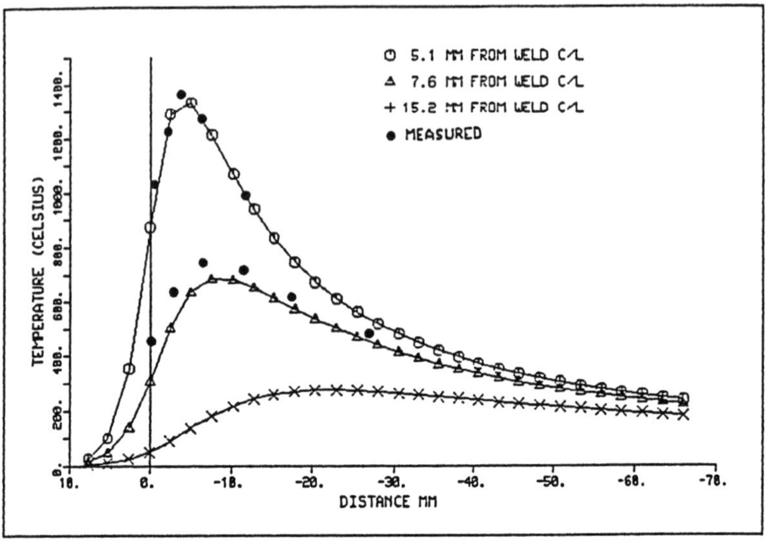

Fig.4. Comparison of predicted and measured temperatures due to a single-pass welding of a stainless steel plate.

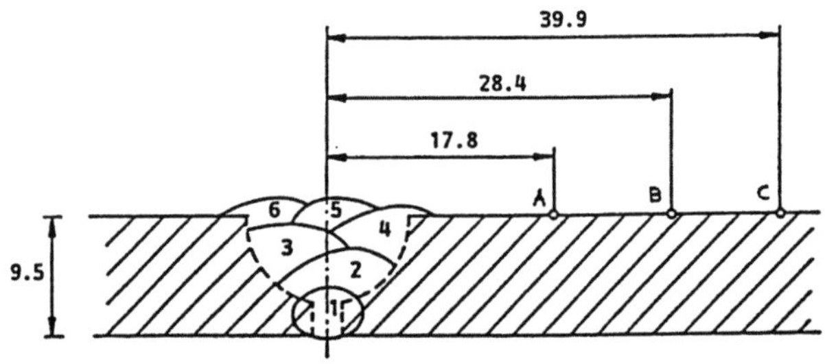

Fig.5. Welding of a multi-pass girth weld on a pipe with six passes. The letters A, B and C refer to the locations of the thermocouples.

Figure 6 compares the predicted and measured values of passes 2 to 4. One may see from Fig. 6 that the increase of minimum interpass temperature (horizontal line) can be predicted rather accurately. However, the calculated peak temperatures remain lower than the measured ones. One of the reasons for this is that the average welding parameters used in the calculation actually vary from pass to pass. Moreover, particularly in the case of weld pass no. 4, a large error results due to the fact that the actual distance from the arc to the thermocouple A is about 20 % smaller than 17.8 mm (see Fig. 5), which increases the difference between calculated and measured results.

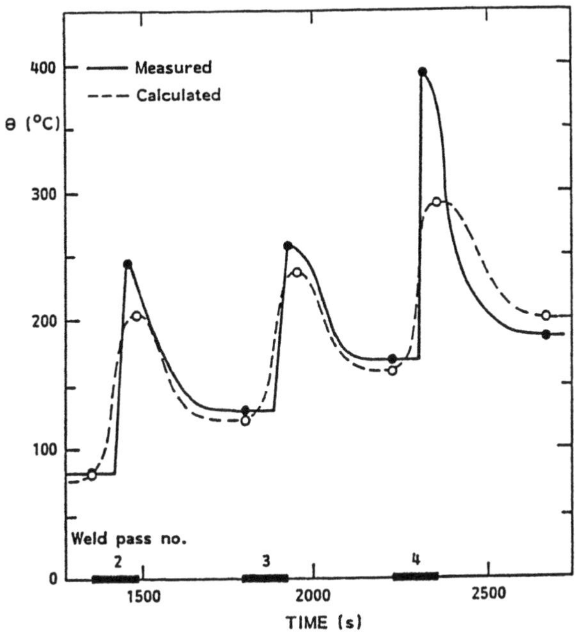

Fig. 6. Comparison of predicted and measured temperatures near the thermocouple A (see Fig. 5) in a multipass girth weld on a pipe.

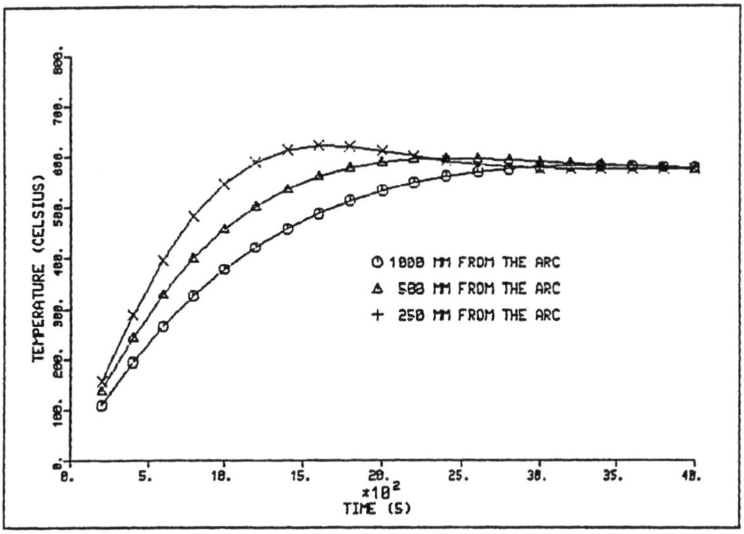

Fig.7. Temperature distribution during overlay welding of a pipe. Each 200 s in time means one rotation and at the time 1000 s, the arc has just finished the pass which locates at the section where the temperatures were evaluated.

The third case study was the analysis of an overlay welding of a pipe where a total of 20 passes were welded resulting to a 100 mm wide overlay. Figure 7 shows temperature time histories in three locations behind the arc at the section located 20 mm from the left end of the overlay weld (Fig. 2). During the time interval from zero to 1000 seconds, the arc becomes closer to the 20 mm section. Therefore, an increase of the temperature is natural. After passing the 20 mm section one could expect that the temperature should decrease in time. However, this is the case only for distances smaller than 250 mm to the arc. In other distances the temperature increases towards an overall temperature level of 550 ^{0}C.

Conclusions

It appears that HAZ temperatures can be calculated quite accurately in single pass welding conditions. In more complicated situations, such as in the case of multipass welding, the results are not as accurate. However, it is expected that the temperature distribution obtained is accurate enough in this case to obtain a reasonable basis for the future analysis of the stress field from the temperature variation. The solution of temperature fields in overlay welding showed that the overall temperature increase in the pipe may be significant if the welding speed is large and if no waiting times are allowed between passes. However, no experimental evidence was available for this analysis. Therefore, the results of this case study are considered to be tentative only.

References

1. Karlsson L.: Thermal Stresses in Welding. In: Hetnarski, R. (ed.) Thermal Stresses I. Elsevier Science Publishers 1986. pp. 300-389.

2. Masubuchi, K.: Analysis of welded structures. Pergamon Press 1980. 642 p.

3. Rosenthal, D.: Mathematical theory of heat distribution during welding and cutting. Welding Journal 20 (1941). pp. 220-234.

4. Rybicki, E., Schmueser, D., Groom, J., Lemcoe, M., Mishler, H. and Rodabaugh, E.: Residual stresses at girth-butt welds in pipes and pressure vessels. Battelle Columbus Laboratories. Progress report BMI-NUREG-1958 (1976). 45 p.

5. Solomon, H., and Levy, S.: HAZ temperatures and cooling rates as determined by a simple computer program. General Electric Report No. 79CRD240 (1980). 15 p.

6. Vaindyanathan, S., Weiss H. and Finnie, I.: A Futher study of residual stresses in circuferential welds. Transactions of the ASME, Paper No. 73-Mat-Z (1973). 5 p.

Design of Weldments in Structures Subjected to Creep

L. A. Samuelson[1], P. Segle[1] and S. T. Tu[2]

[1]The Swedish Plant Inspectorate, Box 49306
S-100 28 Stockholm, Sweden
[2] Department of Materials Technology
The Royal Institute of Technology
S-100 44 Stockholm, Sweden

SUMMARY

Most of the current pressure vessel and piping design codes contain rules for design in the creep range. In practical applications, creep damage is often noted at a life much shorter than the design life. Weldments are particularly susceptible to creep damage and it is thus of importance to investigate the reasons why. Only a few codes recognize that the creep strength of the weld material may be a limiting factor in design and it is not clear what parameters would be important for the designer to consider.
The present paper discusses some aspects of creep life assessment of piping systems with circumferential welds. Analytical solutions and FEM results indicate that large strain concentrations may be expected if the weld material creep rate is significantly higher than that of the parent material. The effect is augmented by external forces and thus by elastic follow up.

1 INTRODUCTION

The current European codes for welded pressure vessels and piping in power plants and other high temperature applications require design for a safe life of the order of 100,000 hours. In most cases the creep design is based entirely on the average rupture stress of the parent material divided by a safety factor of 1.5 [1]. A critical review of several European codes given in [2] revealed that although the analyses are based on the same material data and very similar design criteria, the design life varied by as much as a factor of seven even for a plain pipe under internal pressure. Work in the US has led to a creep design procedure defined in ASME, Code Case N-47-26 [3]. Here a reduction factor is introduced defined as an appropriate ratio of the weld metal creep rupture strength to the base metal creep rupture strength. Recent work shows that it may be reasonable to base the design on such factors [4]. However, there are still parameters to be identified, which may have an influence on the creep life.
Despite numerous studies on creep behaviour of engineering structures little attention has been given to the effect of the weldment. Walters and Cockcroft [5] used the finite element method in analyzing the creep of two material weldments. Later, Browne et al. [6] and Coleman et al [7] incorporated a three-materials model and used a parametric approach to cover a wide range of weld metal/parent metal creep rate ratios. Evans [8] used the finite element method to study a transition joint under creep conditions. By using a finite difference method, see Ball [9] and Samuelson [10], based on thin shell theory,

L. Karlsson, L.-E. Lindgren, M. Jonsson (Eds.)
Mechanical Effects of Welding
IUTAM Symposium Luleå/Sweden 1991

Ivarsson and Sandström [11,12] studied the creep stress redistribution and rupture of welded AISI 316 steel tubes. Ivarsson [13] also studied the creep deformation and rupture of welded 12% Chromium steel tubes by computer simulation.
The present work is intended to examine the applicability and the reliability of the current code methods in creep design for weld components and possible improvements. Both analytical and finite element methods are used in the study of creep of butt welded tubes.

2 SEMI-ANALYTICAL APPROACH

In [14] steady creep of a thin tube under bending and internal pressure is studied. By using this approach and adding; 1) an axial load and 2) a weld i.e. two different material types, a fast running parametric model is given for qualitative study of the influence of a weld in a tube under different load combinations.
It is assumed that the stiffness of the weld is negligible in comparison to that of the pipe in the sense that the circumferential creep strain in the weld is equal to that of the pipe remote from the weld. This approximation is probably reasonable in the case of 'soft' welds whereas 'hard' welds will cause development of a slight neck in the pipe [15].
The compatibility condition may therefore be taken as: $\varepsilon_{\varphi w} = \varepsilon_{\varphi p}$
Equilibrium in the axial direction requires: $\sigma_{xw} = \sigma_{xp} = \sigma_x$
Here, index φ and x represent the hoop and the axial direction in a polar coordinate system, respectively. Index w and p correspond to the weld and the parent material.
The hoop stress remote from the weld is assumed to be $\sigma_{\varphi p}$. The hoop stress in the weld, $\sigma_{\varphi w}$, is not known since the circumferential strain in the weld is forced to equal that of the pipe.
By using the constitutive equation according to Odqvist [16], 2-d strain, with k_p and n_p as the coefficient and exponent in Norton's law for the parent material and k_w and n_w for the weld material, respectively, we get;

$$\dot{\varepsilon}_{\varphi w} = k_w(\sigma_{\varphi w}^2 + \sigma_x^2 - \sigma_{\varphi w}\sigma_x)^{(n_w-1)/2}(\sigma_{\varphi w} - 0.5\sigma_x) =$$
$$= k_p(\sigma_{\varphi p}^2 + \sigma_x^2 - \sigma_{\varphi p}\sigma_x)^{(n_p-1)/2}(\sigma_{\varphi p} - 0.5\sigma_x) = \dot{\varepsilon}_{\varphi p}$$

From this equation $\sigma_{\varphi w}$ can be obtained since all other variables are known.
This equation has to be evaluated for different angles, φ, since the stresses vary around the circumference due to the bending moment.
The analysis thus gives stresses and strain rates both in the parent material and in the weld as a function of the circumferential angle, φ.

For a butt welded tube subjected to internal pressure and axial tension, the results indicate that the axial strain is concentrated to the weld if the weld is 'soft'. The higher the axial tension and the softer the weld, the higher is the concentration, see Fig. 1.
If a bending moment is added, the results indicate that the axial strain in the weld is more sensitive to axial stresses than to bending stresses. This is due to redistribution of axial stresses in the latter case. As in the former case, the axial strain concentration in the weld increases with the 'softness' of the weld, see Fig.2.

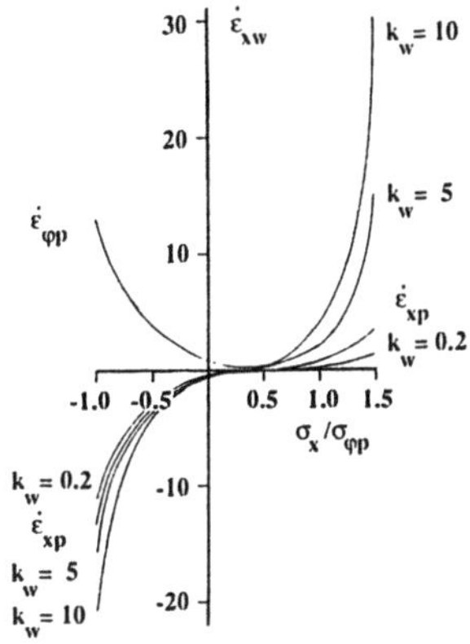

Fig.1 Axial strain rate in the weld as a function of the ratio between axial stress and hoop stress in the parent material. ($\sigma_{\varphi p}$=1MPa, k_p=1, n_p= n_w= 5, M_b=0)

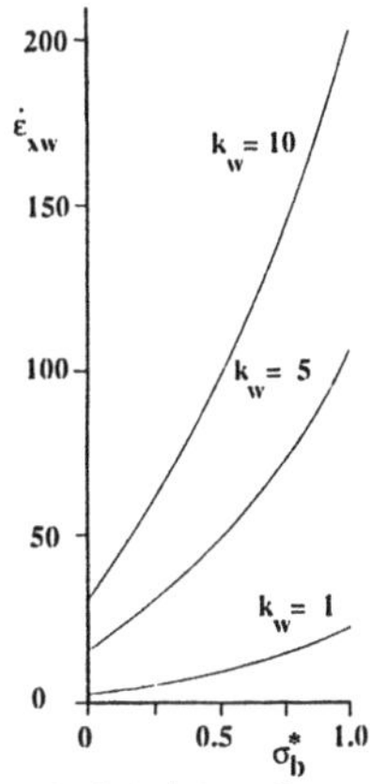

Fig.2 Axial strain rate in the weld as a function maximum elastic stresses due to a bending moment, σ_b^*. Here σ_F is the axial stress due to external axial stress. ($\sigma_{\varphi p}$=1MPa, σ_F=1MPa, k_p=1, n_p= n_w= 5)

In the semi-analytical approach, equilibrium in the axial direction requires that the axial stress, at every circumferential segment, is the same in the parent material as in the weld. The FE simulation of a tube subjected to a bending moment, presented in the next section, shows that this is not the case. The explanation for this is that redistribution in the circumferential direction of the axial stress due to the bending moment, differs between the parent material and the weld. This difference occurs when the exponents in Norton's law differ between the weld and the parent material. Further work will be done to include this effect.

3 FINITE ELEMENT SIMULATION OF CREEP BEHAVIOUR OF WELDED TUBE

Creep tests on welded AISI 316 steel tube were carried out by Ivarsson and Sandström [11]. A tube of 17.8 mm outer diameter and 0.9 mm wall thickness was chosen and was cold stretched to 20% reduction in area. The specimens were made by joining two 70 mm long tubes by TIG welding without filler metal.
The tests were performed at a temperature of 600 °C. Creep strain rate data of each part of the weld joint is presented in Fig.3, where the secondary creep rate is given as a function of stress. It is evident that the weld is 'soft'.

In the FE simulations presented in this paper, the tube dimensions, the parent and the weld material properties and the internal pressures are the same as in [11].

As a first step, an axisymmetric model of a butt welded tube subjected to internal pressure is studied.

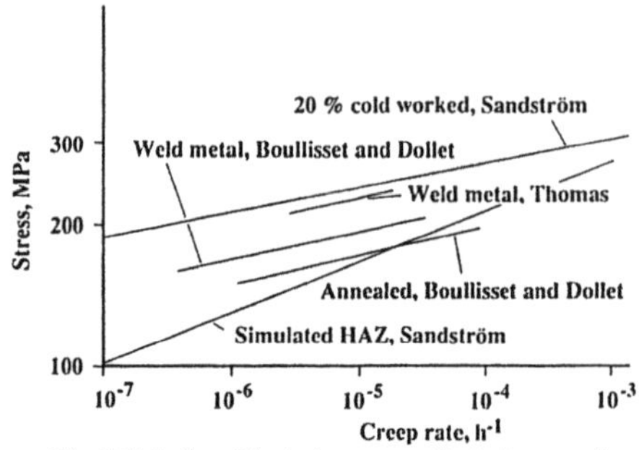

Fig.3 Relationship between applied stress and creep rate for simulated HAZ and 20% cold-worked 316 steel.

The hoop stress distribution at the middle surface of the tube is given in Fig.4 and the hoop strain distribution at the middle surface is plotted in Fig.5, in which a comparison with the finite difference results [12] is made.

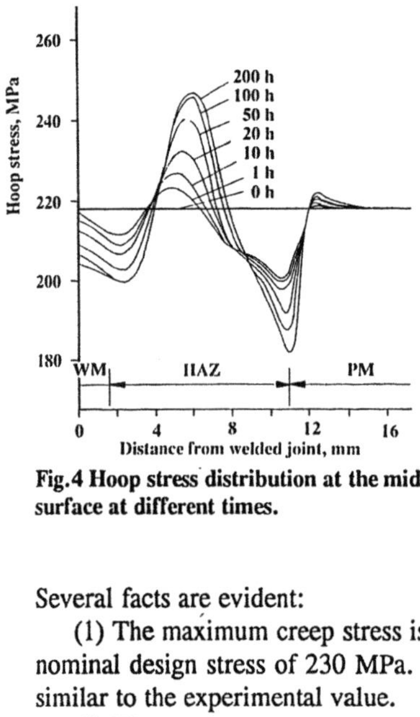

Fig.4 Hoop stress distribution at the middle surface at different times.

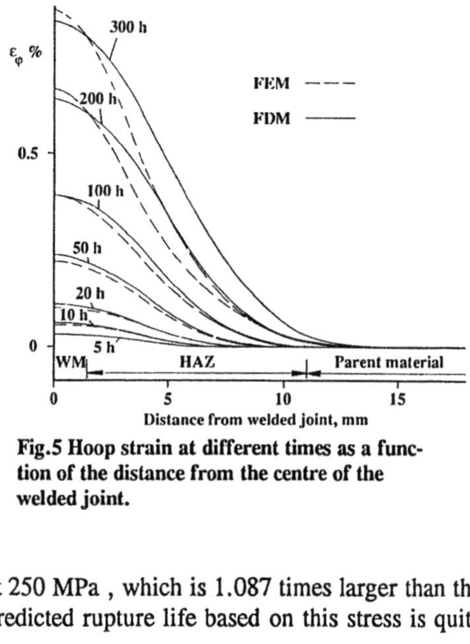

Fig.5 Hoop strain at different times as a function of the distance from the centre of the welded joint.

Several facts are evident:

(1) The maximum creep stress is about 250 MPa , which is 1.087 times larger than the nominal design stress of 230 MPa. The predicted rupture life based on this stress is quite similar to the experimental value.

(2) The maximum stress concentration is first built up at the inner surface at a section 5 mm away from the weld centre and then shifts to a section 6 mm from the centre at the middle of the tube wall rather than at the outer surface. This explains the experimental results where cracks in all cases started at the inside of the tube.

(3) The maximum hoop strain occurs at the centre of the weld rather than at the position with maximum stress. This indicates that it might not be reasonable to use the strain as a criterion in this case. However, the creep ductility of the HAZ-material is not known.

We have also examined some other parameters such as maximum principal stress and equivalent stress in the creep life prediction, but it appears that the maximum hoop stress provides better predictions in this case.

In a second step, different axial stresses are added to the model above. Fig.6 shows the axial strain in the centre of the weld as a function of the added axial stress. It is seen that the axial strain is very sensitive to the increase of the axial stress.
Fig.7 shows the strain evolution with time. In load cases with higher axial forces, the creep strains in the weld exceed those observed in the parent tube by a factor of as much as of 100.

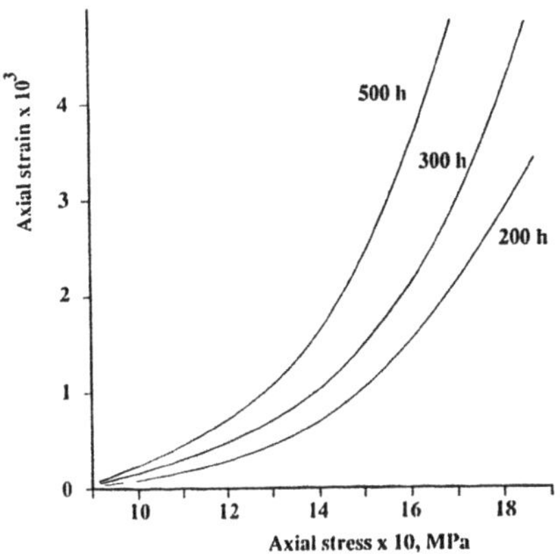

Fig.6 Axial strain in the centre of the weld as a function of the added axial stress.

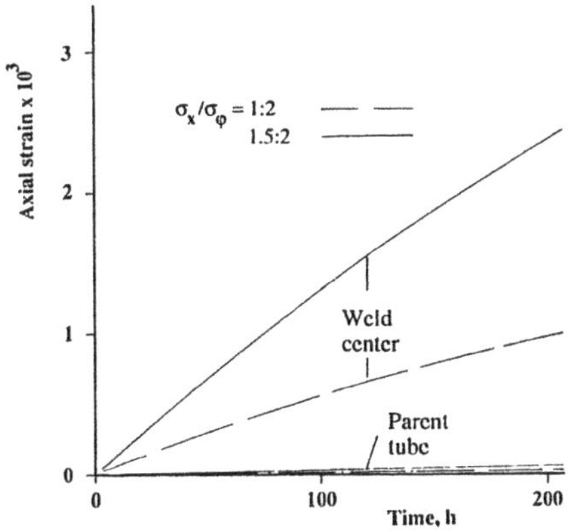

Fig.7 Axial strain as a function of axial stress and time.

According to replica tests on a header weldment carried out at the Swedish Institute for Metal Research [17], microcracks are most frequently found parallel to the fusion line. This may indicate that the high axial strain concentration in the weld found in the FE simulation above, is the governing parameter in cases where large axial forces exist.

In a third step, a shell element model is subjected to a pure bending moment, see Fig 8, in order to verify the results of the semi-analytical solution. Analyses were carried out both for the case of no weld, a) and for a weld at the symmetry line, b).

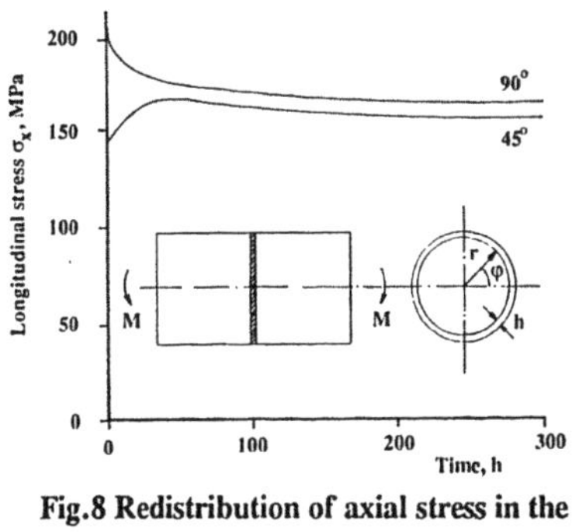

Fig.8 Redistribution of axial stress in the weld centre.

The results of case a) agree well with the analytical results. Some results for case b) are given in Figs. 8, 9 and 10, which show that the stress distribution around the circumference at the weld differs from that obtained in case a).

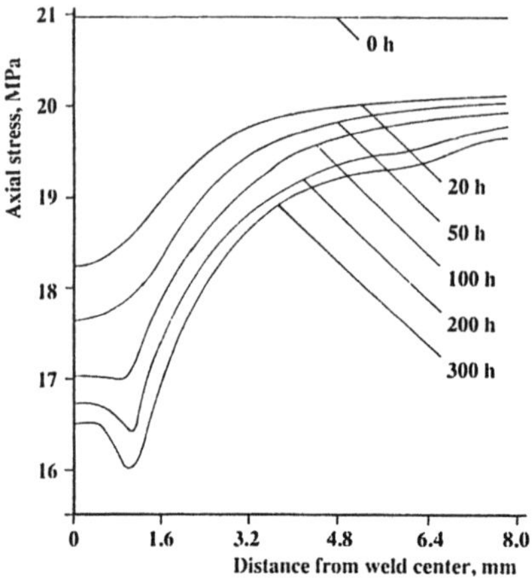

Fig.9 Axial stress along the tube as a function of time, $\varphi = 90$.

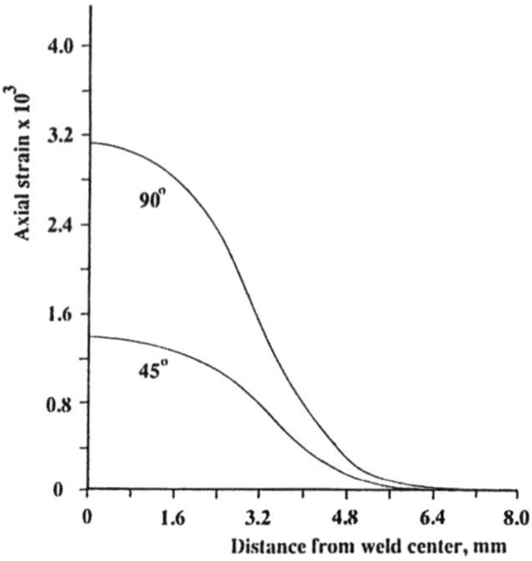

Fig.10 Axial strain distribution along the tube after 200 hours, $\varphi = 45$ and $\varphi = 90$.

4 CURRENT CODE DESIGN PROCEDURES

As was indicated above, most of the current design codes are based on substitution of creep rupture strength of the material for yield strength. Furthermore, the possible reduction of the strength of a component due to the fact that a weld may have creep properties different from those of the parent material is often disregarded. Some codes, though, prescribe a reduction factor on stresses in the structure related to welding of the order of 0.7 to 0.9. There appears to be no specific requirement on the creep properties of the weld material in relation to those of the parent material and it is thus reasonable to believe that 'soft' or 'hard' welds may appear in practice.

The ASME code [3] gives specific reduction factors on stress related to the minimum creep rupture stress of the weld metal. In addition, ASME requires that creep strains are kept within certain limits, such as $\varepsilon_c \leqslant 1\%$ for primary stress fields.

The examples given in section 3 indicate that the following procedure may be used for design. In this case the stress has been raised by a factor 250/230=1.087. If the design is based on the primary stress of the parent tubes, the allowable stress should be reduced by a factor of 0.92.

Internal pressure (MPa)	24.5	22.6	19.6	15.0
Design stress (MPa)	230	212	184	140
Calculated reduction factors	0.920	0.891	0.856	0.791
ASME principle	0.912	0.880	0.830	0.760

Table 1 Creep strength reduction factors

On the other hand, by using the ASME principle, the reduction factor can also be evaluated. The creep strength reduction factors for the tubes subjected to different pressures are listed in Table 1. It may be noted that the lower the stress level(longer life time) the more conservative the code rule.

Since the analyses carried out above were based on very specific creep data, it is not possible to draw any general conclusions. The results indicate, however, that various weld design parameters may have a considerable influence on the creep resistance of a pressure vessel or tube and it is important to evaluate this influence in a comprehensive parametric study.

5 DESIGN CONSIDERATIONS

The analyses of welded tubes indicate that a number of parameters are important for the assessment of the creep life.
If the creep properties, i.e. creep strain rate $\dot{\varepsilon}$, creep exponent n and creep ductility ε_f, of the weld differ from those of the parent material, the safe life of the tube may be considerably reduced. In order to make realistic estimates of the creep life it is necessary to consider the weld material properties in a realistic analysis based on a rigourous theory for nonlinear creep. The effect of external forces must be included in the analysis since they may cause severe strain concentrations in the weldment.
A single design criterion does not seem to exist for creep problems and it may become necessary to check different conditions in order to verify that a specific safe life can be established. Such criteria may be based on stresses or strains where use of effective values or components may apply. It is also possible that a damage parameter may be defined which would provide a more general basis for design, see Sandström [18].

6 DISCUSSION

The present paper discusses some of the related parameters which may influence the life of the weld. The work forms part of a project intended to provide improved methods of analysis and design criteria for creep.
The finite element analyses of butt welded tubes carried out indicate that the models used are sufficiently accurate when comparing results with other numerical results and tests. The method is, however, not particularly suited for design unless the weld and HAZ material properties are known. Parametric studies may, on the other hand, give an insight in the behaviour of the structure under various loading conditions.
Simplified models for analysis of weldments need to be developed. Those studied above appear to provide some useful information about the creep behaviour. A possible result of a parametric study might be a requirement on the material properties of the weld in relation to the parent material. Also, it will be necessary to discuss alternative suggestions for a design criterion. Some kind of a damage parameter, compatible with for instance the cumulative damage in fatigue, might possibly be found.
Finally, application of fracture mechanics methods in cases where cracks are found may give information on the creep crack growth rate and an assessment of the remaining life.

REFERENCES

1. Swedish Pressure Vessel Commission, Swedish pressure vessel code, 1987.
2. CEOC, Assessment of the life/remanent life of the components operating in the creep range, R73/CEOC/CP89 dEf., Paris 1991.
3. ASME Code Case N47-26, Class 1 components in Elevated Temperature Service, ASME, 1987
4. Corum, J.M., Evaluation of weldment creep fatigue strength reduction factors for elevated-temperature design, ASME Trans., J.Pres.Ves.Tech. 112 (1990) 333-339.
5. Walters, D.J., Cockcroft, R.D.M., A stress analysis and failure criteria for high temperature butt welds, Int. Inst. of Welding Colloquium on Creep Behaviour of Welds in Boilers, Pressure Vessels and Piping. Toronto, 1972.
6. Browne,R.J. et al., Creep Failure Analysis of Butt Welded Tubes, Creep and Fracture of Engineering Materials and Structures, eds. B. Wilshire, D.R.J.Owen, Proceedings of the international Conference, Swansea, 24-27 Mar. 1981.
7. Coleman, M.C., Parker, J.D., Walters, D.J., The behaviour of ferritic weldment in thick section 1/2Cr1/2Mo1/4V pipe at elevated temperature, Int. J. Pres. Ves. & Piping, 18 (1985) 277-310.
8. Evans, R.W., Design of welded components for creep, The 2nd Int. Materials Engineering Conference, London, 9-13 May, 1988.
9. Ball, R. E., A geometrically nonlinear analysis of arbitrarily loaded shells of revolution, NASA, CR-909, Jan 1968.
10. Samuelson, L., A, Creep buckling of shells of revolution. User's manual for program DSOR 06. The aeronautical research institute of Sweden, TN HU-1548 part 4, 1975.
11. Ivarsson, B.G., Sandström, R., Creep deformation and rupture of butt-welded tubes of cold-worked AISI 316 steel, Metals Technology, pp.440-448, November, 1980.
12. Ivarsson, B.G., Sandström, R., Creep of Butt-welded Tubes, Creep and Fracture of Engineering Materials and Structures, eds. B.Wilshire, D.R.J.Owen, Proceedings of the International Conference, Swansea, 24-27 Mar. 1981.
13. Ivarsson, B., Creep Deformation and Fracture in Butt-Welded 12% Chromium Steel Tubes, Report IM-1748, Swedish Institute for Metal Research, 1983.
14. Boyle,J.T., Spence,J., Stress Analysis for Creep, Butterworths, 1983.
15. Odqvist, F.K.G., Hult, J., Kriechfestigkeit metallischer Werkstoffe, pp 1 - 303. Springer, Berlin, 1962.
16. Odqvist, F.K.G., Mathematical Theory of Creep and Creep Rupture, Oxford University Press, 1966.
17. Wu, R., Storesund, J., Creep damage in a service-exposed header of 2.25 Cr1Mo steel, Report IM-2391, Swedish Institute for Metal Research, 1988.
18. Sandström, R., Prediction of creep rupture using strain to failure criteria, Application to butt-welded tubes, Department of Materials Technology, Royal Institute of Technology, Sweden, 1983.

Creep Behaviour of 304 Stainless Steel Welded Joints Composed of Two Different 308 Weld Metals

J. KINUGAWA, Y. MONMA, H. HONGO, M. YAMAZAKI and T. WATANABE

National Research Institute for Metals
2-3-12, Nakameguro, Meguroku, Tokyo 153 JAPAN

Summary
A weld metal zone composed of two or more weld metals with different creep properties is proposed as an approach to produce welded joints with well balanced performance in terms of creep lifetime and ductility.
Type 304 steel was welded by depositing a 308L and/or a 308 wire. The resulting weld metal zones are composed of a 'soft' weld metal with high ductility and 0-to-100 vol. % of a 'hard' one with an extended creep lifetime but low ductility.
The creep tests suggest that the creep behavior of a welded joint may be controlled to obtain an optimum balance between its time-to-rupture and elongation.

Introduction

Substantial efforts have been made to control residual elements in weld deposits in order to improve the creep lifetime and ductility of austenitic stainless steel welds [1]. However, for the present, there has been little success in finding a composition which could satisfy both the lifetime and the ductility.

If such incompatibility is unavoidable, we have to choose some balance between the lifetime and ductility from the viewpoint of optimizing the service performance for each expected loading history. For instance, in designing plants to be operated with frequent repetitions of startup-to-shutdown, we had better give the first priority to the guarantee of ductility even at a certain expense of lifetime, because insufficient ductility would give a potential threat of creep-fatigue failure to components subjected to deformation-controlled loading [2].

In regard to welded joints, much depends on the selection of welding materials by which the creep properties of produced weld metals can be fit to our target scope. Unfortunately, the ideal weld metal may not be available. In such a case, another choice may be multi-layer welding in which two or more obtainable welding materials are deposited in proper sequence. That is, the weld metal zone is composed of weld metals with different creep properties. We do not expect the properties of the weld metal of each pass in itself to be ideal. Rather, the whole of the weld metal zone as a composite of different weld metals is optimized.

In the present study, creep tests were conducted for butt-welded joints of 304 stainless steel composed of 308L and/or 308 weld metal. Based on the test results, a discussion is presented about the new possibility of optimizing the creep performance of a weld by composing its weld metal zone with different weld metals.

L. Karlsson, L.-E. Lindgren, M. Jonsson (Eds.)
Mechanical Effects of Welding
IUTAM Symposium Luleå/Sweden 1991

Experimental Procedure

Five welded joints WJ0, WJ35, WJ67, WJ77 and WJ100 of 60 mm thick 304 steel (.04 % C) were prepared by narrow-gap SAW by depositing a 308L wire (.02 % C) and/or a 308 wire (.05 % C). They correspond to the weld metal zone of nothing but the weld metal WMA (308L), WMA plus 35, 67, 77 vol. % of weld metal WMB (308) and all WMB, respectively. **Figure 1** shows the resulting disposition of WMA and/or WMB.

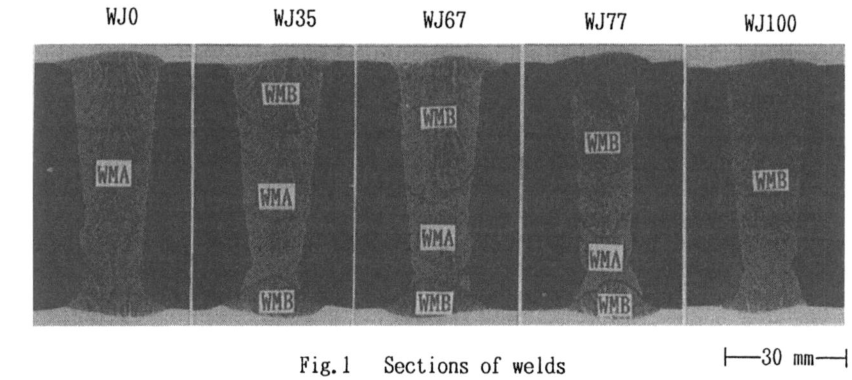

Fig.1 Sections of welds

Specimens for testing 'all-weld-metals' and 'joints' were taken out of the welded joints as shown in **fig. 2.** All-weld-metal specimens of WMA and/or WMB, 6 mm in dia. and 30 mm in gage length, were cut out along the weld center lines at the 1/4-, mid- and 3/4-thickness.

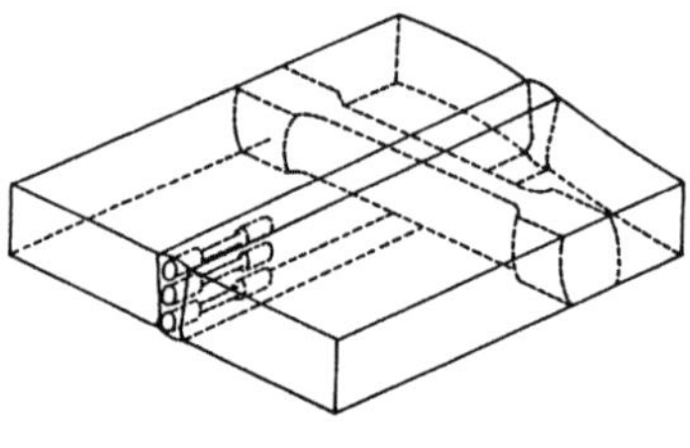

Fig.2 Location of specimens and welded joints

After shaving the weld reinforcement, the joint specimens were prepared with a rectangular cross section of the full thickness × 10 mm and gage length of 160 mm, some 85 % of which traversed the base metal. The loading direction was perpendicular to the weld center line. Over the finished surface of their paralleled parts, heat-resisting grids with a line spacing of about 0.1 mm were photo-etched to measure the local strain distribution during creep tests by interrupting them at specified times for moiré photography. For testing base metal, specimens with 10 mm in dia. and 50 mm in gage length were cut out of the mid-thickness of the plate so that the loading direction agreed with that of rolling.

All these specimens were subjected to creep tests at 550 ℃ under an applied stress of 235MPa. Joint specimens were unloaded every 400 h to check cracking and to measure local strains.

Joint specimens which rupture through the base metal provide little information on the lifetimes and ductility of their weld metal zones. For this reason, especially in testing joints, tests were carried out at 314 MPa in addition to 235 MPA based on data [3], which suggested a greater probability of failure at weld metal zones under higher stresses.

Creep Properties of All-Weld Metals

Figure 3 compares the creep curves of all-weld-metal specimens with 100 % WMA, those with 100 % WMB and a base metal specimen. Great discrepancies in the curves for WMB indicate that this metal is very site-variable. Nevertheless, we surely have a 'softer' weld metal WMA creeping readily and a slightly 'harder' WMB resisting creep more than the base metal. Thus, the resulting joints from WJ0 with 100 % WMA to WJ100 with 100 % WMB are 'undermatched' and 'overmatched' in terms of creep performances.

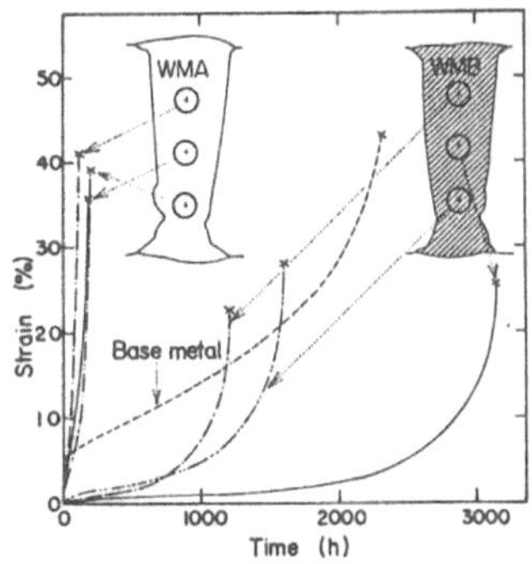

Fig.3 Creep curves of WMA WMB and base metal (235 MPa)

Some of the all-weld-metal specimens from the joints WJ35, 67 and 77 are not composed of WMA or WMB monolithically but of both by sampling them from the vicinities of WMA/WMB boundaries. In **fig. 4,** the time-to-rupture and elongation of all-weld-metal specimens are shown as a

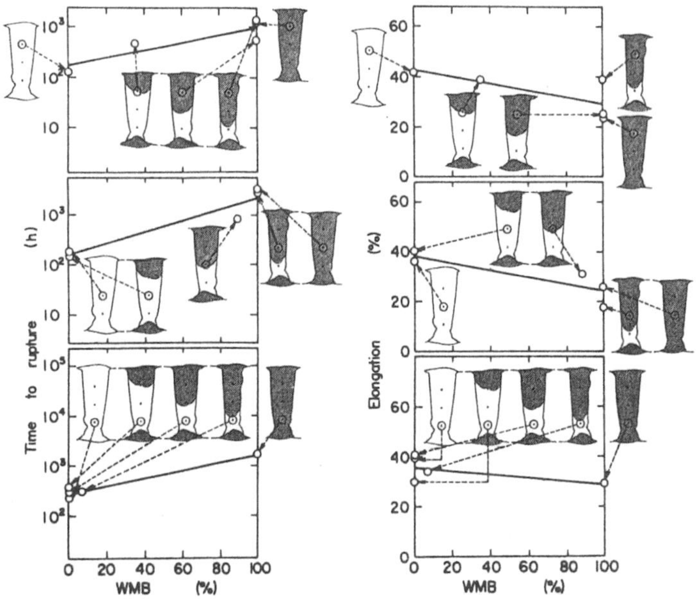

Fig.4 Time-to-rupture(left) and elongation(right) of all-weld-metal vs. vol.% of WMB

function of the vol. % of WMB. For these composites of the two metals, the plotted data lie between those for 100 % WMA and 100 % WMB. That is, their creep properties are never fixed on the level of the shorter time-to-rupture of all WMA's or the lower elongation of WMB's. The 'law of mixtures' for uniformly strained composites explains this as follows:

Any strain distributed evenly to WMA and WMB results in a lowered stress in WMA, since its partner, WMB, being more resistible to deformation, always bears a larger share of the applied load. Thus, even though failure is initiated in WMA, its lowered stress produces longer time-to-rupture for such composites than that for 100 % WMA under the same conditions of loading. The prolonged time-to-rupture is not so long as that of 100 % WMB. As a result, at the time of rupture of the composite, the metallurgical deterioration of WMB is limited. This results in more elongation of the composites than that of 100 % WMB.

Creep Behavior of Welded Joints

Figure 5 shows the creep curves of the joint specimens. From WJ0 to WJ100, increasing the vol. % of the hard metal WMB with a longer lifetime certainly lengthens their time-to-rupture, while it little affects their creep rates probably due to the fact that the base metal occupies the major portion of each specimen. Most of the gain in elongation also comes from the deformations of the base metal.

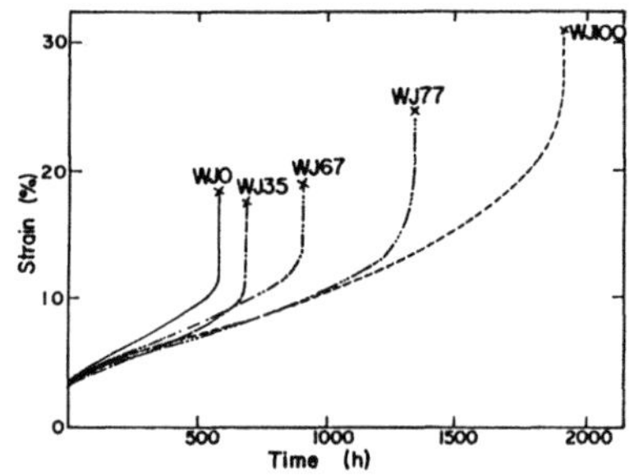

Fig.5 Creep curves of joints (235MPa)

The joint specimens ruptured under an applied stress of 235 MPa and their ruptured sections are shown in **figs. 6** and **7**. The joints WJ0-to-WJ77 containing the soft metal WMA failed at their weld metal zones, while only the most overmatched one WJ100 composed of 100 % the hard metal WMB ruptured through the base metal.

Fig.6 Ruptured joints (235 MPa)

For joints WJ0-to-WJ77, the fracture surface just around the narrowest part between the two bond lines appeared grayish. This proves that this region of the fracture surface has been oxidized heavily during long-term exposure to the atmosphere. Consequently, any failure in the weld metal zones is very likely to have been initiated around their narrow parts, where the soft metal WMA is disposed (See fig.1.). Liquid penetrant in-

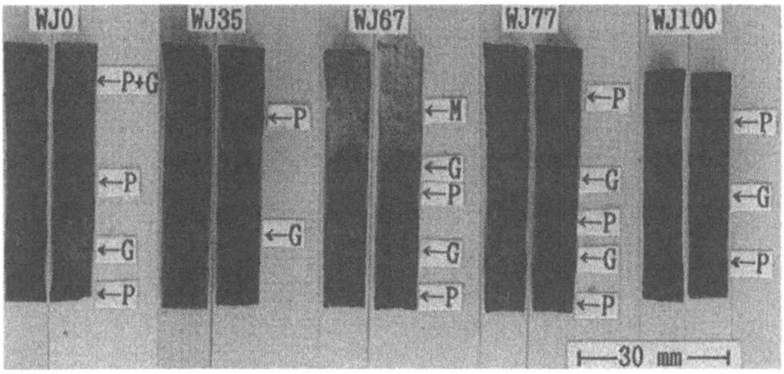

Fig.7 Ruptured sections of joints (G: Grayish, P: Purplish and M: Metallic)

spection of the specimens, which were unloaded every 400 h during creep tests, showed that failure had been initiated at a loading time exceeding 400 h for the joints WJ0-WJ35, more than 800 h for WJ67 and more than 1200 h for WJ77. Meanwhile, for the all-weld-metal specimens with the same metal of 100 % WMA, the time-to-rupture is between 120 h and 210 h under the same applied stress (fig. 3). These facts demonstrate that the metal WMA, while being contained in a weld, has a delayed time-to-failure-initiation compared to the same material cut out of the weld and tested in an uniaxial state of stress. This result could be explained by regarding the 'equivalent stress' as the dominant stress criterion controlling the creep damage of austenitic steels, such as cracking or void formation, and also, by considering the said stress to be lowered with some increase of its hydrostatic component incidental to multi-axial stress systems in the weld. We can certainly anticipate additional delays from increasing the vol. % of the harder metal WMB. The effect of adding WMB is explained by assuming that this metal, being more resistible to deformation, bears more load and lowers the stress to be applied to WMA.

To return to fig. 6, for the joint WJ100, a few macroscopic cracks, which did not lead to the complete rupture of the specimen, are seen on the face of the base metal neighboring its weld toe. These cracks resulted likely from some stress concentration around the toe, which was magnified with the discrepancy of deformations between the 'rigid' weld metal zone of 100 % WMB and the 'soft' base metal adjoining it. At all events, such cracking confirms misgivings that excessive over-matching may accelerate the creep damage of the base metal.

Local Strains in Weld Metal Zones

Figure 8 shows an example of the moiré fringe maps observed on the surface of the joint specimens. In **fig. 9**, a comparison of strain contours after 400 h loading, obtained from the fringe maps, is made for the five joints loaded by a stress of 235 MPa. The drastic change of contours from WJ0 to WJ100 suggests the possibility of controlling the creep deformation of a weld by appropriate disposition of different weld metals.

Fig.8 Typical 'moiré fringe map' (WJ0 after 400 h loading with 235 MPa)

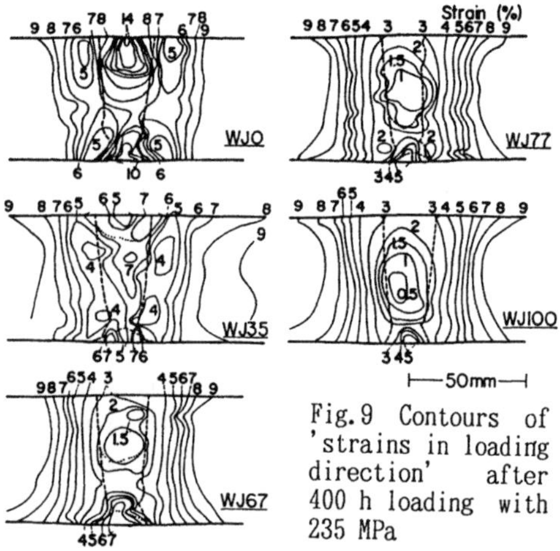

Fig.9 Contours of 'strains in loading direction' after 400 h loading with 235 MPa

As regards weld metal zones, creep strains tend to accumulate on the top and/or bottom side of the specimen. This proves that such outer sides adjacent to their free faces are in more restraint-free states for any deformation. It may be worth noting that strains also seem to concentrate around the narrowest part between the bond lines particularly in the most undermatched joint WJ0. The reason for this could be explained by the following model:

Let us assume the 'weld metal zone of 100 % WMA' to be 'assembled bar specimens', as are drawn in **fig. 10** with fine lines. For each specimen, we take its gage length to be equal to the distance between the two bond lines AOB and A'O'B'. In such an analogy for the undermatched weld metal zone, the 'HAZ-to-base metals' on both sides of it are also analogous to a pair of the 'grips of each specimen' + 'chuck-to-cross-heads of the testing machine', since they are estimated to be more rigid than the soft metal WMA. At a fixed displacement between the cross-heads, the shortest gage length of the specimen between O-O' ought to give the largest strain.

We already referred to the signs showing failure initiation at the weld metal WMA disposed around 0-0'. This is believed to be due to the additive effects of the larger localized strains just discussed and the soft weld metal WMA.

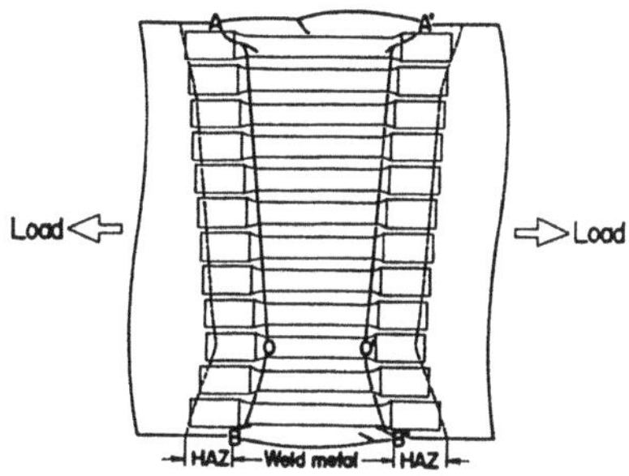

Fig.10 Analogy of weld metal zone to assembled bar specimens

Ductility of Weld Metal Zones

For a weld metal zone in a joint, we have no appropriate measures to evaluate its overall ductility, such as the 'elongation' for ordinary tensile specimens. For this reason, after rupturing of each joint specimen, an additional observation was made of its strain distribution so that we could represent the elongation of its weld metal zone by taking the average of the 'strains at the time of rupture' all over this zone.

The time-to-rupture and the averaged elongation of weld metal zones, after arranging them with the vol. % of the hard metal WMB are shown in fig. 11. Under an applied stress of 235 MPa, their elongation from 100 % WMA to 100 % WMB decreases from 26 to about 20 %, whereas that of all-weld-metal specimens is 36-to-42 % for the soft metal WMA and 23-to-28 % for the hard one WMB (fig. 3). The change of elongation with vol. %, taking the difference between the two all-weld-metals into account, is not as much as would be expected. The reason for this is due to the considerably lessened elongation for the weld metal zone of 100 % WMA (26 %) in comparison with that for the very metal of WMA (36-to-42 %). The discrepancy may be explained as follows:

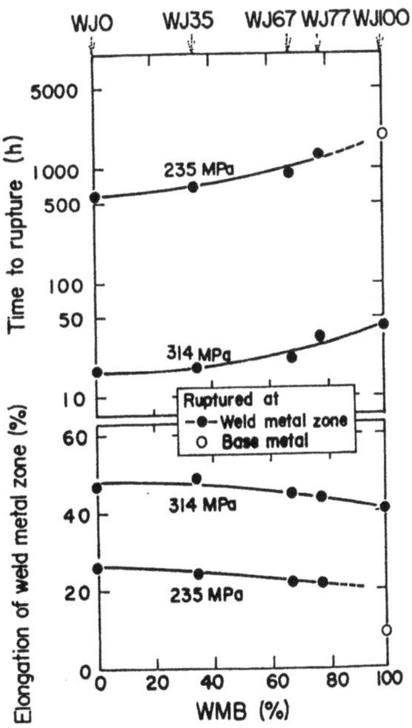

Fig.11 Time-to-rupture and averaged elongation vs. vol. % of WMB in weld metal zones

The time-to-rupture of the WMA zone, 580 h, is longer than that of its constituent metal WMA, 120-to-210 h (fig. 3.). The increased lifetime produces additional metallurgical deterioration and thus results in a loss of ductility.

Nevertheless, increasing WMB lengthens the time-to-rupture of the weld metal zone and reduces its elongation. Thus, we could choose some balance between the time-to-rupture and elongation over their ranges from 100 % WMA to 100 % WMB.

Conclusion

A comparison was made of the creep behavior of 304 stainless steel welded joints composed of a 308L and/or a 308 weld metal with different creep properties.
The results are summarized as follows:
For the so-called 'all-weld-metals' composed of 308L and 308, their creep properties conform to a 'law of mixtures'. This is explained by assuming a uniform strain in both metals.
For the most 'undermatched' weld metal zone filled with the 'soft' weld metal 308L creeping readily, creep strains are localized not only on its front and back sides but around the narrowest part between the two bond lines. Its failure also initiated at this location. The time-to-rupture of this zone is longer than that of the all-weld-metal cut out of it, while the elongation averaged over the zone is less.
'Overmatching' by adding the 'hard' metal 308 produces a considerable change in strain distribution, which results in a longer time-to-rupture with less elongated weld metal zones, and finally, to failure initiation at the base metal near the weld toe.
From these results, we conclude that we could control the creep behavior of welded joints, and also, choose the best balance between their creep lifetimes and ductility to fit the predictable in-service conditions by the rational layout of weld metals with different creep properties.
To realize this possibility, a purely empirical approach, such as the trial-and-error of creep testing for a number of composite welds, does not seem realistic because of the long times involved. Computer-aided approaches are believed to be more or less essential for such a construction. In such an approach, the key to success also may be 'how to simulate the creep rupture behavior of welded joints with high accuracy'. We are now making FEM programs to incorporate the 'damage mechanics'.
The comparison of computations with measurements will be reported before long.
Acknowledgments The authors would like to express their sincere gratitude to Prof. S. Yokoi, at Tokai Univ., Drs. H. Nakamura, Y. Muramatsu and S. Kuroda, at National Research Institute for Metals, for suggestive discussions. They are deeply indebted to Nippon Steel Corp. for preparing the welded joints subjected to the study. Warm-hearted encouragement from Prof. Y. Ueda, at Osaka Univ., and detailed checking of the manuscript by Prof. I. Finnie, at Univ. of California, Berkeley, are gratefully acknowledged.

References

1. See for example, Binkley, N.C. et al.; Effects of slight compositional variations on type E308 electrode deposits, **Weld. J., 53** (1974) 91s-95s.
2. Yamaguchi, K. and Nishijima, S.; Prediction and evaluation of long-term creep-fatigue life, **Fatigue Fract. Engng. Mater. Struct., 9** (1986), 95-107.
3. Yamazaki, M. et al.; Creep-rupture behavior of butt welded joint of 304 stainless steel thick plate using small and large specimens, **J. Soc. Mater. Sci. Jpn., 39** (1990) 509-515 (in Japanese), **Trans. Nat. Res. Inst. Met.**,33 (1991), 64-71.

The Influence of Welding Stresses and Distortions on the Stability of Shells of Revolution

F.G. Rammerstorfer and I. Skrna-Jakl

Institute of Lightweight Structures and Aerospace Engineering,
Vienna Technical University, Vienna, Austria

M.Zelezny

Institute of Applied and Technical Physics,
Vienna Technical University, Vienna, Austria

Summary

This paper deals with the computation of the elastic and elastoplastic stability of welded shells (cylindrical and spherical shells) under mechanical or thermal loading. It is assumed that the residual welding stresses and welding distortions are either known or can be predicted by numerical methods or analytical approximations. The determination of the bifurcation or limit points is performed by numerical discretization methods using finite element or finite difference methods. The influence of welding stresses is simulated either directly by initial stresses introduced into the governing equations or by a fictitious temperature field which leads to a thermal stress field similar to the residual welding stress field. Welding distortions are treated as geometrical imperfections. Specific emphasis lies on circular cylindrical shells with circumferential butt welds. Some fundamental considerations regarding buckling due to fictitious or real thermal loads are presented.

Introduction

Shells are very imperfection sensitive with respect to buckling. In the evaluation of the stability of wélded shells, different sources of imperfections must be considered: geometrical imperfections caused by welding distortions or misalignements on the one hand and load imperfections by residual welding stresses on the other hand. Several design recommendations take these imperfections into account by introducing a knock-down factor by which the theoretical buckling load, i.e. the buckling load of the perfect shell, is multiplied in order to render a so-called practical buckling load. This knock-down facter is chosen to be small enough to ensure safe and conservative estimates of the practical buckling load. In some cases this way is too conservative, i.e. uneconomic, and a direct computation of the influence of weld induced imperfections seems to be appropriate.

L. Karlsson, L.-E. Lindgren, M. Jonsson (Eds.)
Mechanical Effects of Welding
IUTAM Symposium Luleå/Sweden 1991

Estimation of Welding Stresses and Distortions

Residual stresses and distortions of shells, resulting from weldments depend not only on the geometry and material behaviour, but to a large extend on the welding procedure and eventual heat treatment. Nonlinear-Finite-Element computations allow the consideration of the individual influences of the welding process as; for example; the sequence in case of multi-path welding (see e.g. [1,2]), and the complex thermo-visco- elastic-plastic material behaviour (see e.g. [3]). Experimental methods, as e.g. described in [1,4], also give information on residual welding stresses.

Since this paper will not deal with the computation of welding stresses and distortions but on their influence on stability, very simple approximations have been used to determine (i.e. to assume) the residual stresses and distortions in the considered shells. These estimations are based on ideas presented in [5] and applied to circumferentially butt welded circular cylindrical shells.

The approximation of the welding distortions by the so-called "shrinkage-force approach" is based on the assumption that a constant circumferential tensile stress σ_0 acts in a circumferential strip of width $2b_p$ (corresponding to the plastified zone, see Fig.1) along the welding seam. In a conservative approach, σ_0 can be estimated by

$$\sigma_0 = \sigma_Y \tag{1}$$

with σ_Y being the lower value of either the yield stress of the welding material or the structural material. Assuming elastic residual stress states, the distortions can be estimated by the use of linear elastic solutions of the cylindrical shell under an external radial pressure p_0 acting along $2b_p$ with

$$p_0 = \frac{\sigma_0 h}{R} \qquad (2),$$

h is the shell thickness, and R is the radius.

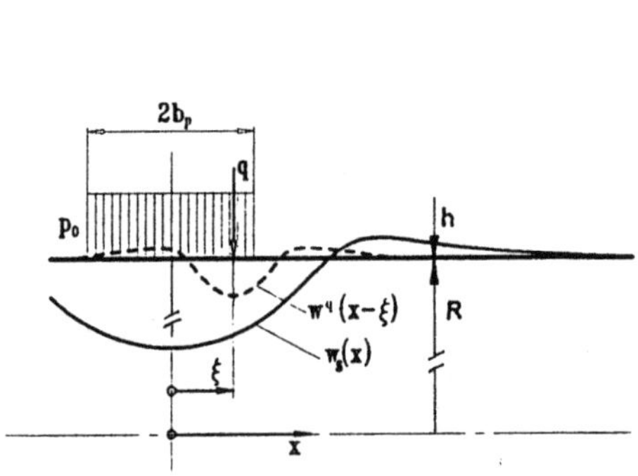

Figure 1. The model for estimating welding distortions

The well known solution for the radial inward deflection under a radial ring load with intensity q (radial force per unit circumferential length) acting at position ξ (see Fig.1) is used as an influence function

$$w^q(x,\xi) = q\frac{e^{-\lambda|x-\xi|}}{8\lambda^3 K}(\sin\lambda|x-\xi| + \cos\lambda|x-\xi|) \quad for \quad x \geq 0 \tag{3}$$

with

$$\lambda = \sqrt[4]{\left[\frac{3(1-\nu^2)}{h^2R^2}\right]}, \qquad K = \frac{Eh^3}{12(1-\nu^2)} \tag{4}$$

Integration along b_p renders an estimation for the residual welding distortions $w_s(x)$ for $x > 0$.

For $x \in [0, b_p]$:

$$w_s(x) = \frac{\sigma_0 R}{2E}\left[2 - e^{-\lambda(x+b_p)}\cos\lambda(x+b_p) - e^{\lambda(x+b_p)}\cos\lambda(x-b_p)\right] \tag{5}$$

and for $x > b_p$:

$$w_s(x) = \frac{\sigma_0 R}{2E}\left[e^{-\lambda(x-b_p)}\cos\lambda(x-b_p) - e^{-\lambda(x+b_p)}\cos\lambda(x+b_p)\right] \tag{6}$$

For $x < 0$, the solution is symmetric to the above one with respect to $x = 0$. The maximum welding distortion is derived from eqn. (5) at $x = 0$:

$$w_s(0) = \frac{\sigma_0 R}{E}\left(1 - e^{-\lambda b_p}\cos\lambda b_p\right) \tag{7}$$

In an analogous way, the residual welding stresses can be estimated by superimposing the elastic solution of the above load case with a properly assumed residual stress field of the undistorted shell (usually the residual welding stress distribution of a corresponding butt welded plate is used). According to [5] the extreme values of the residual stresses in the distorted shell are approximated by:

$$\sigma_{\varphi,s}(0) = \sigma_0 e^{-\lambda b_p}\cos\lambda b_p \tag{8}$$

$$\sigma_{x,s}(0) = \pm 3\sigma_0\sqrt{3(1-\nu^2)}\, e^{-\lambda b_p}\sin\lambda b_p \tag{9}$$

with (+) at the inner and (-) at the outer shell surface, respectively.

In [5] comparisons between measured residual stresses and results obtained by this estimation procedure are shown. In Fig.2, this comparison is presented by normalized values. In this figure results of a simulation by a fictitious temperature field (see following section) are presented, too. These welding residual stresses must fulfill the equilibrium conditions which are written in terms of stress resultants as follows:

$$\frac{dq_s(x)}{dx} + \frac{n_{\varphi,s}(x)}{R} = 0, \qquad \frac{dm_s(x)}{dx} - q_s(x) = 0 \tag{10}$$

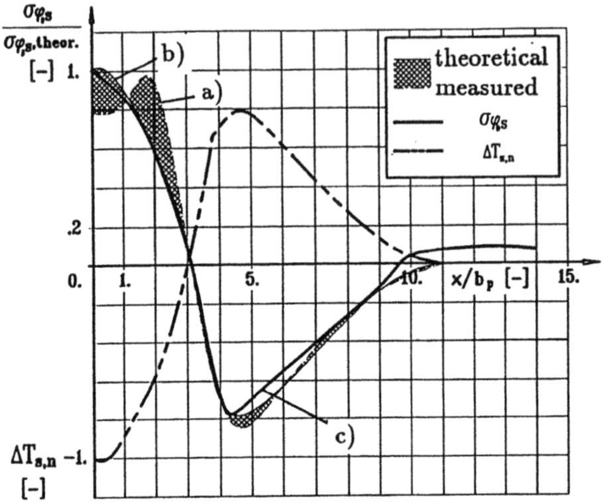

Figure 2. Normalized circumferential welding stress component near the circumferential weld of a circular cylindrical shell: a) measured results, b) approximated by above method and c) simulated by a fictitous temperature field

Consideration of Welding Stresses and Distortions in Discretization Methods

Welding distortions respresent typical geometrical imperfections and can be considered by modelling the geometry of the structure appropriately, i.e. an imperfect geometry is input to the analysis program. Hence, usual geometrical and material nonlinear analyses handle this welding effect. With respect to welding stresses different approaches exist.

A direct approach is the consideration of welding stresses as "initial stresses" $S_{ij,s}$ in the incremental (or incremental-iterative) equations. After discretization - for example by finite element displacement functions - the following linearized incremental equilibrium equation can be derived, e.g. expressed in Total Lagrangean formulation (the index 's' denotes contributions from the residual welding stresses):

$$[{}^m\underset{\approx}{\mathbf{K}}_M + {}^m\underset{\approx}{\mathbf{K}}_u + {}^m\underset{\approx}{\mathbf{K}}_g + {}^m\underset{\approx}{\mathbf{K}}_s](\Delta\underset{\sim}{u})^1 = {}^{m+1}\underset{\sim}{\mathrm{R}} - ({}^m\underset{\sim}{\mathrm{F}} - {}^m\Delta\underset{\sim}{\mathrm{F}}_s) \qquad (11)$$

or

$${}^m\hat{\underset{\approx}{\mathbf{K}}}(\Delta\underset{\sim}{u})^1 = {}^{m+1}\underset{\sim}{\mathrm{R}} - {}^m\hat{\underset{\sim}{\mathrm{F}}} \qquad (12).$$

${}^m\underset{\approx}{\mathbf{K}}_M$ represents that contribution to the tangent stiffness matrix ${}^m\hat{\underset{\approx}{\mathbf{K}}}$ in equilibrium configuration m, which is not explicitly (but implicitly) dependent on the stresses and displacements. ${}^m\underset{\approx}{\mathbf{K}}_u$ is the initial displacement stiffness matrix (explicitly dependent on deformations), ${}^m\underset{\approx}{\mathbf{K}}_g$ is the initial stress or geometric stiffness matrix (explicitly dependent on the stresses), and ${}^m\underset{\approx}{\mathbf{K}}_s$ results from the initial stresses. Also the internal force vector contribution ${}^m\Delta\underset{\sim}{\mathrm{F}}_s$ results from these initial stresses. ${}^{m+1}\underset{\sim}{\mathrm{R}}$ is the external load vector in increment (m+1) and ${}^m\underset{\sim}{\mathrm{F}}$ corresponds to the internal element forces corresponding to the stresses in the elements. Equation (11) is typically applied iteratively to improve the incremental displacement vector $\Delta\underset{\sim}{u}$.

Care must be taken to ensure that $S_{ij,s}$ is a self-equilibrated stress field, i.e. the equilibrium and boundary conditions

$$\frac{\partial S_{ij,s}}{\partial x_j} = 0 \ , \qquad S_{ij,s} n_j = 0 \qquad \forall i \tag{13}$$

must be satisfied. Experimental results of welding stresses are normally not exact enough. Hence, the analysis should start with an initial zero external load increment. Equilibrium iterations should be performed within this zero load increment which modify the initial stress field in order to become a self-equilibrated one. These changes in the residual stress field might be considerable, if the input initial stresses are far from being in equilibrium, and they cause "spurious" distortions in addition to the welding distortions regarded by an imperfect geometry in the input (see Fig. 9 in a following section).

Another approach which may be useful in some cases, is to simulate the welding stress state by a fictitious temperature field applied to the unloaded welded structure and keeping this field constant during incrementation of the real mechanical or thermal loading [7]. The corresponding fictitious temperature field can be determined either by trial and error or by solving an inverse problem. In this way a self-equilibrated initial stress field is automatically achieved, however, distortions due to this temperature field must be expected. An incremental equilibrium equation is obtained which is similar to eqn. (11), where ${}^m\underset{\approx}{\mathbf{K}}_s$ and ${}^m\Delta\underset{\sim}{\mathrm{F}}_s$ are replaced by ${}^m\underset{\approx}{\mathbf{K}}_{th}$ and ${}^m\Delta\underset{\sim}{\mathrm{F}}_{th}$, respectively; see e.g. [8]. Both approaches are applied in this paper.

According to the static stability criterion, in the critical state m^*, i.e. at bifurcation or limit points, the equation

$$^*\,\underset{\approx}{\hat{\mathbf{K}}}\,\delta\underset{\sim}{u} = 0 \tag{14}$$

has a non-trivial solution $\delta\underset{\sim}{u} \neq 0$, and the global tangent stiffness matrix $\underset{\approx}{\hat{\mathbf{K}}}$ becomes singular at the stability limit:

$$det^*\,\underset{\approx}{\hat{\mathbf{K}}} = 0 \qquad (15).$$

This so-called determinant criterion is well estabished in computational stability analyses using discretization methods. Eigenvalue procedures have been derived by starting with eqn. (14), see e.g. [9,10].

Consideration of the Individual Welding Effects on Stability

The critical question now is which of the individual welding effects, i.e. welding residual stresses or distortions, have a more pronounced influence on the stability behaviour of shells. In the case of longitudinally welded, axially loaded cylindrical shells, the influence of the residual welding stresses on the stability is very pronounced [11], whereas for circumferentially welded circular cylindrical shells under axial loads the answer to the above question depends on the radius-to-thickness ratio.

In [12], L. Häfner considered rather thin cylindrical shells ($R/h = 500$) with a circumferential weld. Häfner's computations illustrate that the perfect unwelded shell exhibits elastic nonaxisymmetric buckling and even the welded shell buckles with a wave number larger than zero, i.e. nonaxisymmetrically. His results lead to the conclusion that the weld effect induced reduction of the critical load for thin axially loaded circular shells is mainly due to welding distortions, and the welding residual stresses have a rather small influence, see Fig.3.

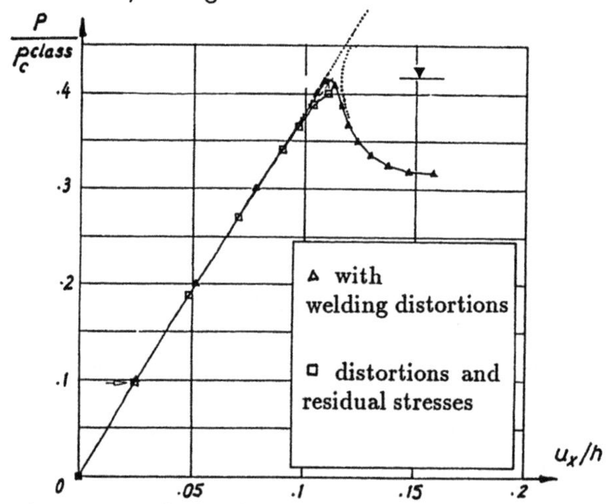

Figure 3. Load displacement paths for the axially loaded butt welded thin shell - taken from [12]

Hence, the drop-down of the buckling load for such thin shells can be estimated by simply considering the geometrical imperfections. This is the reason why Häfner's results agree with the classical results obtained by Hutchinson et al [13], see Fig. 4 in which also results obtained by Amazigo and Budiansky [14] and the well known results by Koiter [15] are presented.

Häfner's computations have shown that the knock-down factor concept of the DASt-013 design recommendation [16] is sufficiently conservative even for determining the stability limit of welded circular cylindrical shells under axial loads, compare Fig. 5.

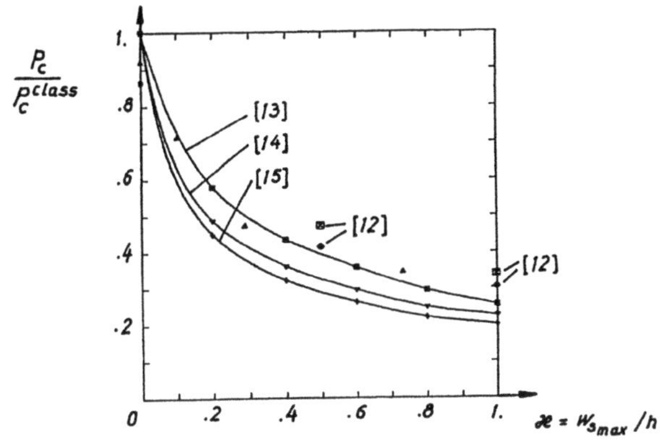

Figure 4. Normalized critical axial load as a function of normalized imperfection amplitude $\varkappa$ - taken from [12]

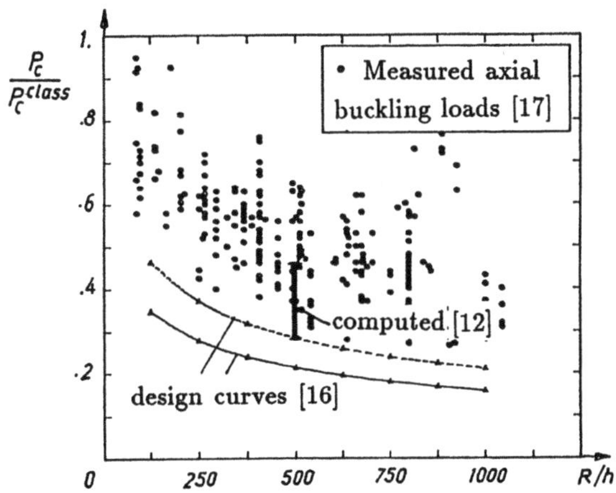

Figure 5. Measured axial buckling loads [17], design curves [16], and results computed by Häfner [12] for axially loaded butt welded thin shells - taken from [12]

As a typical example for a thick walled cylindrical shell a butt welded tube of a buried pipeline is treated, see Fig.6. This shell is rather thick walled as indicated by $R/h \approx 46$. In order to compare the individual influences of the weld effects, the geometrically and materially nonlinear behaviour of this shell under axial compression was computed assuming different conditions: a) perfect unwelded shell, b) shell with assumed welding stresses but without welding distortions, c) shell with assumed welding distortions but without welding stresses, d) shell with both welding stresses and distortions.

Due to the small R/h-ratio, elastic buckling must be excluded, and even in the perfect unwelded case, axisymmetric elastic-plastic collapse is the relevant mode of instability. Hence, an axisymmetrical model is sufficient.

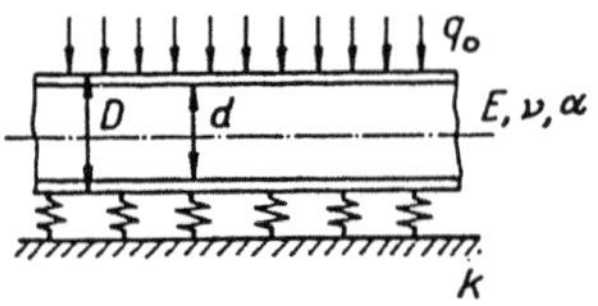

$E = 2.1 \times 10^5$ N/mm^2
$\nu = 0.3$, $\sigma_Y = 287$ N/mm^2
$\alpha = 1.2 \times 10^{-5}$ $^oC^{-1}$
$d = 645.8$ mm
$D = 660.0$ mm
$q_o = 20$ N/mm
$k = 8.3$ N/mm^2

Figure 6. Geometry and material parameters of the buried pipeline

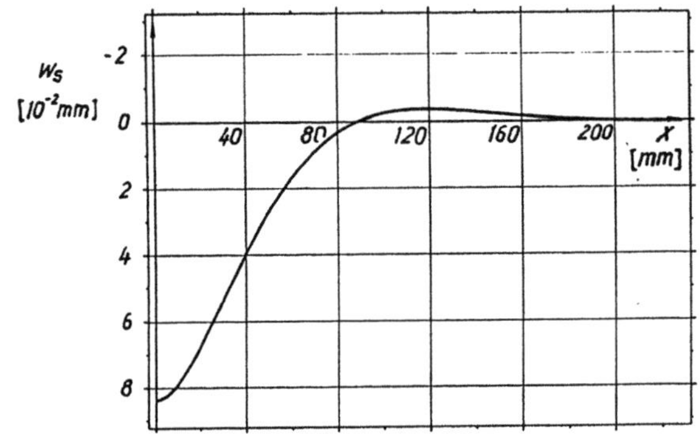

Figure 7. Welding distortions estimated by the above method

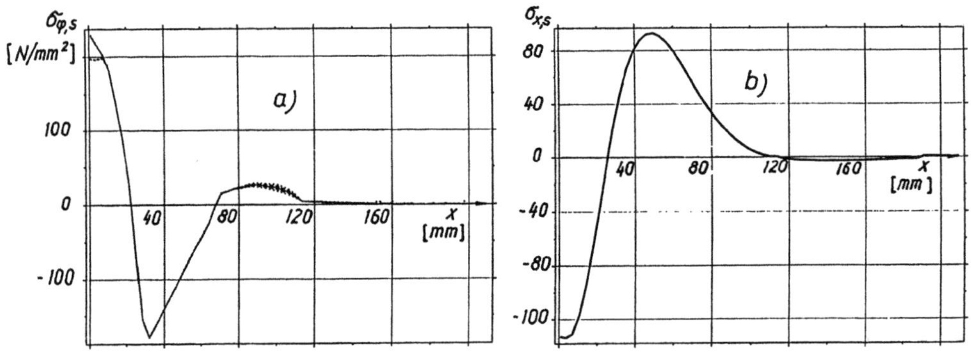

Figure 8. a) Circumferential initial stresses, b) axial initial stresses at the outer surface (for the inner surface multiply by -1)

In this analysis welding distortions and residual welding stresses are estimated according to the corresponding former section. The residual stresses are considered by initial stresses as described in the previous section. The estimated distortions are shown in

Fig. 7. After application of a zero external load increment to the shell under initial stresses assumed in analogy to Fig.2, the resulting initial stresses being in equilibrium are presented in Fig. 8.

As mentioned in a previous section care must be taken regarding the equilibrium of the welding stresses. Figure 9 shows how additional "spurious" distortions are produced by the finite element analysis if the input initial stresses are not in equilibrium. In case a) the complete residual stress field is input as determined by the analytical "shrinkage-force approach"; the small deviations from the analytically determined distortions are caused by a slight plastification in a small region at the weld (which is not considered in the analytical model). For demonstration purposes in case b) only the circumferential residual stress component from the analytical solution is taken and zero axial stresses are used as starting situation. Due to the zero-load equilibrium iterations a self-equilibrated residual stress field is achieved which is very close to the complete analytical solution, however, the additional "spurious" distortions lead to a significant change in the geometry, i.e. the imperfection shape.

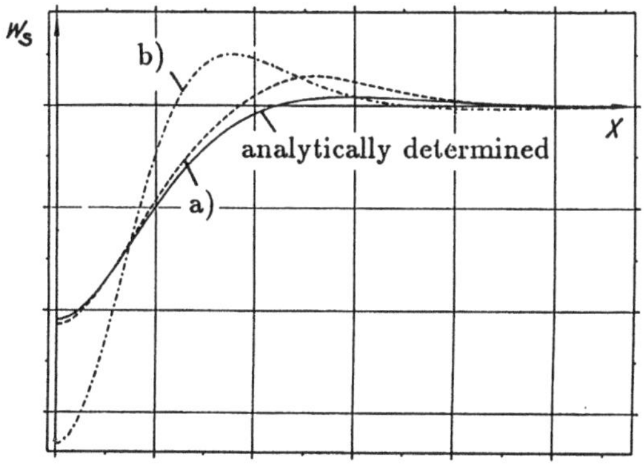

Figure 9. Change of welding distortions by non-equilibrated initial residual stresses: a) complete analytically determined residual stresses used, b) $\sigma_\varphi \doteq \sigma_{\varphi,s}, \sigma_x = \sigma_{x,s}$ used as initial residual stress field

The incremental-iterative finite element analysis using degenerated axisymmetrical shell elements [18] leads to results as shown in Fig.10. In contrast to the aforementioned investigations of welded thin cylinders, the influence of welding distortions (geometrical imperfections) on the limit load of this rather thick shell is very weak. Here the effect of the residual welding stresses is much more pronounced: The onset of plastification is, of course, highly influenced by these welding effects. Hence, the stiffness loss starts at significantly lower load levels. However, the limit load is not too different from the unwelded shell. Nevertheless, the investigations described in the following section will show that in many technical applications welding effects must not be disregarded even

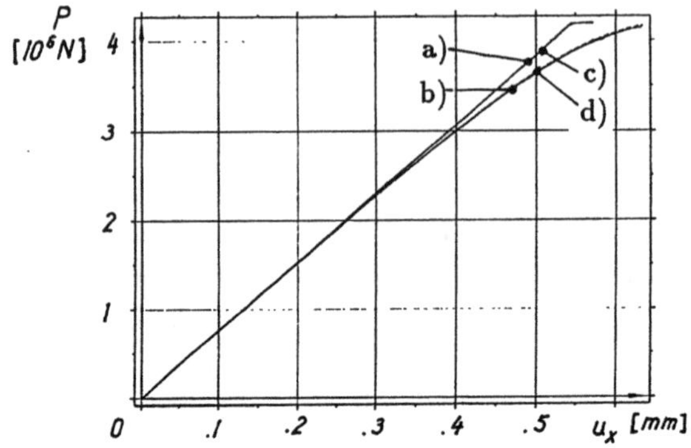

Figure 10. Load-Displacement paths: a) perfect unwelded shell, b) with welding stresses and without distortions, c) with welding distortions and without welding stresses, d) with both welding stresses and distortions

in thick shell analyses.

A more detailed investigation of the postbuckling behaviour shows that in the case under consideration the welding distortions may even have a certain stabilizing effect, see Fig. 11a. The reason for this rather unexpected behaviour is shown in Fig. 11b: The radial welding distortions are inward directed and the postbuckling bulge goes outwards. A complete postbuckling path is shown in Fig. 12.

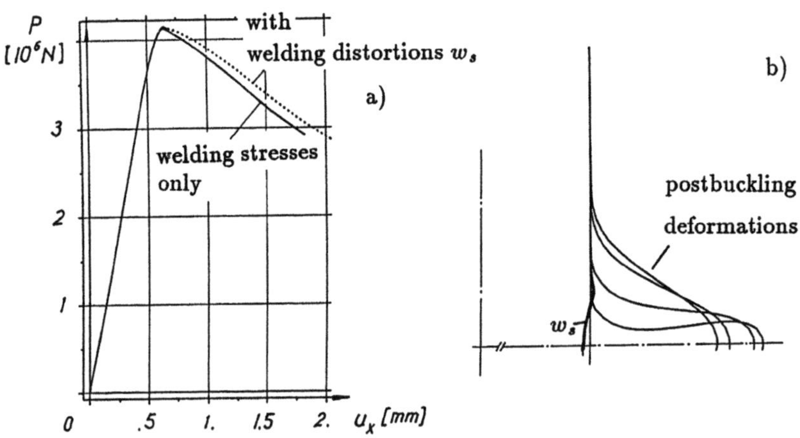

Figure 11. a) The effect of welding distortions, b) postbuckling deformations

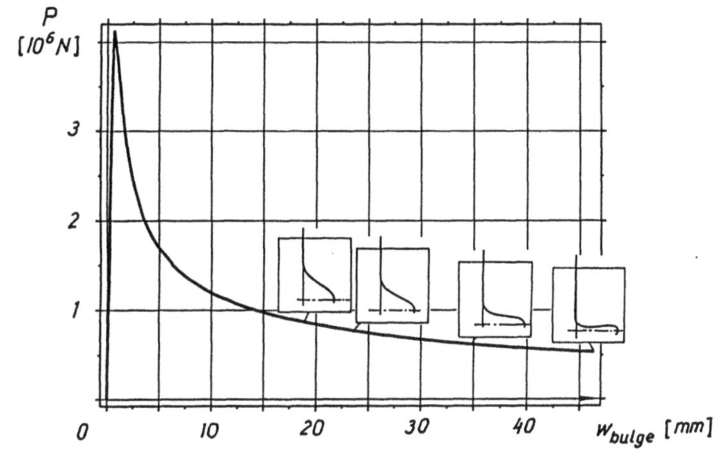

Figure 12. Postbuckling behaviour of the welded shell

Application to the Investigation of the Stability of a Buried Pipeline for Hot Water Supply

Recently, some projects in the field of pipeline construction for the transport of heated media were started in which the pipes are embedded without compensators between the fixpoints. Hence, if the heated medium transported in the pipline causes a temperature rise above the prestressing temperature, i.e. the ambient or stress free temperature, an axial compression of the pipe is induced. The pipeline is constructed by butt welding circular cylindrical shells the geometrical and material parameters of which are given in Fig. 2 (see former section). The thermally induced compression forces may lead to instability and localization effects.

In this respect the complete or overall buckling of the pipeline is considered in [6]. There the pipeline is treated as an elastic beam on an elastic foundation free to uplift in vertical direction and loaded downwards by the dead load of the pipe, the transported medium and the overburden. This buckling problem is very similar to the classical problem of thermal buckling of tracks treated in several papers, e.g. [19].

With respect to this overall buckling, see Fig. 13, the effect of welding stresses and distortions is not directly but indirectly relevant. The localization of axial thermal strains due to welding effects influences the temperature - compressionforce - relation. As will become clear by the following considerations the residual welding stresses and distortions have even a stabilizing effect with respect to overall buckling of the pipeline.

The problem of local thermal buckling of the pipe is treated now. The buckling appers at a misalignement caused by a small sudden change of the pipe dimensions at a weld which may be the consequence of an unfavourable combination of pipe sections still

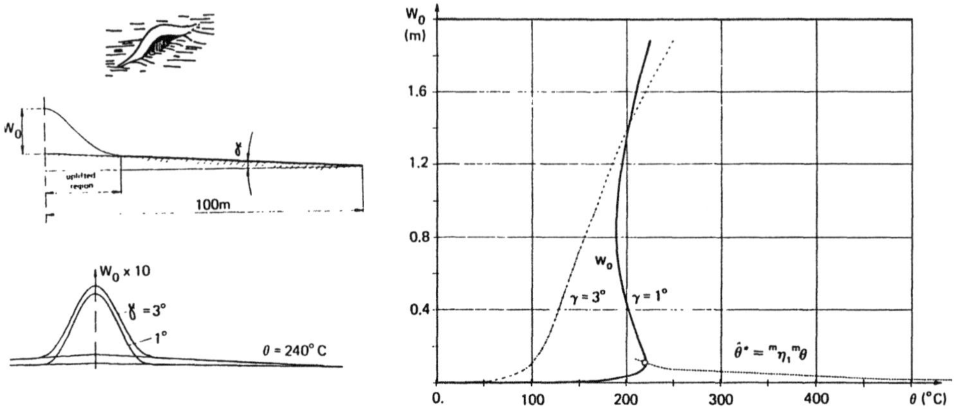

Figure 13. Thermal over-all beam-mode buckling of the buried pipeline, treated in [6,8]

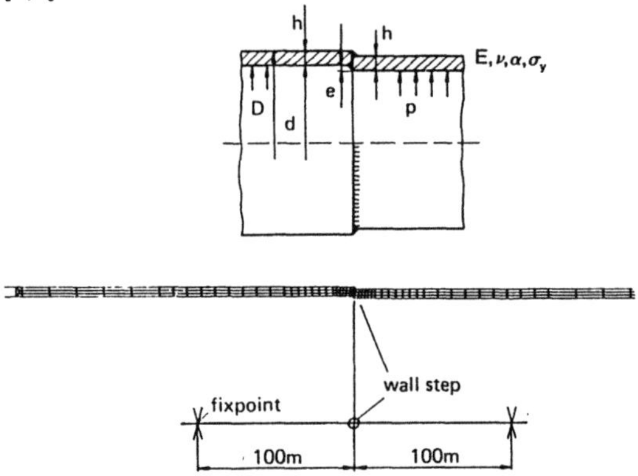

Figure 14. The welded pipe connection with a wall step

within tolerances, see Fig. 14.

This wall step leads to bending deformations in addition to the membrane state if the pipe is compressed due to heating. Consequently, local plastification takes place at a rather low temperature rise. Further heating results in the formation of alocal axisymmetric bulge and, finally, the subsequent thermal strains of the pipe are localized at this bulge, and the increase in axial forces with increased temperature is reduced. On the one hand, this is favorable with respect to the above described overall buckling, but on the other hand, a limit point problem appears locally at the weld which is similar to the necking of ductile specimens in the usual tensile test. In this context the welding stresses plays an important role.

Since the local states of stress and deformation at the wall step are relevant for this thermal buckling problem, 2D-axisymmetric continuum elements rather than shell elements seem to be necessary to model the relevant regime of the pipeline. Figure 15

Figure 15. Localization of the axial thermal strains by formation of a bulge

shows a typical result of such a 2D finite element analysis.

However, our comparsion analysis performed with the BOSOR 5 program [20] has shown that such problems can also be treated with sufficient accuracy by axisymmetrical shell models. Hence, the following considerations dealing with the influence of welding stresses at the wall step are treated by axisymmetric BOSOR models. In this case, the residual welding stresses are accounted for in the model by a fictitious temperature field shown in Fig. 3.

The incrementation of the real temperature (in addition the unchanged fictitious temperature ΔT_s and the constant internal pressure p) leads to the aforementioned bulge formation shown in Fig. 16.

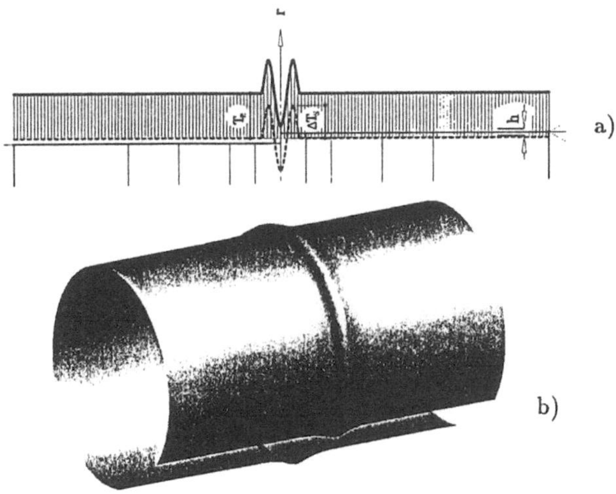

Figure 16. Misaligned pipe connection: a) superimposed fictitious and real thermal loading $\Delta T_s + T_R$, b) computed bulge formation

By consideration of Fig. 17, the individual influences of the butt weld and misalignement can be evaluated. Some interesting features can be observed. The perfect cylinder shows no instability up to the yield limit (at a temperature rise of approximately

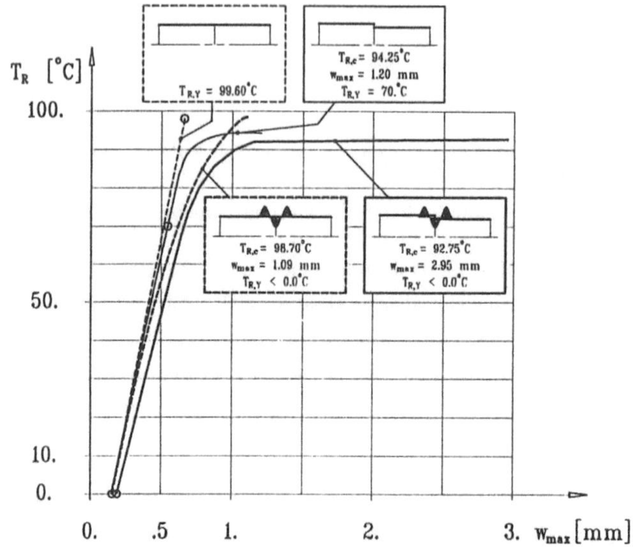

Figure 17. Temperature - displacement path for the apex of the bulge

99°C). Taking the wall step geometry into account first plastification is observed at $T_{R,Y} = 70°C$ and the limit load temperature corresponds to $T_{R,c} = 94°C$.

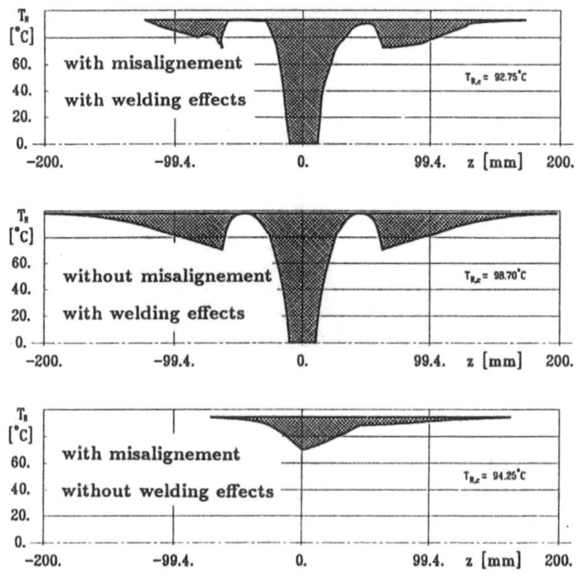

Figure 18. Growth of plastified regions with increased thermal load

Despite the fact that the wall step does not show a dramatic reduction in the thermal load carrying capacity for monotonic loading, it should mentioned that such pipelines

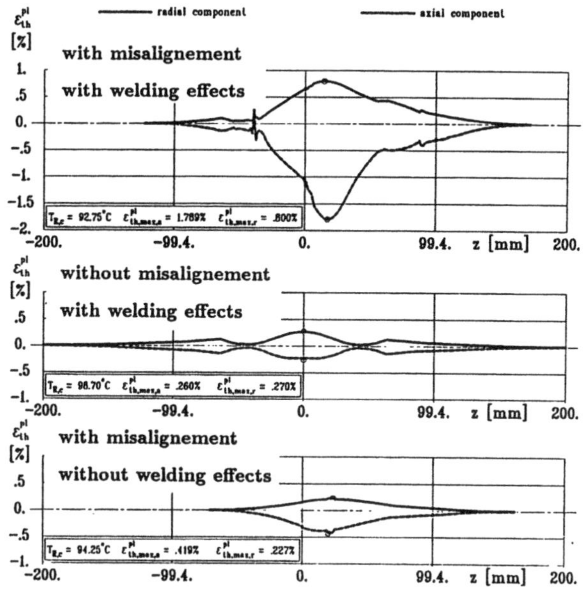

Figure 19. Distribution of the axial plastic strains along the meridian

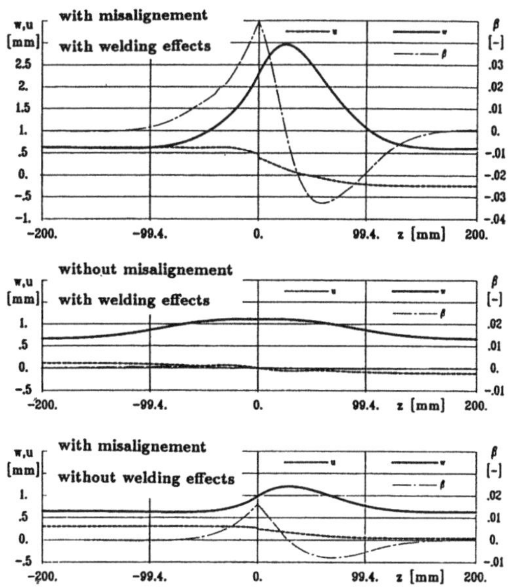

Figure 20. Displacements at the stability limit

are subject to cyclic thermal loads, and hence, plastic damage in terms of low cycle fatigue must be considered. In this respect, the wall step must be considered more carefully. This becomes more pronounced if welding residual stresses are taken into account as indicated in Fig. 18. The growth of the plastified region with increasing temperature load T_R is compared for different cases.

The localization effect becomes obvious if one considers the distribution of the plastic axial strains at the stability limit as shown in Fig. 19. Finally, information about the sharpness of the bulg at the stability limit can be obtained from Fig.20.

Thermoelastic Buckling Effects

In the previous sections the simulation of residual stresses by fictitious temperature fields was discussed. As long as thick walled shells are considered, this approach seems to be appropriate. However, in case of thin walled shells the thermal distortions in conjunction with the stresses due to this fictitious temperature field may lead to unexpected effects with respect to stability. This is shown by the following investigations which can also be useful for the stability analysis of shells during annealing (often performed after welding).

In order to demonstrate the above mentioned effect, it is instructive to consider a somewhat academic problem: the infinitely long cylindrical shell under a thermal step load as illustrated in Fig 21.

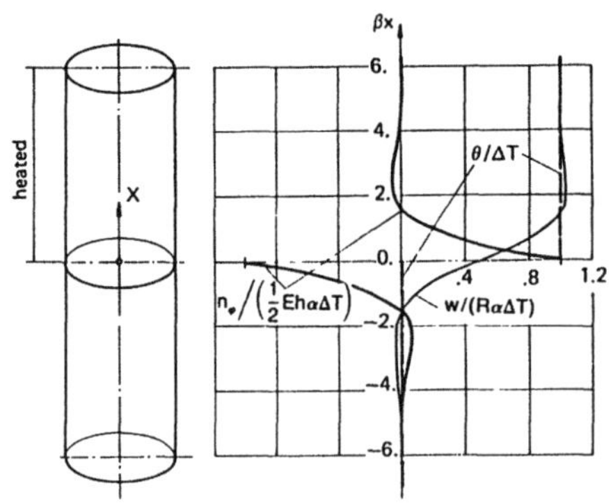

Figure 21. The infinitely long cylindrical shell under a thermal step load

One of the classical solutions of this problem is published by W.Flügge [21] where the influence of prebuckling deformations on the critical temperature rise ΔT_c^{class} is disregarded.

In a more recent paper, D.Bushnell [22] concluded from his nonlinear computational investigations using the determinant criterion (eqn. (15)) that the cylindrical shell hardly buckles under a thermal step load. His conclusion is based on the observation of an increasing determinant with increasing temperature which is a consequence of the stiffening effect due to thermal prebuckling deformations forming the cylindrical shell into a doubly curved shell. Bushnell investigated a circular cylindrical shell of following parameters:

$R = 2540mm\ (100''),\ h = 25.4mm\ (1''),\ E = 2.068 \times 10^5 N/mm^2,\ \nu = 0.3,$
$\alpha = 1.8 \times 10^{-5} K^{-1}.$

In [22] the nonlinear stability analysis is performed under the assumption of a circumferential wave number ($n = n_c^{class} = 14$). The behaviour of the normalized determinant

$$det_n{}^m \hat{\underset{\approx}{\mathbf{K}}} = \frac{det^m \hat{\underset{\approx}{\mathbf{K}}}}{det^0 \hat{\underset{\approx}{\mathbf{K}}}} \tag{16}$$

as a function of temperature rise is shown in Fig. 22 in which the result of [22] is denoted by "BUSH85".

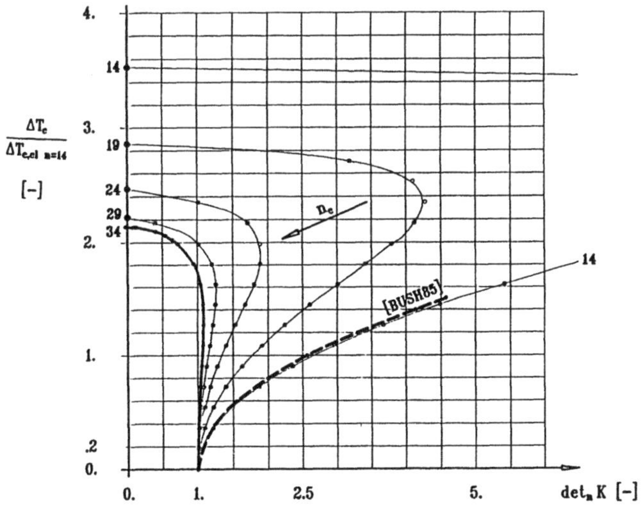

Figure 22. Normalized determinant vs. normalized temperature rise for different buckling wave numbers

If the nonlinear stability analysis is performed with varied n, it is easy to recognize that n_c^{class} does not remain relevant if prebuckling deformations are taken into account. The academic example demonstrated here (which has no physical evidence since unrealisticly high ΔT_c-values are computed under assumptions which are chosen to be comparable with [22]) shows a critical circumferential wave number $n_c = 34$ instead of 14 and a critical temperature rise which is about twice the classical value. Figure 23 shows the prebuckling deformations and the relevant buckling mode.

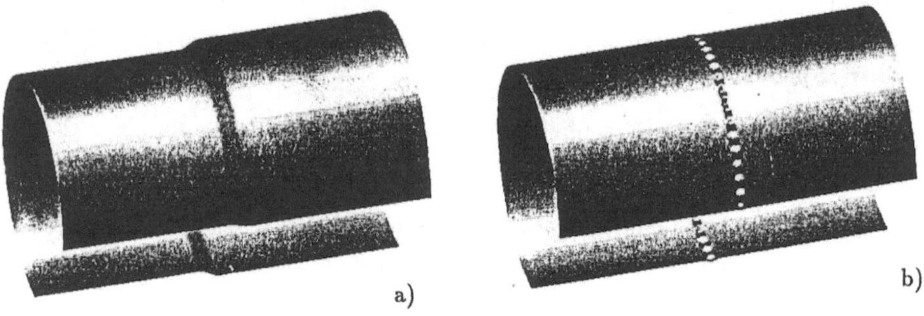

Figure 23. Prebuckling deformations (a) and buckling mode (b) of the cylindrical shell under a thermal step load

This example has shown the significance of prebuckling deformations in stability considerations of thermally loaded shells and shells for which residual welding stresses are simulated by fictitious thermal loads. A more practical example is treated now: the stability of a hemispherical shell during annealing. The shell is clamped at its boundary (e.g. welded to a rather rigid supporting structure) and heated up by an axisymmetrical temperature field

$$T(\varphi) = T_0 e^{-k\varphi^2} \qquad (17).$$

The following geometrical and material data are chosen:
$R = 2500mm, h = 4mm, E = 7.22 \times 10^4 N/mm^2, \nu = 0.34, \alpha = 2.4 \times 10^{-5} K^{-1}$. To demonstrate fundamental features only, the problem is treated as linear thermo-elastic, and initial welding stresses and distortions are not taken into account. Hence, these very local effects at the boundary are not taken into account. Performing a nonlinear incremental analysis analogously to the cylinder example described above we obtain the following results:

The hemispherical shell buckles with a very high circumferential wave number at a critical temperature rise $T_{0c} \approx 344°C$ which may be further reduced due to residual welding stresses and distortions as well as due to plasticity effects! Hence, care must be taken with respect to buckling if thin welded shells are annealed!

In Fig. 24 the temperature field, the circumferential membrane force, and the radial displacement at the stability limit are presented. The buckling mode is also shown in

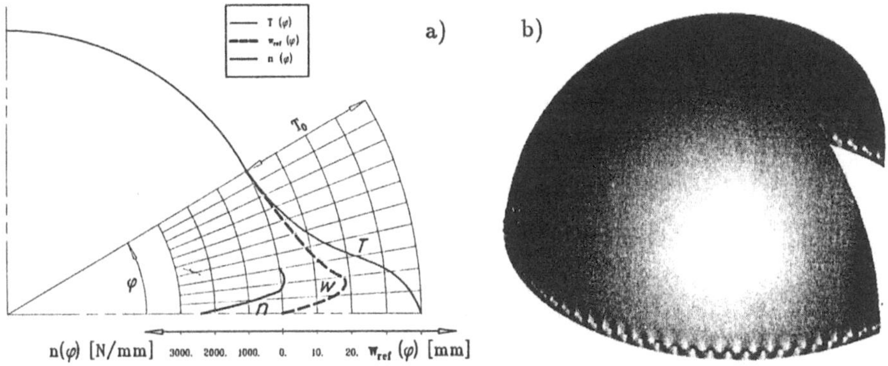

Figure 24. a) Temperature field, radial displacement and circumferential membrane force at the stability limit, b) buckling mode

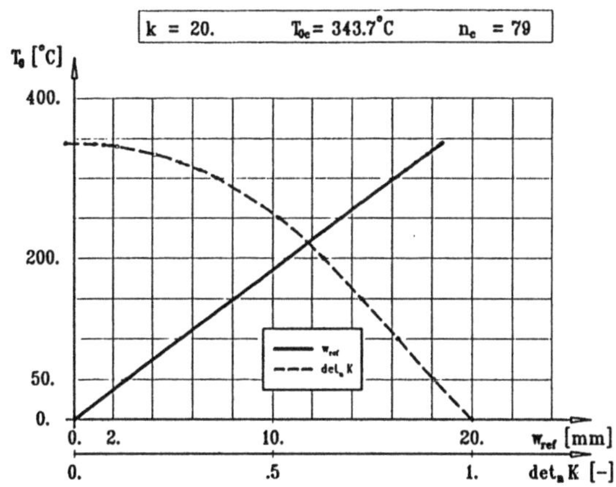

Figure 25. Normalized determinant, circumferential wave number $n = n_c = 79$

this figure and is restricted to a rather narrow regime near the boundary.

The behaviour of the normalized determinant of the tangent stiffness matrix is shown in Fig. 25. In this case, it is not as dramatic as in the previous cylinder example.

Conclusions

The relative thickness of the shell determines whether residual welding stresses or welding distortions have a dominant influence on the stability limit. While thin shells have a more pronounced effect on distortions, in thicker shells residual welding stresses must not be ignored since they cause the onset of early plastification.

It is shown that the consideration of residual welding stresses in discretization methods requires some attention. If the directly input residual stresses are not fully self-equilibrated zero-load equilibrium iterations should be used and "spurious" distortions must be expected. If welding stresses are simulated by fictitious temperature loads self-equilibrated residual stresses are achieved automatically but additional distortions are caused.

If annealing processes are investigated, care must be taken with respect to the influence of thermally induced prebuckling deformations on the stability limit. This influence can only be characterized if geometrical nonlinearities are considered in the buckling analysis.

References

[1] Karlsson, L., 'Thermal Stresses in Welding' in *Thermal Stresses I* (Ed. R.B. Hetnarski), pp. 300–389, North-Holland Publ. Comp. 1986.

[2] Rybicky, E.F. and R.B. Stonesifer, 'Computation of Residual Stresses due to Multipass Welds in Piping Systems', J. Pressure Vessel Techn. **101**, 149–154 (1979).

[3] Wang, Z. and T. Inoue, 'Viscoplastic Constitutive Relation Incorporating Phase Transformation - Application to Welding', Mater. Sci.Technol. **1**, 899–903 (1985).

[4] Macherauch, E. and V. Hauk (Ed.), *Residual Stresses in Science and Technology*, Deutsche Gesellschaft für Metallkunde, Oberursel, FRG, 1987.

[5] Radaj, D., *Wärmeeinwirkungen des Schweißens*, Springer-Verlag Berlin Heidelberg New York London Paris Tokyo 1988.

[6] Rammerstorfer, F.G., K. Dorninger and J. Ployer, 'Nonlinear Stability Problems Concerned with Buried Pipelines for Hot Water Supply', ASME PVP-Vol.169, 57–62 (1989).

[7] Rammerstorfer, F.G. and F.D. Fischer, 'A Method for the Experimental Determination of Residual Stresses in Axisymmetric Composite Cylinders', J. Engng. Mater. Technol. - in press.

[8] Ziegler F. and F.G. Rammerstorfer, 'Thermoelastic Stability' in *Thermal Stresses III* (Ed. R.B. Hetnarski), pp. 107–189, North-Holland Publ. Comp. 1989.

[9] Ramm, E., *Geometrisch nichtlineare Elastostatik und Finite Elemente*, Habilitationsschrift, Univ. of Stuttgart, FRG 1975.

[10] Brendel, B., *Geometrisch nichtlineare Elasto-Stabilität*, Doctoral Thesis, Univ. of Stuttgart, FRG 1979.

[11] Ravn-Jensen, K. and V. Tvergaard, 'Effect of Residual Stresses on Plastic Buckling of Cylindrical Shell Structures', Int. J. Solids Structures **26**, 993–1004 (1990).

[12] Häfner, L., *Einfluß einer Rundschweißnaht auf die Stabilität und Traglast des axialbelasteten Kreiszylinders*, Doctoral Thesis, Univ. of Stuttgart, FRG, 1982.

[13] Hutchinson, J.W., R.C. Tennyson and D.B. Muggeridge, 'Effect of a Local Axisymmetric Imperfection on the Buckling Behavior of a circular Cylindrical Shell under Axial Compression', AAIA Journal **9**, 48–52 (1971).

[14] Amazigo, J.C. and B. Budiansky, 'Asymptotic Formulas for the Buckling Stresses of Axially Compressed Cylinders with Localized or Random Axisymmetric Imperfections', J. Appl. Mech. **39**, 179–184 (1972).

[15] Koiter, W.T., 'The Effect of Axisymmetric Imperfections on the Buckling of Cylindrical Shells Under Axial Compression', Roy. Netherl. Acad. Sci., Ser. B, **66**, 265–279 (1963).

[16] DASt-Richtlinie 013 *Beulsicherheitsnachweise für Schalen*, Deutscher Ausschuß für Stahlbau, Edition July 1980.

[17] Esslinger, M. and B. Geier, *Postbuckling Behavior of Structures*, CISM Courses and Lectures, No. 236, Springer-Verlag, Wien/New York 1975.

[18] ABAQUS User's Manual, HKS Inc., Providence, RI, 1987.

[19] Kerr, A.D., 'Analysis of Thermal Track Buckling in the Lateral Plane', Acta Mechanica **30**, 17–50 (1978).

[20] Bushnell, D., 'Computerized Analysis of Shells - Governing Equations', Comp. & Struct. **18**, 471–536 (1984).

[21] Flügge, W., *Stresses in Shells*, Springer-Verlag, Berlin 1960.

[22] Bushnell, D., *Computerized Buckling Analysis*, Nijhoff Publ., Dortrecht, The Netherlands 1985.

Influences of Welding Imperfections on Stiffness of Rectangular Plate Under Thrust

T. Yao,[1] P. I. Nikolov[2] and Y. Miyagawa[1]

[1] Faculty of Engineering, Hiroshima University, Higashi-hiroshima, Japan
[2] Graduate School of Hiroshima University, Higashi-hiroshima, Japan

Summary

In this paper, stiffness of plates in compression was discussed considering the influences of initial deflection and welding residual stresses. For this purpose, an elastic large deflection analysis was performed in an analytical manner, as well as an elasto-plastic large deflection analysis by the finite element method. It has been found that:
(1) Initial deflection reduces the stiffness of a compressed plate according to its magnitude.
(2) Welding residual stresses do not affect the stiffness itself but buckling load. Consequently, they affect only the load at which the stiffness changes due to the occurence of deflection in the elastic range.

1. Introduction

Recently, High Tensile Steel has become widely used in structures, and this enables to reduce the plate thicknesses and hence the weight of structures. However, to use the High Tensile Steel effectively, the plate thickness may become so thin that the occurence of elastic buckling is inevitable and design allowing plate buckling may be necessary. In this case, precise assessment of the stiffness of plates above buckling load is important considering the influences of initial deflection and residual stresses due to welding. In this paper, series of elastic and elasto-plastic large deflection analyses were performed to estimate the influences of welding imperfections on stiffness of plates.

2. Initial Imperfections Produced in Plates due to Welding

A plate element from a stiffened plate assemblage is considered in this paper. The plate has an initial deflection generally of a complex mode as illustrated in Fig. 1 (a). Assuming that a plate element is simply supported along four edges to represent the assemblage continuity, the initial deflection can be expressed in a general form as:

$$w_0 = \Sigma\Sigma A_{0mn} \sin\frac{m\pi x}{a} \sin\frac{n\pi y}{b} \quad (1)$$

L. Karlsson, L.-E. Lindgren, M. Jonsson (Eds.)
Mechanical Effects of Welding
IUTAM Symposium Luleå/Sweden 1991

The total deflection of a plate under external load may be expressed in the same form as initial deflection for the simply supported condition.

$$w = \Sigma\Sigma A_{mn} \sin\frac{m\pi x}{a} \sin\frac{n\pi y}{b} \tag{2}$$

When a plate is subjected to uni-axial compression in x-direction, many components in the x-direction may increase, but only the first component, $\sin(\pi y/b)$, in the y-direction will increase. In this case, expressions for the initial and total deflections may be simplified as:

$$w_0 = \Sigma A_{0m} \sin\frac{m\pi x}{a} \sin\frac{\pi y}{b} \tag{3}$$

$$w = \Sigma A_m \sin\frac{m\pi x}{a} \sin\frac{\pi y}{b} \tag{4}$$

The shape of initial deflection was measured on 42 panels at the deck of 100,000DWT bulk carrier. Applying the method of least squares on the measured results of deflection along the centerline of a panel in longitudinal direction, the coefficients, A_{0m}, were evaluated taking m up to 11. It was found that one half-wave component was the largest and the coefficients of odd terms were greater than those of even terms when a heat treatments were not carried out to remove welding distortions in the plates. These features are the same as those observed in the previous measurements [1], and result in a so-called hungry-horse mode of initial deflection.

In Ref.[1], an idealized mode of initial deflection was proposed, which consists of a flat portion in the middle and a quarter of a sine wave at both ends. This mode consists of only odd terms. In this paper, some modifications were made introducing small even terms. The coefficient of the $2n$th even term was taken as 1/5 of the coefficient of $(2n+1)$th term. The coefficients of the proposed idealized initial deflection are given in Table 1. These modes represent the general features of the measured ones.

The residual stresses in the plate are produced by either fillet welding of stiffeners to plates or butt welding of plates. At the middle portion of the plate, the former produces compressive residual stresses, whereas the latter tensile residual stresses. In both cases, rectangular distribu-

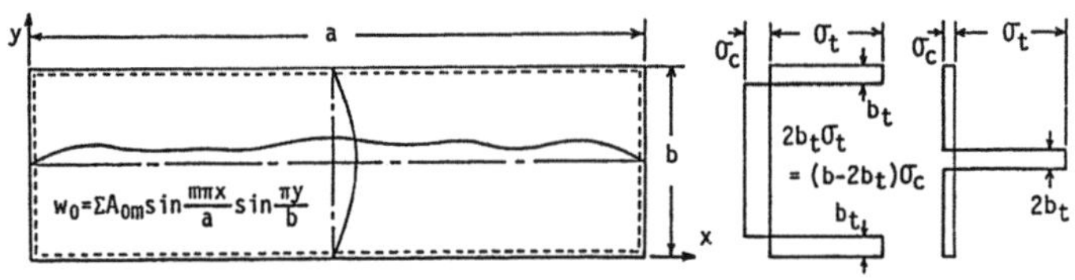

(a) Initial deflection (b) Residual stresses

Fig. 1 Initial imperfections produced by welding

Table 1 Coefficients of component modes of idealized initial deflection

a/b	A_{01}/t	A_{02}/t	A_{03}/t	A_{04}/t	A_{05}/t	A_{06}/t	A_{07}/t	A_{08}/t	A_{09}/t	A_{010}/t	A_{011}/t
$1 < a/b < \sqrt{2}$	1.1158	-0.0276	0.1377	0.0025	-0.0123	-0.0009	-0.0043	0.0008	0.0039	-0.0002	-0.0011
$\sqrt{2} \leq a/b < \sqrt{6}$	1.1421	-0.0457	0.2284	0.0065	0.0326	-0.0022	-0.0109	0.0010	-0.0049	-0.0005	0.0027
$\sqrt{6} \leq a/b < \sqrt{12}$	1.1458	-0.0616	0.3079	0.0229	0.1146	-0.0065	0.0327	0.0000	0.0000	-0.0015	-0.0074
$\sqrt{12} \leq a/b < \sqrt{20}$	1.1439	-0.0677	0.3385	0.0316	0.1579	-0.0149	0.0743	0.0059	0.0293	-0.0012	0.0062
$\sqrt{20} \leq a/b < \sqrt{30}$	1.1271	-0.0697	0.3483	0.0375	0.1787	-0.0199	0.0995	0.0107	0.0537	-0.0051	0.0256

tions were assumed as indicated in Fig. 1 (b) From the self-equilibrating condition of welding residual stresses, the following equation is derived.

$$2b_t\sigma_t = (b - 2b_t)\sigma_c \tag{5}$$

In the analysis, the four edges of a plate were assumed to remain straight while subjected to in-plane movement.

3. Method of Theoretical Analysis

3.1 Elastic large deflection analysis

When a flat plate is loaded by a compressive in-plane force, the plate undergoes buckling. If a plate is initially deflected, lateral deflection increases from the beginning of loading, but the increase of deflection is small until the load reaches near the buckling load. This behaviour exhibits the geometrical nonlinearity. To analyze such behaviour, an elastic large deflection analysis was performed in an analytical manner with the initial and total deflections expressed by Eqs.(3) and (4).

3.2 Elasto-plastic large deflection analysis

When a plate is further loaded, the yielding takes place. The behaviour after initial yielding up to collapse exhibits the material nonlinearity as well as the geometrical nonlinearity. To analyze such behaviour, an elasto-plastic large deflection analysis was performed applying the finite element method.

4. Influence of Initial Deflection on Stiffness of Compressed Plates

4.1 Short plates

When a short plate is compressed, only the term, A_1, in Eq.(4) increases with an increase of compressive load. In this case, elastic large deflection behaviour of a plate can be accurately approximated by Eqs.(A1), (A2) and (A3) in Appendix taking m as 1. The results of elastic large deflection analysis for the plate of $a/b = 0.8$ are plotted by dashed curves in Figs. 2 (a), (b) and (c). The magnitude of initial deflection are taken as 0.0, 0.01, 0.25, 0.5 and 1.0 times the plate thickness.

These figures indicate that the in-plane rigidity abruptly changes at the buckling load when a plate is flat. When a plate has small initial deflection, the in-plane rigidity rapidly decreases as the compressive load approaches the buckling load. When initial deflection is large, the plate shows low in-plane rigidity from the beginning of loading. The rigidity decreases with an increase in compressive load.

To check the influence of yielding on in-plane rigidity, a series of elasto-plastic large deflection analysis was also performed. Two cases were considered either elastic or plastic buckling takes place when the plate is initially flat. The results are shown by solid curves in Fig. 2.

It is known that the in-plane rigidity rapidly decreases with the spread of plastic zone after yielding has started especially when the plate is thick.

It may be said that the in-plane rigidity of a plate with initial deflection never goes below that of a flat plate after buckling until yielding takes place.

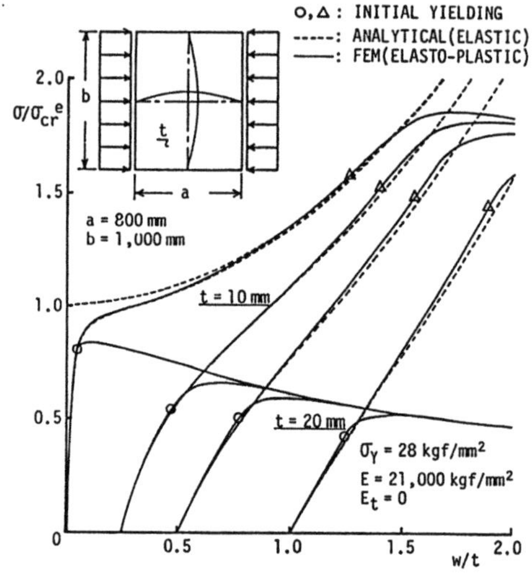

(a) Average stress - deflection relationships

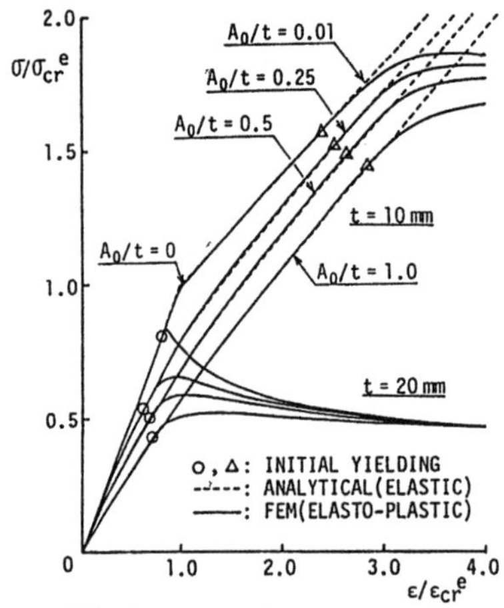

(b) Average stress - strain relationships

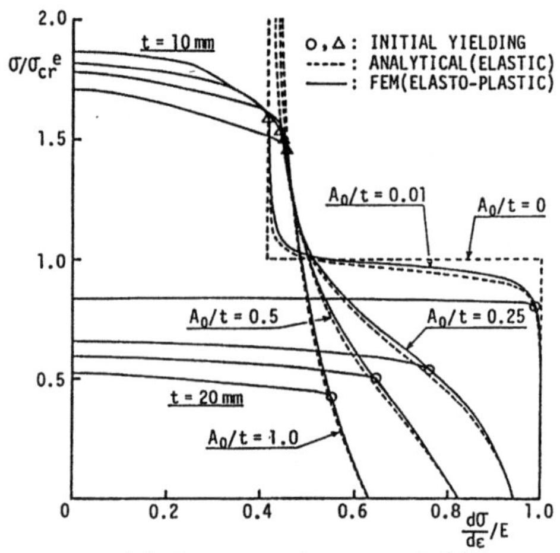

(c) Average stress - rigidity relationships

Fig. 2 Behaviour of short plate under thrust

3.2 Long plates

A series of elastic large deflection analysis was performed on the plates of which shapes of initial deflection were measured in Chapter 2. One of the typical results are shown in Fig. 3 (a), where the relationships between the average compressive stress and deflection components are plotted.

All component modes of initial deflection increase until the load approaches the buckling load, but all except one mode ($m = 5$) diminish above the buckling load. Considering such behaviour of a long plate under increasing load, uni-modal analysis was performed taking only the stable deflection component, A_5, giving only the corresponding component of initial deflection, A_{05}. The magnitude of initial deflection was taken the same as that of the multi-modal analysis. The calculated result is plotted in fig. 3 (a) by a dashed curve. It is known that the relationship between the average stress and the deflection component, A_5, is almost the same for both cases.

For these two cases, the average stress-strain relationships are compared in Fig. 3 (b), and the changes in rigidity with an increase of load in Fig. 3 (c). These figures indicate that the behaviour of a long plate with complex multi-

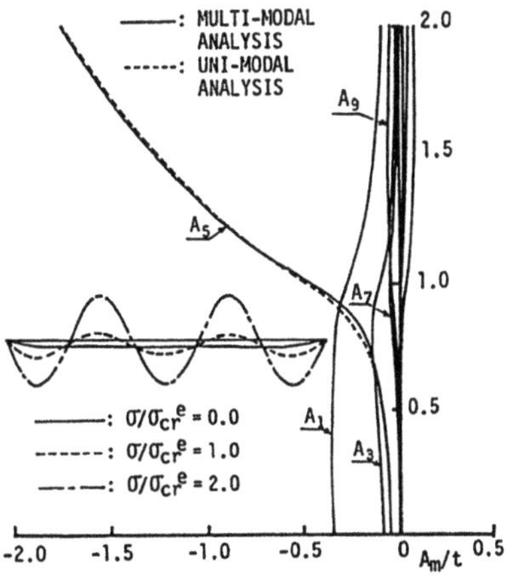

(a) Average stress - deflection component relationships

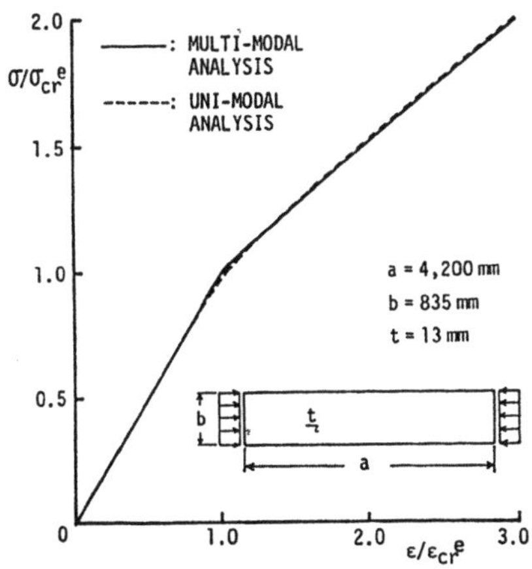

(b) Average stress - strain relationships

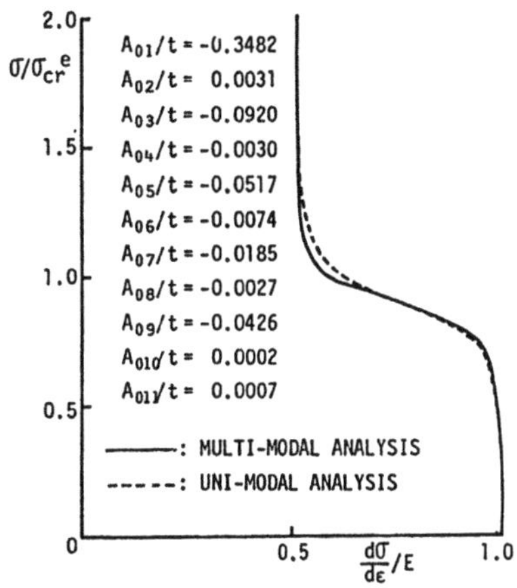

(c) Average stress - rigidity relationships

Fig. 3 Behaviour of long plate under thrust

modal initial deflection is approximately known by performing uni-modal analysis with initial deflection of which mode coincides with the stable mode above the buckling load.

A series of elastic large deflection analyses was performed on a rectangular plate with idealized initial deflection in Table 1 to know the stable deflection mode above the buckling load. The aspect ratio of the plate was changed up to 5.0. The stable deflection mode for various magnitudes of initial deflection is indicated in Fig. 4 together with the in-plane rigidity above the buckling load.

It is well known that the buckling mode changes from *m* half-waves to *m*+1

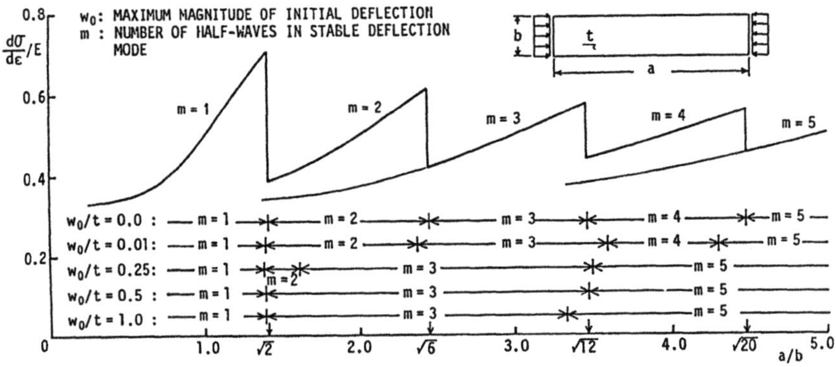

Fig. 4 Stable deflection mode and in-plane rigidity of a compressed plate above buckling load

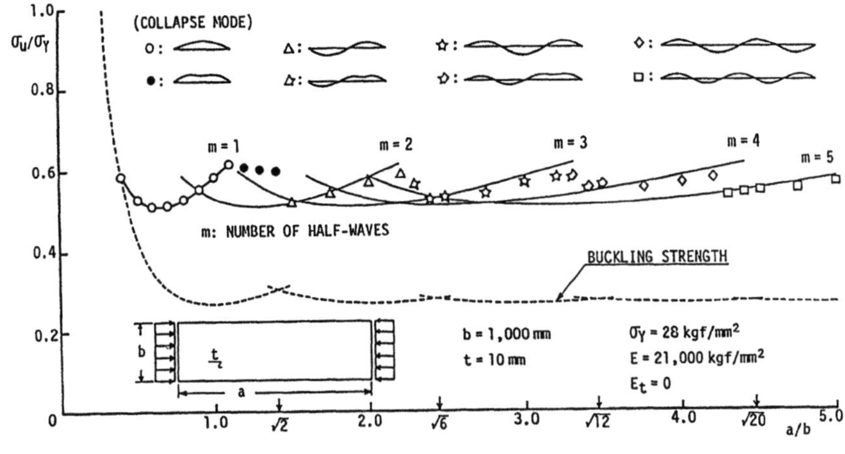

Fig. 5 Compressive ultimate strength of rectangular plate with initial deflection

half-waves in the loading direction at the aspect ratio of $\sqrt{m(m+1)}$. However, the stable mode does not necessarily coincide with the buckling mode when the plate is accompanied by initial deflection. Figure 4 indicates that the stable mode of even half-wave numbers does not appear when a large initial deflection exists in a hungry-horse mode.

A series of elasto-plastic analysis was also performed on the same rectangular plate in Fig. 4. The magnitude of initial deflection was taken as 0.01 times the plate thickness. The evaluated ultimate strengths are plotted in Fig. 5. The solid curves represent the ultimate strength when completely uni-modal collapse takes place in a long plate. For certain ranges of aspect ratio, the collapse mode is not uni-modal. This may be because yielding has started before the stable deflection mode is attained.

It should be noted that both the in-plane rigidity and ultimate strength show abrupt changes at the aspect ratio where the stable deflection mode terminates.

4. Influence of Welding Residual Stresses on Stiffness of Compressed Plates

The influence of welding residual stresses was also examined by performing elasto-plastic large deflection analysis. The results for a short plate are shown in Figs. 6 (a) and (b). It is observed that the in-plane rigidity in the elastic range above buckling load is not affected by residual stresses. Only the buckling load is affected, and hence the start of reduction in rigidity. After yielding has started, the behaviour is influenced by residual stresses especially when compressive stresses exist in the middle portion of the plate.

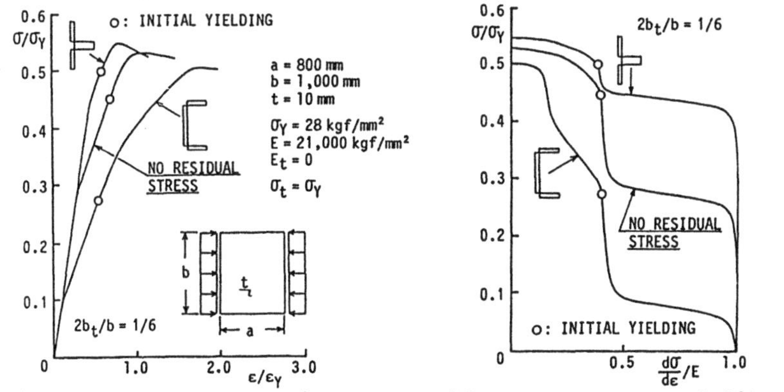

(a) Average stress - strain relationships

(b) Average stress - rigidity relationships

Fig. 6 Influence of welding residual stresses on behaviour of compressed plate

5. Conclusions

(1) The in-plane rigidity of a compressed plate decreases with the increase in the magnitude of initial deflection, but it never goes below that of a flat plate after buckling in an elastic range.

(2) As the compressive load increases above the buckling load, the in-plane rigidity approaches to that of a flat plate asymptotically after buckling.

(3) When a plate is under thrust in the direction of its shorter side, the in-plane rigidity is reduced according to the magnitude of maximum initial deflection.

(4) When a plate is under thrust in the direction of its longer side, the in-plane rigidity is not so much reduced even if the magnitude of maximum initial deflection is large.

(5) Welding residual stresses do not affect the in-plane rigidity itself but have influence on buckling strength in the elastic range.

References

1. Ueda, Y. and Yao, T.: The influence of complex initial deflection modes on the behaviour and ultimate strength of rectangular plates in compression. J. Construct. Steel Research. 5 (1985) 265-303.
2. Yao, T. and Nikolov, P.I.: Stiffness of plates after buckling. J. Kansai Soc. Naval Arch. 215 (1991) 137-146.

Appendix: Fundamental Equations for Uni-nodal Analysis

When only A_{0m} and A_m are taken as the initial and total deflection, the equilibrium equation in the elastic range is expressed as follows [2]:

$$\alpha(A_m^2 - A_{0m}^2)A_m + \beta(A_m - A_{0m}) + \{\gamma_x + (a/mb)^2\gamma_y - \sigma\}A_m = 0 \qquad \text{(A1)}$$

where

$$\alpha = \frac{m^2\pi^2 E}{16a^2}\{1 + (a/mb)^4\}, \qquad \beta = \frac{\pi^2 E}{12(1-\nu^2)}\left(\frac{t}{b}\right)^2 (a/mb + mb/a)^2$$

$$\gamma_x = \int_0^b f(y)\cos\frac{2\pi y}{b}\,dy, \qquad \gamma_y = \int_0^a g(x)\cos\frac{2m\pi x}{a}\,dx$$

$f(y)$ and $g(x)$ represent the residual stress in x- and y-directions, respectively. The average compressive strain in this case is expressed as:

$$\varepsilon = \gamma(A_m^2 - A_{0m}^2) + \sigma/E \qquad \text{(A2)}$$

where $\gamma = \pi^2/8a^2$. The in-plane rigidity is then represented as:

$$d\sigma/d\varepsilon = E/\{1 + 2\gamma E/(2\alpha + \beta A_{0m}/A_m^3)\} \qquad \text{(A3)}$$

Influence of Residual Stresses and Deformation on Buckling Behaviour of Plates

Delan Yin

PH.D., Lecturer of Bridge and Underground Engineering Department
Southwest Jiaotong University,
Chengdu, Sichuan, 610031, P.R.CHINA

Shizhong Qiang

PH. D., Professor of Bridge and Underground Engineering Department
Southwest Jiaotong University,
Chengdu, Sichuan, 610031, P.R.CHINA

Summary

In this paper, multi-layer/DR combination method is used to study the buckling behaviour of plates , giving a detailed discussion for the influence of the parameters, such as loading and boundary conditions of the plate , the slenderness of the plate , the mode of the residual deformation and the distribution of the residual stresses.

1. Introduction

Welding is a main connection method of steel components. Residual stresses and deformation will occur in part of the components because of non-uniform expansion with heat and contraction with cooling during welding process. They are initial imperfection of components. Initial imperfection has a certain effect of the buckling behaviour of a plate. It makes the practical panel not matching with the ideal flat panel which is a basic assumption in the theory of linear elastic buckling. Therefore , the theory of classic linear buckling does not always predict the ultimate capacity of plates correctly. But before 1970s , there were only few series of solutions and classical finite difference solutions for the nonlinear analysis of plates with initial imperfections. And they were not enough to be used in engineering practice.

Since 1970 , researchers in Great British had studied the elasto-plastic large deflection behaviour of plates with initial imperfections. The typical results of them can be seen in Refs. [1] and [2].

L. Karlsson, L.-E. Lindgren, M. Jonsson (Eds.)
Mechanical Effects of Welding
IUTAM Symposium Luleå/Sweden 1991

It is complex to research the influence of residual stresses and deformation on ultimate capacity of a plate , for there are many parameters concerned , such as loading and boundary conditions of plates , the slenderness of plates , the mode of the residual deformation and the distribution of the residual stresses. In the paper, multi-layer/dynamic relaxation method is used to study the buckling behaviour of plates , giving a detail discussion for the influence of the parameters referred above.

2.Method of Analysis Used in the Present Paper

The method used in this paper is a multi-layer/DR combination. The basic equation is Von Karman's large deflection equation in the incremental form. The displacements are taken as unknown quantities. Governing equations are dispersed by finite difference technique. The plate is divided into five layers toward its thickness to simulate the elasto-plastic behavior. Ideal elastic-perfectly plastic material is assumed , although a form of isotropic hardening has been added to the routine.

The multi- layer/DR combination analysis followed a step-by-step procedure with continuously updated stiffness matrix to allow for plasticity. In the main iterative loop of the DR program, at the first , the forced displacement increments are given on the loading edges of the plate. After the equilibrium state has been attained at each load increment , the incremental strains and curvatures in each layer are evaluated using previous multi-layer stiffness and total stresses in each layer. Based on Von Mises's yield criterion , the elasto-plastic condition of each discrete element is judged. For un-yielding element , elastic stiffness matrix is remained. And for yielding element, the plastic strain rate multiplier , λ , is calculated according to Prandtl-Reuss flow rule. If $\lambda < 0$, it means un-loading from yielding surface, and the stiffness is reset at their elastic value. If $\lambda > 0$, elasto-plastic stiffness matrix is calculated. The average in-plane rigidities and average bending rigidities of the plate can then be given from a numerical integration of the in-plane moduli of five layers through the plate. They will be used in the next loading increment. Then another forced displacement is given.

The detailed description of DR program is in Ref.[3].

3. Details of Present Analysis

3.1. Residual stresses.

Two types of distributions of residual stresses are considered in this paper. One is taken as an idealized distribution of residual stresses consisting of stress blocks with tensile yield zones at the panel edges and a uniform compressive stress block across the remainder of the panel. This form is used in the plates of which slenderness is smaller than 80. Slender panels are unable to sustain any significant uniform compressive stresses , so an improved distribution of residual stresses is used to establish a more meaningful initial state. The idealized block is first applied to the panel in the presence of the required geometric imperfections. Iterations of the main DR loop are then performed. The residual stresses tend to magnify the initial deformation , so the magnitude of initial deformation before starting calculation is factored down to produce the required deflection level (equal to the required deformation), associated with a set of in-plane and bending stresses which are in self- equilibrium. This improved self-equilibrating stress distribution is then used as the starting point for the main loading.

The compressive residual stress levels have been restricted to $0.3\sigma_0$ and $0.1\sigma_0$ in x and y directions,respectively.

3.2. Residual Deformation

The initial deformation patterns discussed in the paper are assumed in double sinusoidal shapes of magnitudes varied with slenderness of the plate, which can be expressed as follow:

$$w_0 = w_c \operatorname{Sin}(m\pi\ x/a) \operatorname{Sin}(n\pi\ y/b) \tag{1}$$

while $w_c = a/165\ (\sigma_0/355)^{1/2}$, as defined in BS5400. It can also be seen in Fig. 1.

3.3. Boundary and Loading Conditions

The boundary displacements and the loading condition of the plate are dealt with as follows. For the loading edges , forced displacements are given according to the loading conditions. The loading conditions considered in this paper are pure shear and uniform compression in one direction (see Fig. 2). The displacement increment given at each step is

$$\Delta v=\alpha\ (\gamma\ /\gamma\ _{0})a \tag{2-a}$$

$$\Delta u=\alpha\ (\varepsilon\ /\varepsilon\ _{0})a \tag{2-b}$$

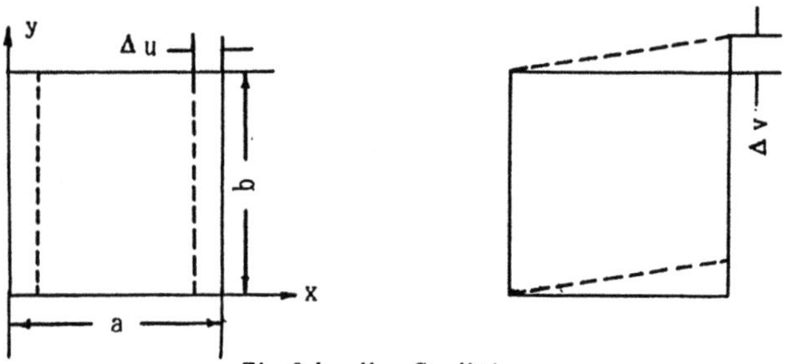

Fig.2 Loading Conditions

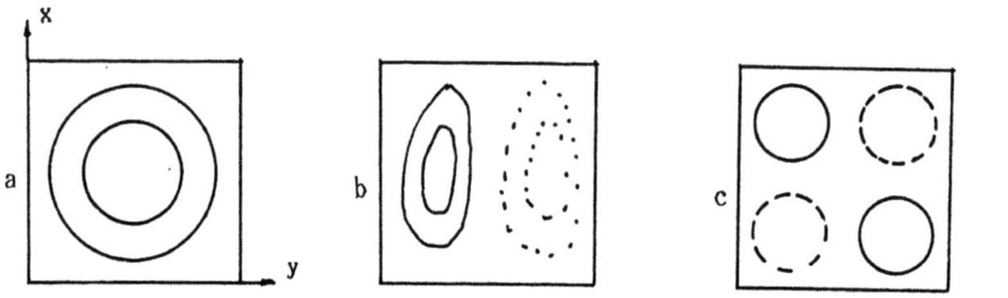

$w_0=w_c\mathrm{Sin}(\pi\ x/a)\mathrm{Sin}(\pi\ y/b)$ $w_0=w_c\mathrm{Sin}(2\pi\ x/a)\mathrm{Sin}(\pi\ y/b)$ $w_0=w_c\mathrm{Sin}(2\pi\ x/a)\mathrm{Sin}(2\pi\ y/b)$

Fig.1 Initial Deformation Modes

where , mean yield strains, and a is the length of the plate, see Fig.2 , is a parameter to control the magnitude of forced displacement increments. In an elastic range, is large; while in elasto-plastic range, is taken as a smaller value.

Regarding the out-of-plane restraints along the four edges of the plate , a simply supported condition is assumed. As for the in-plane displacements along the non-loading edges , three types of boundary conditions are considered, which are: no displacement in perpendicular direction to loads , Eq. (3-a); free displacement under stress free condition , Eq. (3-b); and uniform displacement in perpendicular direction to loads, Eq.(3-c).

$\Delta u=0.0$ (under pure shear) (3-a1)

$\Delta v=0.0$ (under uniform compression) (3-a2)

$\Delta N_y=0.0$, $\Delta u, \Delta v$ are free (3-b)

$\int_0^a \sigma_y dx = 0.0$, (on top and bottom edges) (3-c)

4. Results of the Analysis and Discussions

4.1. The Influence of Residual Stresses on Compressive Panels

Figure 3 shows the effects of residual stresses and the initial imperfection on the plate which is loaded by uniform

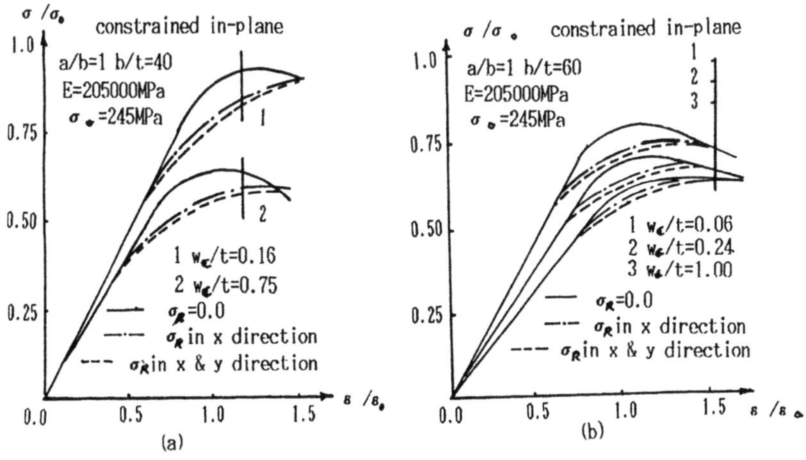

Fig.3 Initial Imperfection VS The Behabiours of Compressed Plates

compression. The in-plane movements of the bottom and the top of the plate are uniform (the third boundary condition). The figure explains the following things. For a thicker plate (b/t=40), severe initial deformation decreases the ultimate capacity of the plate to a high degree; the effect of residual stresses is making the yielding earlier. The ultimate capacity is reduced by about 4% for the case of the plate with residual stresses. For a slender plate (b/t=60), more severe initial deformation does not affect on the plate in same degree as on a thicker plate. Both in- plane rigidity and ultimate capacity are reduced according to the magnitude of initial deflection, but the decrease of ultimate capacity is not so much comparing to the thicker plate.

4.2. The Effect of Residual Stresses on Plates Loaded by Shearing

Figure 4 shows typical average stress-strain curves for shearing plates with residual stresses in longitudinal

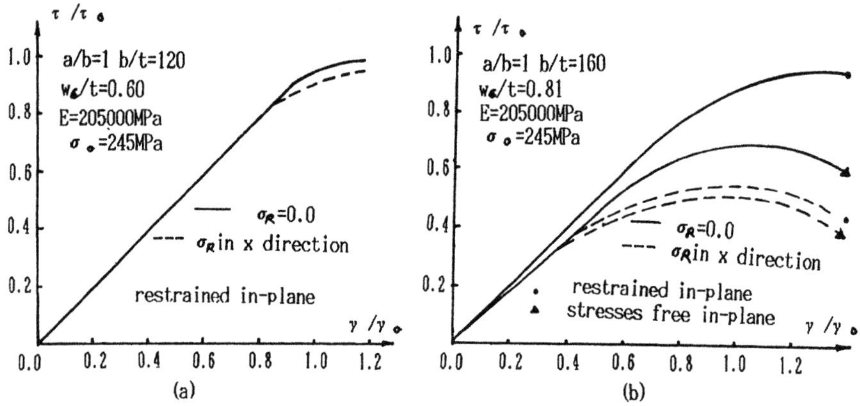

Fig.4 Residual Stresses VS The Behaviours of Shearing Plates

direction. It demonstrates that the effect of residual stresses on a plate of which slenderness b/t is 160 is more remarkable than on a plate of which b/t is 120. In the former , the ultimate capacity of the plate is reduced by 20% , while in the latter , that is reduced by about 8%. Corresponding to two types of boundary conditions of the plate , the effect of residual stresses is not different seriously judging from the results in Fig.4b.

4.3. Effects of Initial Deformation Modes on the Behaviour of Plate

Figure 5 gives average stress-strain curves of plates loaded by shearing. The results shows that initial deformation modes do not exert a remarkable effect on plates loaded by shearing. This is because after it is buckled by shearing , a thinner plate will form a inclined tensile field (see Ref. [4]) to bear more shear forces. The inclined membrane tensile force tends to decrease the initial deformation , so its effect can be reduced to a certain degree.

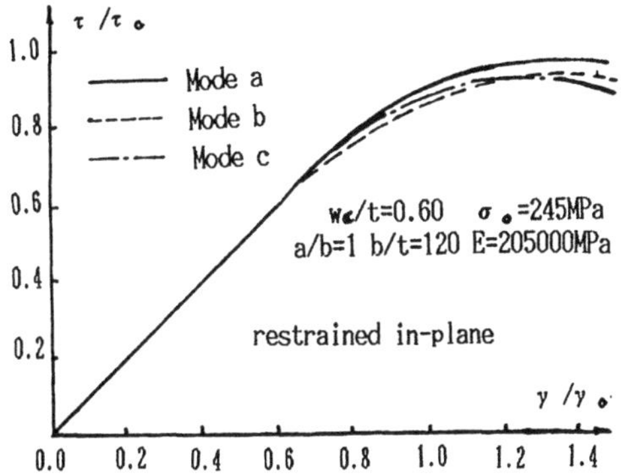

Fig.5 Initial Imperfection modes VS The Behaviours of Shearing Plates

5.Conclusion

5.1. When compressive force is applied for a thicker plate (b/t= 40) , severe initial deformation decreases the ultimate capacity of the plate to a high degree; the effect of residual stresses is making the plate into yielding earlier. The ultimate capacity is reduced by about 4% for the case of the plate with assumed residual stresses. For a slender plate (b/t=60) , more severe initial deformation does not exert effect on the plate in the same degree as on a thicker plate.

5.2. When shear force is applied, the effect of residual stresses on a plate of which slenderness b/t is 160 is more remarkable than on a plate of b/t=120. In the latter, the ultimate capacity of the plate is reduced by 8% , while in the former , that is reduced by about 20%.

5.3. When shear force is applied, the effect of residual stresses on the plate which is restrained at non-loading edges is not different seriously from that on the plate of which top and bottom edges are stress free.

5.4. Initial deformation modes do not exert a remarkable effect on the behaviour of plates loaded by shearing.

6.Acknowledgment

The financial support of National Education Commission of China for the work in the paper is appreciated.

References

1. Crisfield , M. A. 'Large-deflection elasto-plastic buckling analysis of plates using finite elements', Department of the Environment, TRRL Report LR 593 , Crowthorne , 1973, (Transport and Road Research Laboratory).

2. Harding J. E. , Hobbs R. E. and Neal B. G. , 'Ultimate load behavior of plates under combined direct and shear in-plane loading' , International Conference on Steel Plates Structures, Edited by Dowling, P.J., Harding, J.E. and Frieze, P.A., Grosby Lockwood Staples, 1977.

3. Shizhong Qiang , 'An adaptive dynamic relaxation method for non-linear problems' , Computer and Structures , vol.30, No.4, pp855-859, 1988, Printed in Great Britain.

4. Rockey , K. C. , 'An ultimate load method of design for plate girders', International Conference on Developments in Bridge and Construction, Cardiff, 1971.

Harmonic Response of a Spot Welded Box Beam – Influence of Welding Residual Stresses and Deformations

K. S. ALFREDSSON, and B. L. JOSEFSON

Division of Solid Mechanics,
Chalmers University of Technology,
S-412 96, Göteborg, Sweden

Summary

The transverse point mobility for a straight spot welded box beam loaded with a transverse end force is calculated numerically by use of the FE-code ABAQUS. The influence of the residual stress field caused by the spot welding and (approximately) by the cold forming on the harmonic response is studied in particular. The residual deformations and stresses caused by the spot welding are estimated using a simplified approach (employing FEM) where the temperature field present during welding is not modelled in detail. This approach is calibrated against calculated results from a detailed 3D-analysis of welding of one spot. The lowest eigenfrequencies of the box beam are found to decrease about 10 % when residual stresses from the manufacturing processes are considered.

Introduction

Resistance spot welding has been used for over 100 years as a manufacturing process for joining parts of thin-walled structures. Today the method dominates fabrication in the automobile industry. There are, for example, more than 5000 spot welds in the body and frame components of a car.

One major difficulty (from a mechanical point of view) experienced with spot welds is the low fatigue strength compared with that of the unaffected base metal. This is due to the crack-like geometry (normally) created at the weld nugget where large tensile residual stresses are formed, and where the material may have been embrittled by the temperature history during welding. This local lowering of the fatigue strength has been demonstrated in numerous experiments and fracture mechanics analyses, several references have been collected in Radaj [1]. Fatigue of spot welded box beams, which may be enhanced by large deflections of the flanges between the spots, have been studied analytically and experimentally by Oshima and Kitagawa [2]. The present paper focusses, however, on another problem where residual stresses caused by the manufacturing will affect the mechanical properties of the complete structure, namely the harmonic response due to vibrating design loads. It is well known that large tensile residual stresses in a small region near the spot weld are balanced by compressive stresses in remote regions of the structure.

L. Karlsson, L.-E. Lindgren, M. Jonsson (Eds.)
Mechanical Effects of Welding
IUTAM Symposium Luleå/Sweden 1991

For the box beam geometry studied here the regions which deflect in the lowest vibrational eigenmodes are found to have compressive welding residual stresses. Thus one may expect a lowering of the eigenfrequencies of a spot welded box beam. This effect has been observed for stiffened butt welded plates, see Kaldas and Dickensson [3].

Welding residual stress field By use of FEM it is possible to follow, in detail, the temperature- and corresponding stress history during welding and during cooling after welding. Complex nonlinear relations for the constitutive behaviour may also be employed. Such analyses to obtain the shape and magnitude of the residual stress field at one spot weld have been performed by, for example, Nied [4] and Schröder and Macherauch [5].
However, for built-up spot welded structures, like the present box beam, it is not practically feasible to carry out a detailed FE-analysis of the welding of each spot, since the computer time needed will be prohibitively high. Furthermore, insufficient knowledge of the material behaviour and the necessity of use of several simplifying assumptions regarding the geometry underline that a simplified approach is needed. Therefore, we propose that the welding residual stress field is obtained as proposed in Josefson [6] by assigning to each spot (only) an idealized temperature history. Thus, the development of the temperature field during welding and subsequent cooling is not followed in detail.

Present study The harmonic response of a spot welded box beam is studied numerically by use of the FE-code ABAQUS [7]. The particular box beam studied models a straight part of a beam that extends (longitudinally) from the front to the passenger compartment in a car. Figure 1 shows the box beam studied. The transverse end force applied in the harmonic response analysis is also included in this Figure. The wall thickness of all parts is 1.5 mm. Ten spot welds, with the spacing 50 mm, are located at each of the two flanges, thus connecting two halves (one small at left and one larger at right) of the box beam. The material in the box beam is a C-Mn steel which contains less than 0.1 % C and about 0.2 % Mn. The virgin yield stress of the base metal is 250 MPa.
The influence of the residual stress field from the spot welding and from the cold forming on the harmonic response is studied in particular. The analysis is divided into three parts: First, the history of the local temperature-, strain- and stress fields present during welding of one spot are calculated in detail using a 3D-model. This detailed analysis is divided into two subsequent parts: First, a heat transfer analysis is performed where the history of the temperature field is determined. Thereafter, a thermo elasto-plastic analysis is performed and the history of the stress field is calculated from the results of the heat transfer analysis. Figure 2 shows the time history for the clamping force and for the heat supply used here for the spot welding process. A simplified local model employing shell elements and a rudimentary temperature history is created and adjusted to give qualitatively the same results as in the detailed 3D-model.

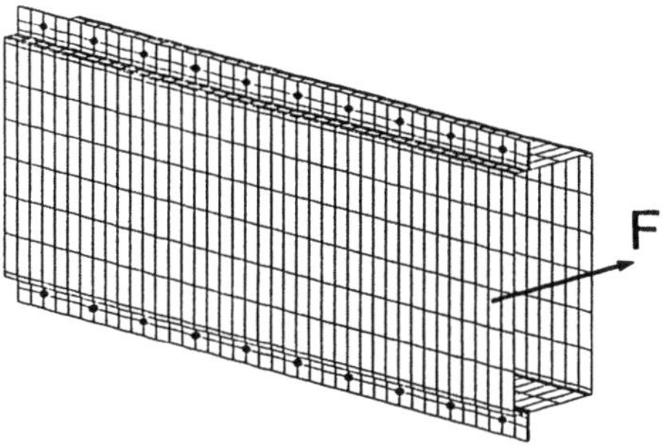

Fig. 1 Geometry for box beam studied. Spacing between spots is 50 mm. Figure is drawn to scale. Total length of box beam is 500 mm. Location of transverse end force for dynamic analysis

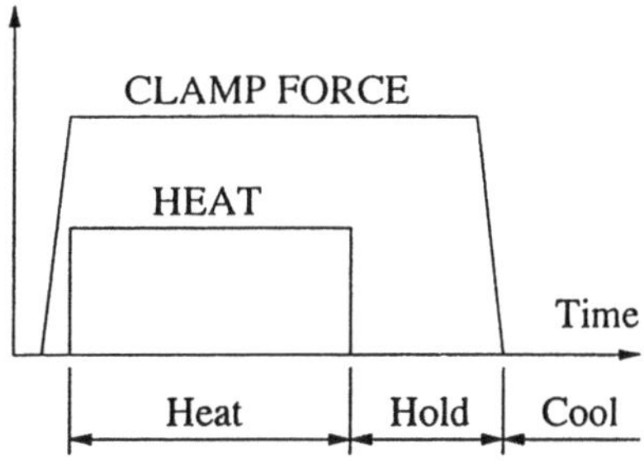

Fig. 2 Assumed time history for clamp force and heat supply in local analysis of one spot weld

This simplified method is then used for the calculation of the welding residual stress field (from each spot) in the box beam model. A simplified method to simulate residual stresses from cold forming using a temperature field is also proposed. Finally, based on the estimated residual stresses from the manufacturing process, and on assumptions on local damping in the flanges of the box beam, the harmonic response is calculated.

FE-models used

Local 3D-model of one spot If all spots are formed simultaneously, which is assumed here for simplicity of calculations, we only need to study one half of a spot weld spacing. This is because symmetry conditions can be used to model neighbouring parts of the local model, since a small part of the beam far away from the ends of the beam is studied. The in plane rotations are deleted at the nodes of the two planes of symmetry that are perpendicular to the longitudinal direction of the beam. At one of these two planes the out of plane displacements are also deleted, while at the other plane the out of plane displacements are equal at all nodes. At the symmetry plane parallel to the longitudinal direction, both the in plane rotations and the out of plane displacements are deleted. Furthermore, one of the nodes of the model is constrained in the vertical direction to prevent rigid body translation.

The flanges of the beam are modelled with eight-node solid elements to include effects of stresses and deformations perpendicular to the flanges due to the electrode forces. Since other parts of the beam are not subject to such loadings they are modelled with four-node shell elements. The left and right halves of the box beam share nodes at the flanges at the electrode positions only. Uni-axial contact elements are added between other nodes of the flanges so that they are not allowed to penetrate, but are free to separate from each other. In the thermal analysis these parts of the flanges do not interact. The electrodes are not included in the mechanical FE-model since they can be regarded as very stiff compared to the heated flanges. Instead they are assumed to be rigid and in full contact over the contact region where a resulting mean pressure of 200 MPa is developed.

Local simplified shell model of one spot In the simplified local model employing four-node shell elements, the heat transfer during the welding is not followed. Instead, the temperature is increased 1400 °C at nodes located in the weld nugget. All other nodes are kept at ambient temperature. Since shell elements can not deform in the thickness direction, local transverse deformations caused by the electrodes can not be modelled. The local 3D and shell FE-meshes are displayed in Fig. 5 below. The number of degrees of freedom (dofs) for these FE-models are about 6300 and 1300 respectively.

Global shell model of box beam The FE-model of the complete box beam is displayed in Fig. 1. Also here four-node shell elements are employed. At both ends of the box beam, a rigid massless membrane is attached. One of the membranes is connected to the ground through six springs, one for each global degree of freedom, with stiffnesses chosen such that the corresponding "rigid body" eigenfrequencies fall well below the eigenfrequencies for the box beam. For all FE-models, that is for both solid and shell elements, thermal strains are evaluated at the centroid point to ensure consistency between thermal and mechanical strains.

Material data To simplify the numerical calculations, and to account for uncertainties in values for the material parameters, constant values for all material parameters are employed in the heat transfer analysis. Thus, the thermal conductivity λ = 45 W/mK, the specific heat c = 450 J/kg °C and the density ρ = 7800 kg/m^3. In the mechanical analysis, the material is taken as thermo elasto-plastic employing the von Mises yield function and an associated flow rule. The material forming the weld nugget between the electrodes will reach austenitization temperature when heated. During the subsequent rapid cooling phase transformations to ferrite and subsequently to bainite will take place. The temperature variation for the yield stress of this material is calculated, at each temperature, by linear volume fractioning from values of the different phases present taken from the literature. The temperature variation of the hardening is more difficult to predict. Hardening is high at lower temperatures (below 600 °C) while plastic strains accumulate mainly at higher temperatures. Furthermore, it is argued that the heat affected material will loose part of its memory during the final phase transformations, see Leblond [8].

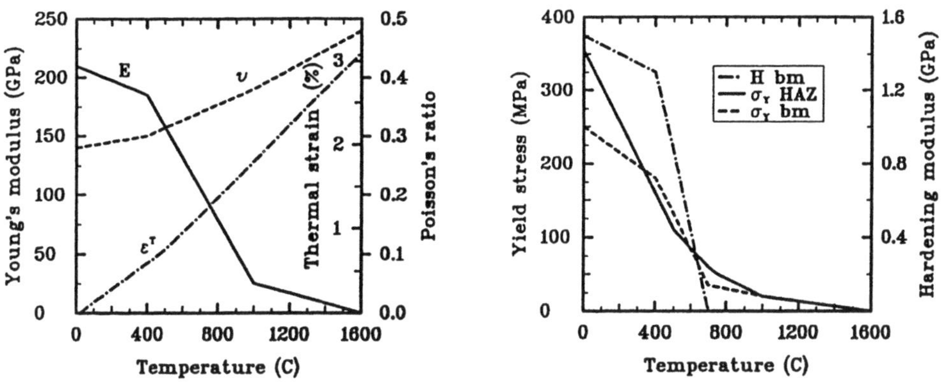

Fig. 3 Temperature dependence of material parameters employed in mechanical analysis for low alloyed C-Mn steel

Due to these uncertainties of the hardening behaviour, the heat affected material is taken as elastic ideally plastic. No effects of latent heat production in the thermal analysis nor volume changes and transformation plasticity in the mechanical analysis are included in this material model. The temperature variation of the yield stress and the hardening modulus for the material outside the spot weld is taken from the literature (this material does not experience phase transformations). Figure 3 summarizes the values used for the material parameters in the mechanical analysis.

Local temperature field at one spot

In resistance spot welding the workpieces are heated due to the Joule heating process when the weld current passes through the workpieces between the electrodes. Here, the heat power supplied is assumed to be uniformly distributed over the cylindric material volume between the electrodes with the magnitude adjusted to give a correct maximum temperature as compared to values from literature.

The heat conduction from the flanges to the electrodes (during the weld cycle) is simulated as convective heat transfer with a temperature dependent heat transfer coefficient, see [5]. Other surfaces of the FE-model are assumed to be cooled by the surrounding air through natural convection. The heat transfer analysis shows that cooling is very fast in the weld nugget region, $\Delta t_{8/5} < 1$ s, thus one may anticipate a fully bainitic structure in this region. Figure 4 shows the temperature distribution at the end of the heat input phase, and during cooling after welding when the highest temperature in the flanges has decreased below 100°C.

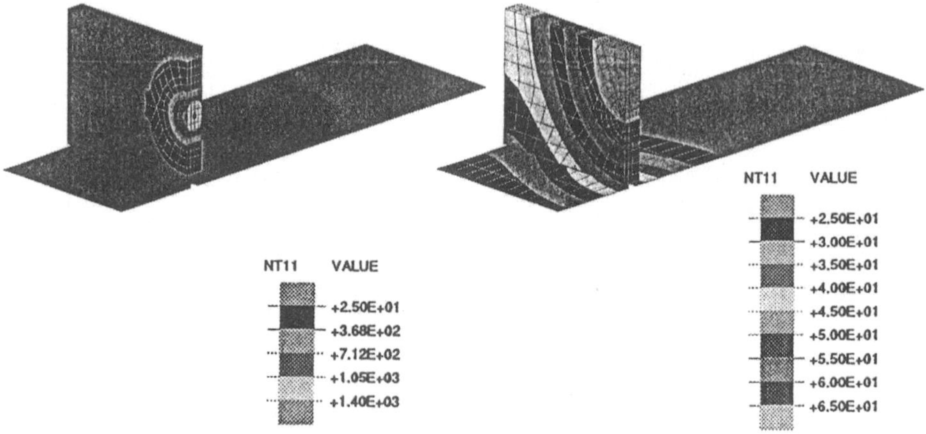

Fig. 4 Temperature distribution (3D-model) in flanges at the end of heat input phase, and when highest temperature in box beam is below 100°C

Local stress field at one spot

The distribution of the longitudinal stresses determined with the detailed 3D-model and with the simplified model are shown in Fig. 5. As can be seen in Fig. 5 the stresses in the longitudinal direction determined with the simplified model agree qualitatively well with those from the more detailed analysis. The agreement is sufficiently good to justify use of a simplified model. Thus, both the 3D-model and the shell model predict the longitudinal stress 200 MPa on the flange surface at the spot. However, below the surface, in the weld nugget, an almost tri-axial residual stress develops with a resulting hydrostatic pressure higher than 200 MPa. The longitudinal stresses on the lateral surfaces are found to be compressive with a maximum magnitude of about 10 MPa.

Global stress field in spot welded box beam

Cold-forming process An attempt is also made to consider the residual stresses caused by the cold forming process that proceeds the spot welding. When a corner is cold-formed, yielding in the circumpherential direction takes place. This will result in a decrease in thickness and the formation of residual stresses in the longitudinal direction, see Ingvarsson [9]. The longitudinal stresses in the plane parts of the cross section of the beam will be compressive to balance the tensile longitudinal stresses in the cold-formed parts. Here, these stresses are introduced by lowering the temperature locally at the nodes along each of the cold-formed corner lines.

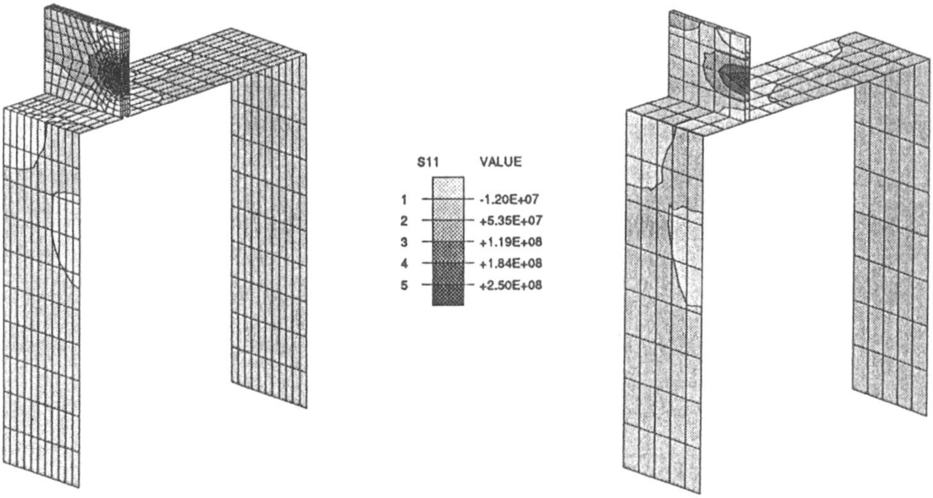

Fig. 5 Comparison of calculated welding residual longitudinal stress field (in Pa) in flange part of local model after use of 3D FE-analysis (considering temperature history) and after use of simplified shell FE-analysis (where temperature history is not followed)

Thus, the elements that share these nodes will contract. The size of these elements are chosen to coincide with the extension of the cold-formed area to give these stresses a proper distribution.

Simplified stress field The welding process for each spot is simplified as described above in the local shell model. This is done simultaneously for all twenty spots. During this welding simulation, and also during the subsequent loading, the left and right flanges are not allowed to penetrate each other. This is achieved through the use of gap elements available in the FE-code used (ABAQUS). One may note that second order effects caused by the membrane residual (welding and forming) stresses are also considered. The total number of (real parts of the) dofs for this model is about 11100.

One finds that the longitudinal residual stresses will reach the yield stress level in the flanges, and that they will be compressive with the maximum magnitude 40 MPa on the left lateral surface of the box beam. This value differs considerably from the value -10 MPa obtained for the same location with the local model. However, since the heated flanges are located outside the neutral surface of the box beam, the spot welding process will induce an additional bending moment (and curvature) not accounted for in the local model. This resulting curvature of the box beam is displayed in Fig. 6 where also the separation of the flanges between the spots is displayed (which is demonstrated in both the local and the global model). The separation of the flanges is caused by the constrained membrane thermal expansion in the flanges, and this effect is found to be stronger than the compression caused by pressure from the electrodes.

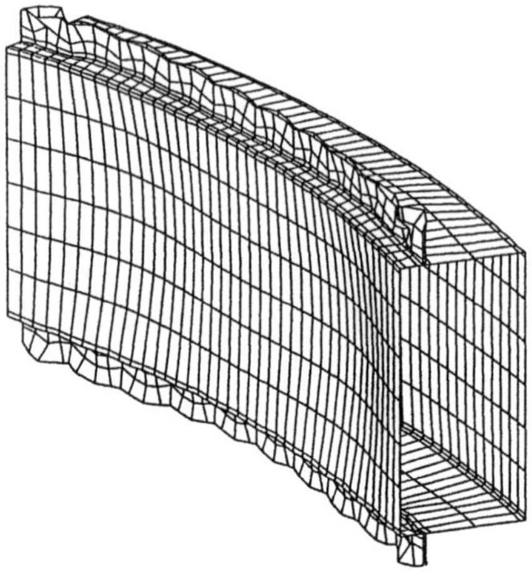

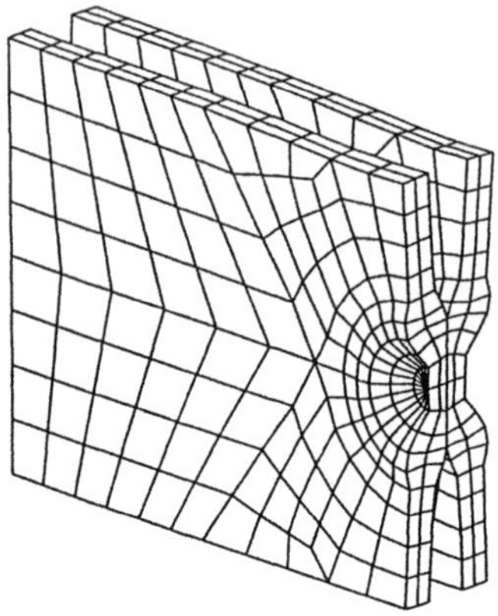

Fig. 6 Residual deformation field for box beam as calculated by global shell model, considering both forming and spot welding, and residual deformation of flange at one spot as calculated by the 3D-model (magnification factors used are 400 and 50 respectively)

Dynamic analysis of spot welded box beam

Eigenfrequencies The present box beam has about eight non-zero eigenfrequencies below 500 Hz. Considering first an ideal box beam where the flanges are continuously connected, the lowest eigenmodes are found to excite the lateral surfaces only and they correspond, approximately, to the lowest eigenmodes for simply supported plates. The global bending eigenmodes for the box beam are found to have considerably higher frequencies. The flanges deflect noticeably first in eigenmode no 13 (with the eigenfrequency 760 Hz).

The values of the lowest eigenfrequencies change only marginally when the geometrical constraints at the spots are introduced since the flanges are much stiffer than the lateral surfaces. However, when residual stresses from the welding and from the cold forming are considered (together with the geometrical constraints) a noticeable decrease, about 10%, for the lowest eigenfrequencies are calculated. Figure 7 shows the calculated eigenmode 8 for the box beam with the geometrical constraints at the spots introduced, with f_8 = 283 Hz (f_8 = 284 Hz for an ideal box beam), and for the same box beam where residual stresses have also been considered, with f_8 = 253 Hz. This reduction can be explained by the presence of compressive longitudinal residual stresses on the lateral surfaces of the box beam, and the figures correspond well with previous results for stiffened welded plates, see [3], and with handbook estimations of eigenfrequencies for simply supported plates loaded with inplane forces.

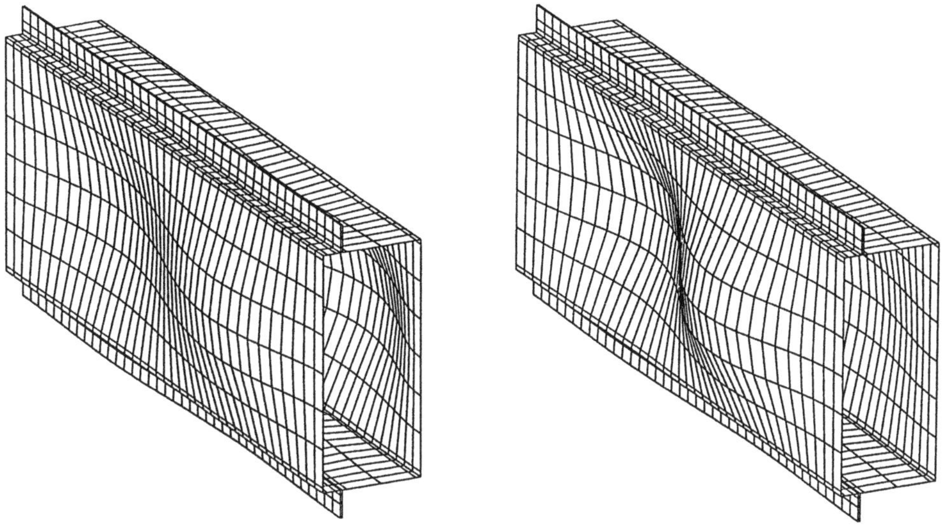

Fig. 7 Calculated undamped eigenmodes (no:s 8) for box beam with geometrical constraints at the spots introduced (left) and for the same box beam with residual stresses from manufacturing also included (right)

Harmonic response The box beam will now be subject to a harmonically varying transverse point load applied at one of the membrane ends, see Fig. 1. This loading case excites low eigenmodes of the box beam. The corresponding magnitude of the transverse point mobility in the point of load application is chosen as output (the real part of the mobility is a measure of the input of mechanical power). The damping in spot welded (or bolted) structures is believed to be caused mainly by pumping of air between the flanges at locations between the spots, see Cremer, Heckl and Ungar [10]. In the present analysis, air pumping is modelled primarily as viscous Rayleigh damping with the loss factor η in the range 1% to 5% at the frequencies of interest, see [10]. Since the webs and the lateral surfaces of the box beam can be considered as undamped, the structure will be nonproportionally damped. Formally, a direct solution of the resulting system of FE-equations by use of complex mathematics must then be employed. Figure 8 shows the influence of the manufacturing process on the transverse point mobility. Beside the lowering of resonance frequencies, the presence of residual stresses in the lateral surfaces is also seen to give higher mobilities at frequencies below 270 Hz. The lowest elastic eigenmode (f_7 = 222 Hz for the ideal box beam) is symmetric and is not excited by a transverse force located at the end of the beam.

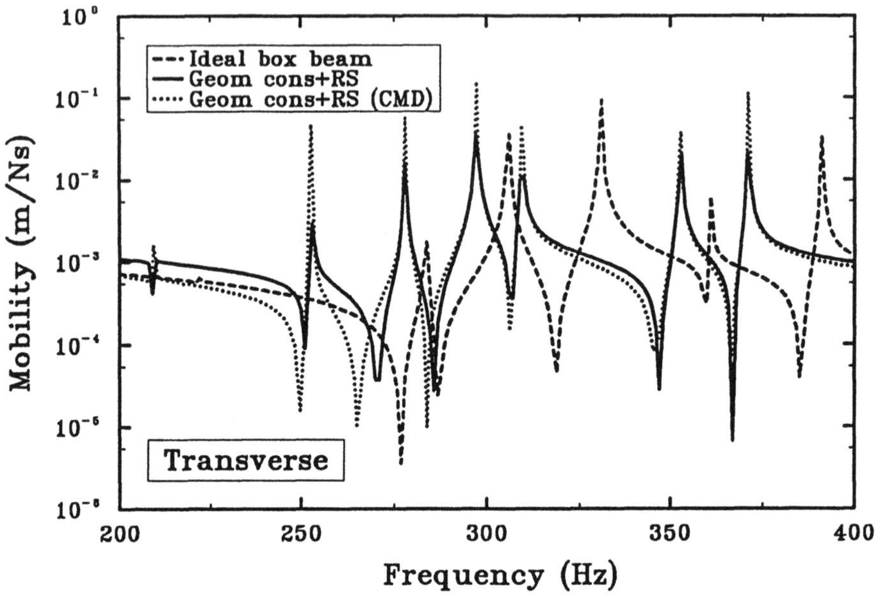

Fig. 8 Calculated magnitude of transverse point mobility for point of load application. Influence of residual stresses (RS) and of modelling of damping in flanges are displayed

The numerical calculations have also shown that qualitatively the same results can be obtained (see Figure 8), with a drastic reduction in CPU-time, using a so called composite modal damping (CMD). In CMD modal superposition with a weighted frequency independent fraction of critical damping, ξ, for each mode is used. The corresponding factors for each part of the structure are then weighted by the mass matrix for each part, and by the (global) eigenmodes to give the average value of ξ for each mode. For the flange parts the loss factor η is here taken as 2.5% which is believed to be a reasonable average value in the frequency range studied (200-400 Hz), see [10]. Due to difficulties in locating resonance frequencies with the direct solution of the FE-equations, the calculated results will differ somewhat at these points. However, Fig. 8 shows that the CMD results also differ from the results obtained from the correct direct solution at frequencies below 250 Hz. Here, the CMD computed response is too stiff. This is probably due to the finite number of eigenmodes (30) employed in the CMD approach.

Acknowledgements

This work was financially supported by grants from the Volvo Research Foundation and Volvo Educational Foundation. Parts of the FE-computations were performed on the Cray X-MP/48 at the National Supercomputer Centre in Sweden at Linköping University.

References

1 Radaj D.: Design and analysis of fatigue resistant welded structures, Cambridge, England, Abington Publishing, 1990.

2 Oshima, M.; Kitagawa, H.: Buckling assisted fatigue of spot-welded box beam under bending. SAE Paper 860605, Society of Automotive Engineering, Warrendale, Pa, USA, 1986.

3 Kaldas, M.M.; Dickinsson, S.M.: The flexural vibration of welded rectangular plates. Journal of Sound and Vibration 75 (1981) 163-178.

4 Nied, H.A.: The finite element modeling of the resistance spot welding process. Welding Journal 63 (1984) 123s-132s.

5 Schröder, R.; Macherauch, E.: Berechnung der Wärme- und Eigenspannungen bei Widerstandspunktschweissverbindungen unter Zugrundelegung unterschiedlicher mechanisch-thermischer Werkstoffdaten. Schweissen und Schneiden 35 (1983) 270-276 (in German).

6 Josefson, B.L.: Prediction of residual stresses and distortions in welded structures. Proceedings of The 10th International Conference on Offshore Mechanics and Arctic Engineering 1991, Volume III-Part A. The American Society of Mechanical Engineers, New York, NY, USA (1991) 25-30.

7 ABAQUS User's manual - Version 4.8. Hibbit, Karlsson and Sorensen Inc, Providence RI, USA, 1988.

8 Leblond, J.B.: Mathematical modelling of transformation plasticity in steels II: coupling with strain hardening phenomena. International Journal of Plasticity 5 (1989) 573-591.

9 Ingvarsson, L.: Cold-forming residual stresses and Box columns built up by two cold-formed channel sections welded together. Bulletin No 121 of The Department of Building Statics and Structural Engineering, The Royal Institute of Technology, Stockholm, Sweden 1977.

10 Cremer, L.; Heckl, M.; Ungar, E.E.: Structure borne sound, Second Edition. Berlin, Germany, Springer Verlag, 1988.

MIX
Papier aus verantwortungsvollen Quellen
Paper from responsible sources
FSC® C105338

If you have any concerns about our products,
you can contact us on
ProductSafety@springernature.com

In case Publisher is established outside the EU,
the EU authorized representative is:
Springer Nature Customer Service Center GmbH
Europaplatz 3, 69115 Heidelberg, Germany

Printed by Libri Plureos GmbH
in Hamburg, Germany